Studies in Big Data

Volume 82

Series Editor

Janusz Kacprzyk, Polish Academy of Sciences, Warsaw, Poland

The series "Studies in Big Data" (SBD) publishes new developments and advances in the various areas of Big Data- quickly and with a high quality. The intent is to cover the theory, research, development, and applications of Big Data, as embedded in the fields of engineering, computer science, physics, economics and life sciences. The books of the series refer to the analysis and understanding of large, complex, and/or distributed data sets generated from recent digital sources coming from sensors or other physical instruments as well as simulations, crowd sourcing, social networks or other internet transactions, such as emails or video click streams and other. The series contains monographs, lecture notes and edited volumes in Big Data spanning the areas of computational intelligence including neural networks, evolutionary computation, soft computing, fuzzy systems, as well as artificial intelligence, data mining, modern statistics and Operations research, as well as self-organizing systems. Of particular value to both the contributors and the readership are the short publication timeframe and the world-wide distribution, which enable both wide and rapid dissemination of research output.

The books of this series are reviewed in a single blind peer review process.

Indexed by zbMATH.

All books published in the series are submitted for consideration in Web of Science.

More information about this series at http://www.springer.com/series/11970

Pardeep Kumar · Amit Kumar Singh
Editors

Machine Learning for Intelligent Multimedia Analytics

Techniques and Applications

Springer

Editors
Pardeep Kumar
Department of Computer Science and Engineering and Information Technology
Jaypee University of Information Technology
Solan, Himachal Pradesh, India

Amit Kumar Singh
Department of Computer Science and Engineering
National Institute of Technology
Patna, Bihar, India

ISSN 2197-6503 ISSN 2197-6511 (electronic)
Studies in Big Data
ISBN 978-981-15-9491-5 ISBN 978-981-15-9492-2 (eBook)
https://doi.org/10.1007/978-981-15-9492-2

This Springer imprint is published by the registered company Springer Nature Singapore Pte Ltd.
The registered company address is: 152 Beach Road, #21-01/04 Gateway East, Singapore 189721, Singapore

Preface

Recently, multimedia such as text, image, audio, video, graphics stands as one of the most demanding and exciting aspects of the information era. Surveys established that every second a lot of multimedia information are created, published and transmitted explosively, becoming an indispensable part of today's big data. Such large-scale multimedia data has disclosed new challenges and opportunities for intelligent multimedia analysis. It provides innovative opportunities to address several potential research problems, e.g., enabling comprehensive visual classification to fill the semantic gap by exploring large-scale data, offering a promising frontier for detailed multimedia understanding, as well as extract patterns and making effective decisions by analyzing the large collection of data. The processing of multimedia large-scale data has emerged as a most demanding area for the application of machine learning techniques. Therefore, the machine learning for intelligent multimedia analytics is becoming an emerging research area for multimedia research community.

Outline of the Book and Chapter Synopsis

In view of above, this book presents state-of-the-art intelligent techniques and approaches, design, development and innovative use of machine learning techniques for demanding applications of intelligent multimedia analytics. We have provided potential thoughts and methodologies that help senior undergraduate and graduate students, researchers, programmers and industry professionals in creating new knowledge for the future to develop intelligent multimedia analytics-based novel approach for multimedia applications. Further, key role and great importance of machine learning techniques as a mathematical tool is elaborated in the book. A brief and orderly introduction to the chapters is provided in the following. The book contains fifteen chapters.

Chapter "Secure Multimodal Access with 2D and 3D Ears" presents a multi-modal technique that uses 2D and 3D ear images for secure access.

A two-stage, coarse and fine alignment to fit the 3D ear image of the probe to the 3D ear image of the gallery. The probe and gallery image keypoints are compared using feature vectors, where very similar keypoints are used as coarse alignment correspondence points. Once the ear pairs are matched fairly closely, the entire data is finely aligned to compute the matching score. A detailed systematic analysis using a large ear database has been carried out to show the efficiency of the technique proposed.

Chapter "Efficient and Low Overhead Detection of Brain Diseases Using Deep Learning-Based Sparse MRI Image Classification" presents efficient and low overhead detection of brain diseases using deep learning-based sparse MRI image classification for four different brain diseases like edema, necrosis, enhancing and non-enhancing tumor. The main contribution of the this book chapter lies in utilizing the sparsity of the set of images that reduces the training and inference time of the CNN.

Chapter "Continual Deep Learning Framework for Medical Media Screening and Archival" introduces continual deep learning framework for medical media screening and archival. The primary section of the proposed piece of work helps us to automatically diagnose tuberculosis (TB) from chest X-rays in minimal execution time and memory footprint. Secondarily, medical media screening and analysis would drive the process of automated archival strategy, tagged to self-generated metadata. Gradual self-improvement of the AI engine narrowing down the fuzzy zone between confident positive and confident negative of diagnosis would be the key achievement of proposed continuous deep learning framework.

Chapter "KannadaRes-NeXt: A Deep Residual Network for Kannada Numeral Recognition" presents KannadaRes-NeXt: a deep residual network for Kannada numeral recognition framework. The proposed framework is based on deep residual network ResNeXt to correctly classify the images of handwritten Kannada numerals. The proposed method offered high accuracy on different datasets. These results also indicate efficiency of the proposed framework is superior to other similar approaches.

Chapter "Secure Image Transmission in Wireless Network Using Conventional Neural Network and DOST" introduces a secure image transmission in wireless network using conventional neural network and discrete orthonormal Stockwell transform (DOST). Experimental results demonstrate that the method offered better performance in terms of mean error, PSNR and entropy, to other similar approaches. Chapter "Robust General Twin Support Vector Machine with Pinball Loss Function" recognizes that twin support vector machines (TWSVM) with hinge loss suffer from noise sensitivity and instability. To overcome these issues, pinball loss-based general twin support vector machines (Pin-GTSVM) was recently proposed in the literature. However, TWSVM and Pin-GTSVM implement the empirical risk minimization principle. Also, the matrices in their dual formulations are positive semi-definite. To overcome these issues, authors propose pinball loss-based robust general twin support vector machines (Pin-RGTSVM). Pin-RGTSVM implements the structural risk minimization principle which embodies the marrow of statistical learning and pinball loss function makes it more robust for noisy datasets. Numerical experiments and statistical evaluation on the real-world benchmark datasets show the efficacy of the proposed scheme.

Chapter "Noise Resilient Thresholding Based on Fuzzy Logic and Non-linear Filtering" proposes a novel image thresholding technique resilient to noise based on the fuzzy logic and the nonlinear filtering. The rationale of the proposed technique is to use adaptive Kuwahara filtering, which essentially suppresses the noise while retaining the image texture information. Finally, an optimized threshold selection process is formalized using the fuzzy logic properties.

Chapter "Deep Learning Methods for Audio Events Detection" focuses on deep learning methods for audio events detection. This chapter describes how to identify audio events in a complex sound scenario using convolutional neural networks. The variability of sound events determines continuous changes in frequency content that can hardly be traced with the instruments commonly used in traditional acoustics. Authors throw some light on the possibility to develop automatic systems to recognize audio events that could inform us of possible threats to the safety of people or things or could offend the sensitivity of people.

Chapter "A Framework for Multi-lingual Scene Text Detection Using K-means++ and Memetic Algorithms" presents a framework for multilingual scene text detection using K-means++ and memetic algorithms. This framework is experimentally evaluated on two standard datasets, namely MLe2e and KAIST, and on an in-house dataset of 400 images, all having multilingual texts. The results obtained are comparable with some state-of-the-art methods.

Chapter "Recent Advancements in Medical Imaging: A Machine Learning Approach" discusses the importance of machine learning in the field of medical imaging for reconstructing medical images from the measured raw data. Besides discussing the use of machine learning for medical image reconstruction, a general overview is also provided on all the existing techniques of medical imaging. Mathematical models are provided in order to understand better the use of machine learning for reconstruction purposes. Unsupervised learning techniques such as dictionary learning (DL) and K-sparse autoencoders (KSAE) for reconstructing medical images have also been discussed.

Chapter "Solving Image Processing Critical Problems Using Machine Learning" discusses various algorithms like neural network, support vector machine, genetic algorithm, convolutional neural network, etc., to solve different critical problems like denoising, image compression, image color enhancement, segmentation, etc., in image processing.

Chapter "Spoken Language Identification of Indian Languages Using MFCC Features" describes a framework for spoken language identification of Indian languages using MFCC features for any multilingual automatic speech recognition (ASR) process. Support vector machine (SVM) is applied on the extracted features. The proposed framework is tested on a standard Indic TTS speech database. SVM is trained for 13 (static), 26 (static + delta) and 39 (static + delta + delta-delta) MFCC features. It is found that the best results are achieved using only 13 static features, and addition of delta and delta-delta features decreases the performance.

Maximum accuracy obtained is 60.33% for the above-mentioned six languages, and 89.33% for only Bangla, English and Hindi.

Chapter "Performance Evaluation of One-Class Classifiers (OCC) for Damage Detection in Structural Health Monitoring" describes the performance evaluation of one-class classifiers (OCC) for damage detection in structural health monitoring. An array of classifiers is executed on a benchmark structure dataset (IASC-ASCE) both from supervised and unsupervised machine learning domain and a comparison between their success rates has been in determining damage in civil structures.

Various classical approaches are tested over International Association for Structural Control (IASC)—American Society of Civil Engineers (ASCE) SHM benchmark problem at varying noise levels and force intensities to cover wide variations in the generated dataset.

Chapter "Brain Tumor Classification in MRI Images Using Transfer Learning" presents brain tumor classification method in MRI images using transfer learning. The proposed method focuses on a binary classification problem that seeks to differentiate between normal and tumorous MRI images by using the idea of deep transfer learning based on pre-trained VGG16 model. Results show that the proposed method performed better accuracy than the other state-of-the-art techniques. Further, the method uses pre-trained model to reduce the training time.

Finally, Chap. "Semantic-Based Vectorization Technique for Hindi Language" discusses semantic-based vectorization technique for Hindi language. This book chapter sheds light on how to embed words from a Hindi corpus scrapped from various newspapers and other literary sources. Using bidirectional LSTM, a neural model is trained to embed entire sentential contexts and target words in the same low-dimensional space, which is optimized to reflect inter-dependencies between targets and their entire sentential context as a whole. It uses the preceding and succeeding parts of sentences to predict the missing word in a sentence.

We especially thank the Studies in Big Data Series Editor, Prof. Janusz Kacprzyk, for his continuous support and great guidance.

We would also like to thank publishers at Springer, in particular Aninda Bose, Senior Editor, Springer, for their helpful guidance and encouragement during the creation of this book.

We are sincerely thankful to all authors, editors and publisher whose works have been cited directly/indirectly in this manuscript.

The editors believe that our book will be helpful to the senior undergraduate and graduate students, researchers, professionals and providers working in the area demanding state-of-the-art machine learning solutions for multimedia applications.

Special Acknowledgements

The first author gratefully acknowledges the authorities of Jaypee University of Information Technology, Waknaghat, Solan, Himachal Pradesh, India, for their kind support to come up with this book.

The second author gratefully acknowledges the authorities of *National Institute of Technology Patna*, India, for their kind support to come up with this book.

Solan, India — Dr. Pardeep Kumar
Patna, India — Dr. Amit Kumar Singh

Contents

About the Editors

Dr. Pardeep Kumar is currently working as an Associate Professor in the Department of Computer Science & Engineering and Information Technology at Jaypee University of Information Technology (JUIT), Wakanaghat, Solan, Himachal Pradesh, India. He has been associated with his current employer since 2008. Prior to joining Jaypee Group, he was associated with Mody University of Technology & Science (Formerly known as Mody Institute of Technology & Science) Laxmangarh, Sikar, Rajasthan. He has completed PhD (Computer Science and Engineering) from Uttarakhand Technical University, Dehradun, India, M.Tech (Computer Science & Engineering) from Guru Jambheshwar University of Science & Technology, Hisar, Haryana, India and B.Tech (Information Technology) from Kurukshetra University, Kurukshetra, Haryana, India. He has served as Executive General Chair of 2016 Fourth International Conference on Parallel, Distributed and Grid Computing (PDGC), Guest Editor of Special Issue on "Robust and Secure Data Hiding Techniques for Telemedicine Applications", Multimedia Tools and Applications: An International Journal, Springer (SCI Indexed Journal, IF = 1.346), Lead Guest Editor of Special Issue on "Recent Developments in Parallel, Distributed and Grid Computing for Big Data", published in the International Journal of Grid and Utility Computing, Inderscience (Scopus Indexed), and Guest Editor of Special Issue on "Advanced Techniques in Multimedia Watermarking", published in the International Journal of Information and Computer Security, Inderscience (Scopus Indexed). Dr. Kumar has been appointed as an Associate Editor of IEEE Access (SCI Indexed, IF = 3.5) Journal. His area of interest includes machine learning, medical image mining, image processing, health care informatics, etc.

Dr. Amit Kumar Singh is currently an Assistant Professor with the Computer Science and Engineering Department, National Institute of Technology Patna, Bihar, India. He received his PhD from National Institute of Technology Kurukshetra, Haryana, India in 2015. He has authored over 100 peer-reviewed journal, conference publications, and book chapters. He has authored three books and edited five books with internationally recognized publishers such Springer and

Elsevier. Dr. Singh have been recognized as "WORLD RANKING OF TOP 2% SCIENTISTS" in the area of "Biomedical Research" (for Year 2019), according to the survey given by Stanford University, USA. He is the associate editor of IEEE Access (Since 2016), IEEE Future Directions (Since 2020), IET Image Processing (Since 2020), Telecommunication Systems, Springer (Since 2020), Journal of Intelligent Systems, De Gruyter (Since 2020), and former member of the editorial board of Multimedia Tools and Applications, Springer (2015–2019). He has edited various international journal special issues as a lead guest editor such as ACM Transactions on Multimedia Computing, Communications, and Applications, ACM Transactions on Internet Technology, IEEE Consumer Electronics Magazine, IEEE Access, Multimedia Tools and Applications, Springer, International Journal of Information Management, Elsevier, Journal of Ambient Intelligence and Humanized Computing, Springer. He has obtained the memberships from several international academic organizations such as ACM and IEEE. His research interests include multimedia data hiding, image processing, biometrics, & Cryptography.

Secure Multimodal Access with 2D and 3D Ears

Iyyakutti Iyappan Ganapathi, Surya Prakash, and Syed Sadaf Ali

Abstract This chapter introduces a multimodal technique that uses 2D and 3D ear images for secure access. The technique uses 2D-modality to identify keypoints and 3D-modality to describe keypoints. Upon detection and mapping of keypoints into 3D, a feature descriptor vector is computed around each mapped keypoint in 3D. We perform a two-stage, coarse and fine alignment to fit the 3D ear image of the probe to the 3D ear image of the gallery. The probe and gallery image keypoints are compared using feature vectors, where very similar keypoints are used as coarse alignment correspondence points. Once the ear pairs are matched fairly closely, the entire data is finely aligned to compute the matching score. A detailed systematic analysis using a large ear database has been carried out to show the efficiency of the technique proposed.

Keywords Biometrics · 3D ear · Verification/identification · Secure access · Feature keypoints · 3D Descriptor · Co-registered ear images

1 Introduction

Biometrics is a computer vision sub-field that utilizes an individual's physiological or behavioral features in authentication. Classic password, Identification card, etc. authentication methods have some drawbacks, such as lost passwords and stolen or fake identity cards [1–3]. Biometrics provides users with a secure means of authen-

I. I. Ganapathi · S. Prakash (✉) · S. S. Ali
Department of Computer Science and Engineering,
Indian Institute of Technology Indore, Indore, MP 453552, India
e-mail: surya@iiti.ac.in
URL: http://iiti.ac.in/people/~surya/

I. I. Ganapathi
e-mail: phd1501101002@iiti.ac.in

S. S. Ali
e-mail: phd1301101006@iiti.ac.in

P. Kumar et al. (eds.), *Machine Learning for Intelligent Multimedia Analytics*,
Studies in Big Data 82,
https://doi.org/10.1007/978-981-15-9492-2_1

tication which can be used to overcome conventional methods of authentication [4–17]. Since each individual owns his/her biometric which is used for authentication and is difficult to steal or forge. Many biometric features have been explored in the past, and in recent years the ear has gained significant attention due to its successful recognition [18–21]. "There's real power in using the appearance of an ear for computer recognition, compared to facial recognition. It's roughly equivalent if not better," said computer scientist Kevin Bowyer of the University of Notre Dame. Moreover, unlike the face, the ear does not change shape under various expressions; not influenced by makeup, ageing wrinkles, scarf occlusion, etc. Besides, the ear size is greater than the fingerprint, iris, retina, etc., and smaller than the face; it can be collected and processed easily. In addition, Ear biometry is non-intrusive, so users need less cooperation to obtain data. Few examples of co-registered samples of different subjects from the UND-J2 dataset are shown in Fig. 1. It is clear from the samples that the dataset is subject to illumination and also to a slight hair occlusion.

Researchers initially focused on 2D ear images [22, 23] and moved to 3D ear recognition [24] as performance-degrading variables like posture, lighting and scaling affect 2D ear images. The reason for choosing 3D is that the above factors have no effect on 3D ear models and information on geometric shapes of 3D ear models have greatly improved recognition performance compared to 2D. A comprehensive ear detection and recognition survey can be found in [25–35]. This chapter focuses on 2D and 3D co-registered ear recognition techniques and emphasizes related works on ear recognition that uses 2D and 3D images. The reason why we are particularly interested in co-registered ear images is that (i) 3D models overcome the above-mentioned challenges (viewpoints and light changes) caused by 2D images by capturing the shape variations irrespective of light changes; (ii) the richness of 3D geometric properties are highly distinctive, and (iii) using 2D textures along with 3D would help to identify potential keypoints. We introduce a few human recognition approaches using co-registered 2D and 3D ear images. Approaches are focused on registration, feature detection and description. We follow two recognition methods. (i) Since keypoint detection in 2D images is well researched, we rely on the keypoints detected by 2D algorithms and map them to 3D images. Followed by a two-step registration is used to match pairs of ears to identify similarities. First, the mapped keypoints are aligned roughly to obtain the transformation matrix. Second, a fine alignment of complete 3D ear data with the transformation matrix obtained is performed to measure registration errors and is used as the matching score. (ii) In contrast to the previous step, 3D images do not explicitly use the marked keypoints for alignment. Since all keypoints in one image may not match the other, a filtering method is used to find the best matching keypoints. Before coarse alignment, each keypoint is described using a descriptor to compute a vector feature. Similarity is obtained between these feature vectors, where a pair of feature vectors showing the least distance are selected as correspondence. In the same way, we find all possible pairs for the image of the probe and gallery to find all correspondence points. Since we use the geometric properties of each 3D point together with 2D, the detection of keypoints has improved. A local 3D descriptor called rotational projection statistics (RoPS) [36] is used to compute the feature vector. RoPS is a recently developed

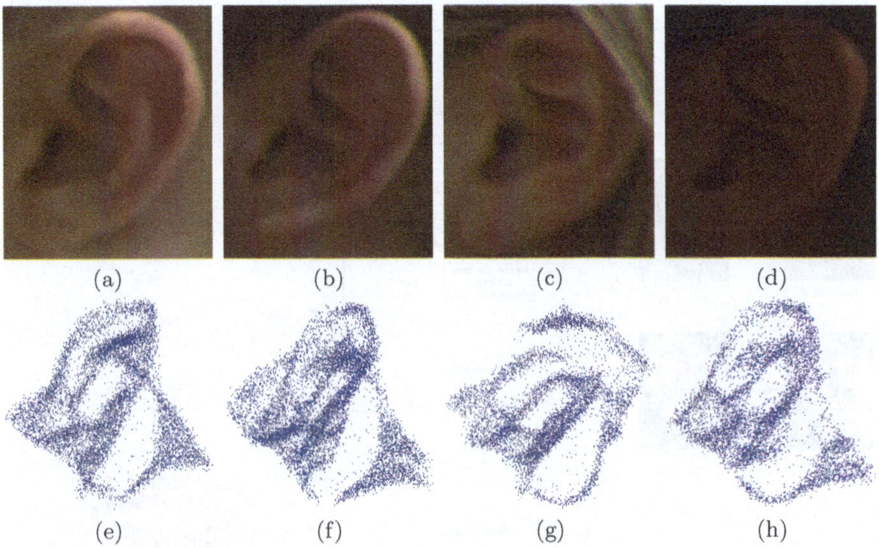

Fig. 1 Few samples of 2D-3D co-registered from UND-J2 database. **a–d** 2D ear image, **e–h** co-registered 3D ear image

3D descriptor that shows satisfactory results in matching inter-class objects. We use RoPS to identify the correct ear pair keypoints using 3D information at each keypoint. We proposed the use of a two-stage combination to significantly improve efficiency by combining RoPS with the iterative closest point (ICP) [37]. A similar approach is followed in technique [38], however, keypoints are found based only on 2D image information and no 3D image information contributes. In the contrary, the proposed method used both 2D and 3D image information. Outline of the proposed technique is shown in Fig. 2.

1.1 Main Contributions

1. 2D and 3D ear data is used to recognize human for secure access.
2. To find keypoints in 3D, We utilize the versatility of 2D keypoint detection algorithms.
3. Also, we include 3D information along with 2D information to find robust keypoints.
4. The recognition performance is analysed using registration algorithms; and 3D descriptor based algorithms.
5. The performance of the proposed approach is encouraging in presence of noise and occlusion.

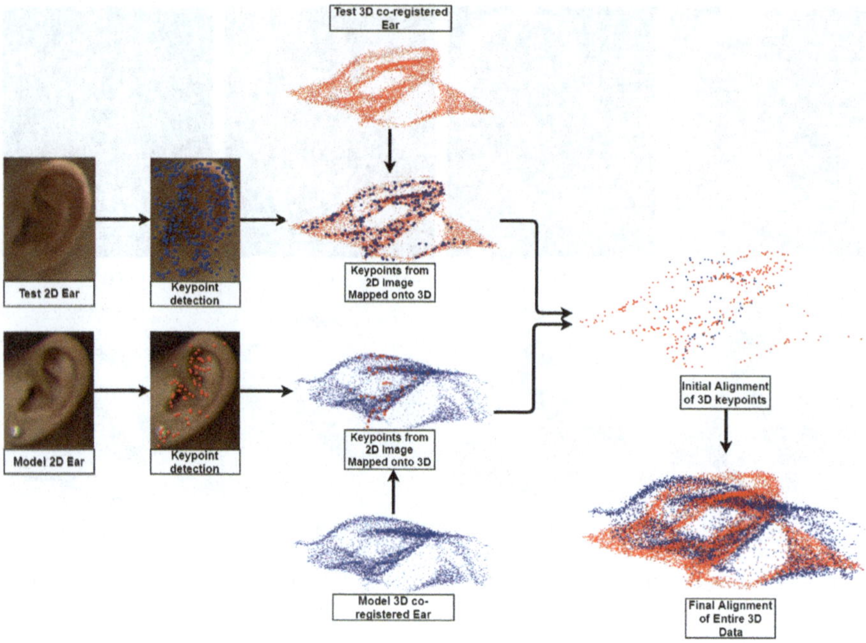

Fig. 2 Outline of the proposed technique

The remainder of the paper is organized as follows. A brief overview on 3D ear recognition is discussed in Sect. 2 and the use of co-registered ear images in human recognition is given in Sect. 3. In Sect. 4 preliminaries on a few strategies are discussed in the recognition context. The proposed technique is explained in detail in Sect. 5 and the evaluation of the technique proposed in Sect. 6. Finally, the chapter concludes in Sect. 7.

2 Related Work

Different methods for 2D, 3D, and 2D+3D ear recognition exist in literature. Chen and Bhanu developed a generalized method of 3D object recognition, tested using 3D ears [39]. Extreme shape indices detect the feature keypoints, and each keypoint is defined by Local Surface Patch (LSP) [40]. The LSP includes a 2D histogram, surface type, and centroid patch. To reduce the computational complexity, the 2D LSP histogram is mapped using an embedding algorithm. The similarity of the probe to gallery is measured in the space of the lower dimension. Further, in [41], Chen and Bhanu introduced a 3D ear recognition system. Local surface patch descriptors are described for pairs of images. Similar surface descriptors are used to match gallery and test for initial rigid transformation. Then, the ICP algorithm is used to optimize transformation to match the probe and gallery images with the least registration error possible. In the UND-F report, the technique obtained a 2.30% equal error rate. A two-step ICP algorithm is proposed to fit two 3D ears in [42]. Next, the initial rigid transformation obtained via the ICP algorithm is to align the ear helix of the pairs of images. Therefore, the initial transformation accomplished was considered to put about the best final alignment of the two ear images. With 6.70% EER in a small 3D ear sample of 30 participants, this technique has achieved rank-1 recognition of 93.30%.

The technique [43] employed an annotated generic ear model (AEM) to register 3D data and generated a biometric signature with encoded 3D geometric data. Biometric signatures measure similarity among pairs of ears. This technique received a 93.90% recognition from 525 subjects on a dataset of 1031 images. Videos applied to an ear recognition framework by Cadavid and Abdel-Mottaleb in [44]. Several frames were extracted, and the ear segmented from each frame. The 3D ear is further developed from shape from shading of each segmented image. Compared to other models, the built 3D ear model uses a cost similarity feature. Rank-1 accuracy of 95.00%, an EER of 3.30%, was reported using the gallery's 402 video clips and 60 video clips. Ding et al. [45] have proposed a sparse, 3D ear recognition based dictionary learning. First, they extracted the PCA-based features for each image in the gallery and a dictionary was created using those features. Next, it presented a minimum residual optimization problem to identify the image of the probe. To improve performance, more samples were suggested per subject and 10 samples per subject reported rank-1 accuracy of 95.23%.

Very few techniques with local and global features have been employed for 3D ear recognition. Zhou et al. [46] proposed a 3D ear-recognition method using patches of the shape index, curve, and surface around a point to create the local feature vector. Using a standard voxel grid, 3D gallery and probe images are voxelised, and the voxel image is used to build the global feature vector. Eventually, local and global feature vectors are combined to evaluate the model and the matching score is calculated using cosine similarity. Islam et al. [47] have made use of a keypoint neighborhood to create local features. This feature sets the initial alignment between the image test and the gallery, and uses ICP for fine alignment. In the UND database, the technique achieved a recognition rate of 90.00%. For a given 3D ear image a shape-indexed image is generated in [48] and keypoints are detected on the shape-indexed image obtained. Local coordinate systems define each keypoint, and the support region aligns with the local coordinate system. Additionally, transformed local 3D points into a range image. Finally, in order to produce a feature vector, a 3D center-symmetric local binary pattern is applied to the range image. For the probe and gallery images, the features are used to find the corresponding keypoints. Combinations of few local descriptors together with a global descriptor to recognize ears is proposed in [49]. Local descriptors are carefully selected and a weighted combination of descriptor scores used to find similarities among the ears. Similarly, a local descriptor based multistage local 3D descriptor is proposed to identify ears in [50]. In 3D ear recognition, this technique has shown encouraging and promising results.

A person's ear-parotic face angle has been used as a special feature for 3D ear recognition in [24], where the ear-parotic angle is the angle between the normal ear-plane vector and the normal face-plane vector. This technique produced EER 2.80–2.30% in its in-house database. Sun et al. [51] uses Gaussian mean curvature to measure each point's salience in the 3D ear point cloud. Using saliency values, key points and Poisson disk sampling were created and each keypoint has a quadratic manifold to define the local feature. For a 4.00% EER in the UND-J2 database, a 95.10% recognition accuracy is achieved. Dan and Guo [52] introduced 2.5D ear recognition based on SIFT. 3D ear data is rotated along x, y, and z axes for multiple images and transformed into range images. These images are used to match the 3D image, where SIFT [53] algorithm is used to find the vector of the images obtained. SIFT features' closest combination decides the similarity between two ear images. They tested 415 subjects and 830 UND-J2 samples, achieving 98.87% accuracy in identification. In [54], three measures matched a 3D image. A quadratic surface with neighborhood information is constructed around each keypoint, thus used as a feature vector. Basically, the similarity of 3D ear pairs is computed using α-entropy and minimum spanning tree. A minimum spanning tree is generated for matched keypoints, and α-Entropy is utilized to find the similarity.

3 Co-registered 2D and 3D Ears in Recognition

This section reviews the techniques that use 3D ear images along with co-registered 2D images for ear recognition. In the literature, the data of 2D and 3D ears has been associated in different ways.

1. The combination is used as a multimodal biometric in some works such as [55–57], where 2D and 3D ears are used as separate modalities. These studies show that the accuracy of recognition is enhanced by combining matching scores from 2D images of ears and 3D ear models.
2. 2D ear images have been used in a couple of other works to get keypoints in 3D ear images. The reason is that 2D keypoint algorithms are well studied in contrast to early-stage 3D keypoint detection algorithms. For example, keypoints detected in co-registered 2D ear images in [38, 58] have been used to find keypoints in corresponding 3D ear images.
3. There are a few works such as [41, 59, 60], where co-registered 2D ear images are used to detect and segment ear in 3D profile images.

In [61], Yan and Bowyer use multi-modal, multi-algorithms, and multi-instances to examine the efficiency of ear recognition. The methodology [61] is similar to [62] and [63] where, for a detailed analysis, 2D data and algorithms as well as 3D data and algorithms are used. The technique concludes that the multimodal approach which uses PCA for 2D ear images and ICP for 3D ear images provides better performance compared to the other two approaches. This indicates that the weighted sum of scores with a 0.2 and 0.8 PCA and ICP weight gave 202 subjects a high 93.10% recognition and 302 subjects a 90.40% recognition. Woodard has proposed a technique of human recognition in [64] by integrating the ear with the 3D face and the finger. The technique uses an ICP algorithm to register images of the ear and face in 3D. It integrates all the modalities used to improve score-level fusion recognition efficiency, and reports a 97% recognition rate on 85 subjects using all three methods.

In [65], a concept of multimodal biometric signatures is introduced using the 3D ear and 3D face. The technique uses the average class image shape as [66] to construct an annotated model for each class. To create a geometric image that distinguishes the salient features, it matches the model obtained on the ear and the face. The procedure gives more weight to the rigidity of the annotated regions, relative to less rigid regions. For recognition, computed features are used to match face and ear samples, where the same subject samples are used to evaluate a fusion score. The technique reported improved performance by matching the fusion-based score as opposed to the same obtained by using separate modalities. The technique requires a sufficient number of samples from each subject for satisfactory results to accurately identify the annotated ear and face model. In [38], a combination of 2D and 3D ear images to recognize a human being is proposed. The keypoint features in the 2D image are first identified and then mapped to the co-registered 3D image. These keypoints are first used to coarsely align two 3D ears and then a fine ear alignment is obtained using a modified ICP algorithm for full 3D ear data. In [58], a technique is presented for detecting keypoints in 2D ear images based on 2D curvilinear structure. As in [38], the keypoints obtained in 2D are mapped into the co-registered 3D ear image to obtain the keypoints in it. A descriptor is further used in the technique to describe each keypoint in 3D. The technique measures the distance between the feature vectors of the probe and gallery ear images to match a pair of ears. By evaluating the distance ratio of the next closest neighbor, we determine the similarity of the match. The match is considered to be accurate if the ratio is less than or equal to the threshold in the range (0, 1).

4 Preliminaries

In this section, a brief introduction to ICP and RoPS algorithms are given. The proposed technique uses both of these algorithms in the recognition framework.

4.1 *Iterative Closest Point*

Let probe P and gallery G are finite sets with m and n points where, $P \in \mathcal{R}^{m\times 3}$ and $G \in \mathcal{R}^{n\times 3}$. For every point in set P, we calculate the nearest neighbor in set G. These points are known as correspondence points and a translation and rotational matrix is determined using these points, to minimize the error E.

$$E(\phi, t) = \sum_{i=1}^{m} \sum_{j=1}^{n} ||p_i - (\phi(g_i) + t)||^2 \tag{1}$$

where, $p_i \in P$ and $g_i \in G$. The matrices ϕ and t are computed using a covariance matrix $H = \tilde{P}\tilde{G}^T$, where $\tilde{P}$ and $\tilde{G}^T$ are the mean subtracted matrices. Further, by decomposing the matrix $H = U \Lambda V^T$, we get the rotation matrix $\phi = VU^T$ and the translation matrix $t = P - \phi(G)$. The first iteration is performed using these matrices. Subsequently, for the transformed gallery the error is computed. If the error obtained exceeds a predefined threshold, iteration continues.

4.2 Rotation of Projection Statistics

This section explains how to combine a local 3D descriptor with an ICP. Local characteristics depend on invariant local neighborhood information, with rigid translations and rotations. There are many techniques in the literature which focus on local features. We chose to combine RoPS [36] with 2D keypoint detectors and ICP. A detailed review of the 3D descriptors given in [67] clearly shows that the descriptor chosen outperforms most existing local descriptors in the literature. Through this feature description technique, a local reference frame (LRF) is measured at each keypoint p. On 2D planes of xy, yz and zx, neighbors of p within a radius r are projected for the local reference frame. For every 2D plane, the projected points are used to create a D distribution matrix. Later on, μ_{11}, μ_{21}, μ_{12}, μ_{22} and e entropy are determined from the D matrix. The moment is implemented in the following way.

$$\mu_{mn} = \sum_{i=1}^{L} \sum_{j=1}^{L} (i - \bar{i})^m (j - \bar{j})^n D(i, j) \tag{2}$$

where, $\bar{i} = \sum_{i=1}^{L} \sum_{j=1}^{L} iD(i, j)$, $\quad \bar{j} = \sum_{i=1}^{L} \sum_{j=1}^{L} jD(i, j)$

In above equations, $D(i, j)$ represent a bin in the 2D distribution matrix D. Shannon entropy e for matrix D is defined as follows:

$$\mathbf{e} = -\sum_{i=1}^{L} \sum_{j=1}^{L} D(i, j) \log D(i, j) \tag{3}$$

Computed statistical properties for xy, yz, and zx 2D planes are concatenated to form a sub-feature vector. For different rotation angles, the above procedure is repeated, and vectors are obtained from each rotation angle. Concatenation of all sub-feature vectors give p, the final feature vector.

5 Proposed Keypoint Detection and Description

This section discusses the keypoint detector and descriptor for co-registered 2D and 3D ear images. In Section 5.1, different 2D keypoint detectors are used to find the keypoints. Further, using these keypoints along with ICP, we discuss the recognition performance. Followed by in Sect. 5.2, the performance of the 3D descriptor is discussed. Outline of the proposed technique is shown in the Fig. 2.

5.1 Feature Keypoint Detection

The purpose of a feature detector is to identify which keypoints are stable, repeatable and informative in an image. We use co-registered 2D ear image to detect these keypoints in a 3D ear image. We have selected six keypoint detectors [68–73] to detect keypoints in 2D images. The chosen keypoint detectors detect keypoints based on texture variations, edges and corners. The other properties such as curvilinear structure of ear [58] can also be used as keypoints.

The coordinates of the detected keypoints of the co-registered 2D ear image are mapped to the 3D image to compute the keypoints of the 3D ear image. Figure 3 shows the keypoint detection using three [71, 72] and [68] detectors. It is clear from the figure that [71] has detected more keypoints than the other two detectors. The detected points are also uniformly distributed over the complete image of the ear. However, while applying the registration algorithm to these detected 3D points, the recognition performance of the combination of [71] and ICP has shown poor results. The other two detectors have detected keypoints in the inner structure of the ear. Although fewer keypoints have been detected as shown in Fig. 3b [72], it has shown improved recognition performance as compared to [71]. In [68], the keypoints are detected in the inner structure of the ear and these points used as keypoints in 3D, help the inner structure of the ear to properly register and therefore improved the recognition performance compared to [69–72].

Figure 4 shows keypoint detection using three other [69, 70, 73] detectors. Similar to the [71] detector, the [69] detector also showed more uniform keypoint detection. The detection performance of this detector shows a slight improvement, but is the same as [71]. Using [70], the keypoints are detected in the inner structure but result in a very small number of keypoints. Finally, in [73], the detected keypoints look similar to [68], but the recognition performance is much better than the other detectors. The keypoints detected helped the registration algorithm to achieve perfect alignment may be the reason for improved performance. Finally, it should be noted that keypoint detection in the inner ear structure has helped to improve the initial coarse alignment of the registration algorithm. This leads to the closest alignment of two 3D ear images and also improves the converging time while entire data is finely aligned. The summary of the recognition performance of the combination of detectors and ICP is shown in Table 1.

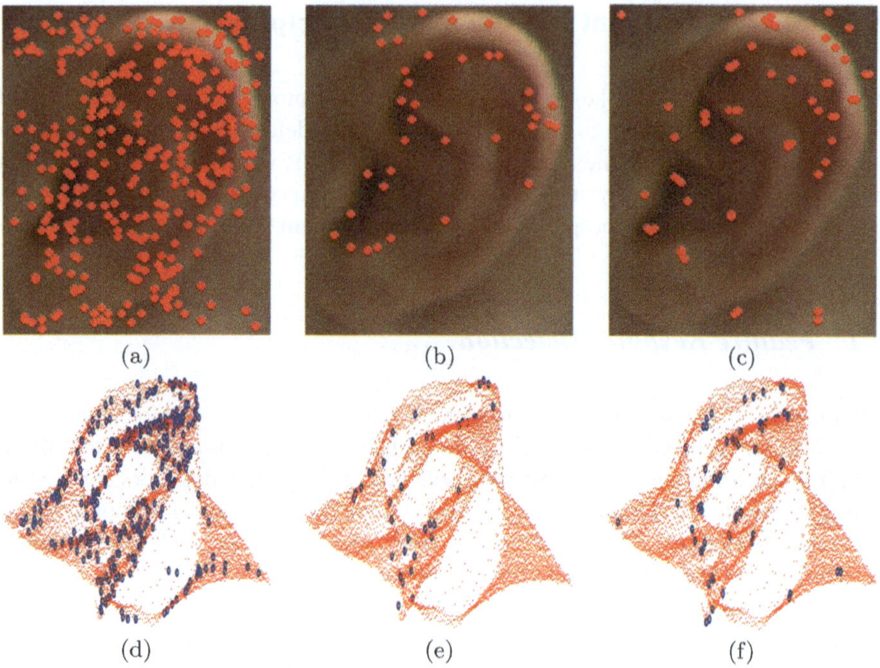

Fig. 3 Mapped keypoints onto 3D ear image found using co-registered 2D ear images using three different descriptors. **a–c** detected keypoints (highlighted in red) in 2D images using [71, 72] and [68] detectors, **d–f** mapped keypoints (highlighted in blue) onto 3D images

5.2 Feature Keypoint Description

In the previous approach, the detected 2D image keypoints are directly mapped to 3D images and used for initial alignment, where 3D information is not used for the detection of keypoints. In this approach, each mapped keypoint is described using a local 3D descriptor to include 3D information. The local descriptor uses the neighborhood information of each keypoint to describe and contributes to the use of 3D information in the selection of keypoints. The descriptor provides a vector function for each main point. We use a popular descriptor called rotational projection statistics (RoPS) to describe every keypoint to obtain a feature vector. As discussed, the close initial alignment of the data will help the registration algorithm to converge quickly, and feature vectors are used to find the corresponding keypoints to enhance this initial alignment. After the keypoints have been mapped to 3D, the local descriptor describes each keypoint using a feature vector. In the case of ear pairs, the feature vectors are compared using distance metrics to find the best keypoints from the detected keypoints in 2D. The best keypoints obtained in both ear images are used for initial alignment to align the ear pairs coarsely. This allows the fine registration process to be effective and increases the recognition efficiency. Also, while describing a key-

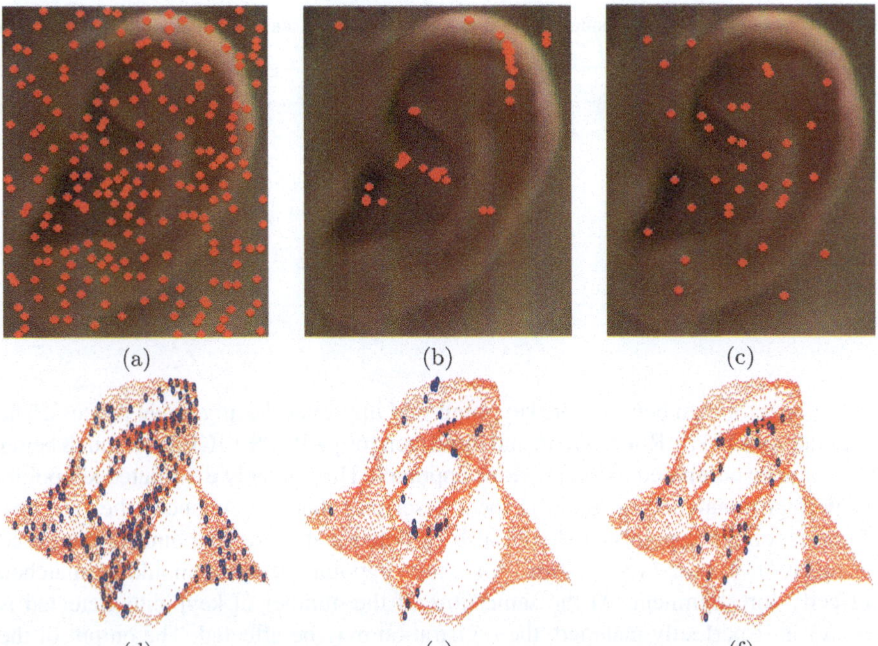

Fig. 4 Mapped keypoints onto 3D ear image found using co-registered 2D ear images using three different descriptors. **a–c** detected keypoints (highlighted in red) in 2D images using [69, 70] and [73] detectors, **d–f** mapped keypoints (highlighted in blue) onto 3D images

Table 1 Recognition performance of different keypoint detectors combined with ICP

Approach	Accuracy (%)	EER (%)
Brisk [71] + ICP	82.69	22.44
Harris [72] + ICP	84.98	18.87
Kaze [68] + ICP	89.02	12.08
minEigen [69] + ICP	83.62	17.10
MSER [70] + ICP	87.74	13.19
SURF [73] + ICP	98.68	1.56

point, the descriptor chooses a support radius to find the neighbors. The information provided by these neighbors is used to find local reference axes and also to construct a feature vector. Unlike the 2D pixel structure, the points in 3D pointcloud are not equally dense, particularly the areas not accessible to the scanner. As a consequence, a keypoint mapped from 2D to 3D may be invalid in a few cases.

The summary of the recognition performance of keypoint detector combinations, 3D descriptor and ICP is shown in Table 2. Few combinations show better performance than the previous approach. The explanation is that the descriptor typically

Table 2 Recognition performance of different keypoint detectors combined with 3D descriptor and registration

Approach	Accuracy (%)	EER (%)
Brisk [71] + RoPS + ICP	96.59	6.25
Harris [72] + RoPS + ICP	74.71	31.13
Kaze [68] + RoPS + ICP	96.87	6.12
minEigen [69] + RoPS + ICP	96.40	6.26
MSER [70] + RoPS + ICP	74.41	30.71
SURF [73] + RoPS + ICP	98.82	1.41

finds a better match between the keypoints and improves the alignment process. The combinations [71] + RoPS + ICP; and minEigen [69] + RoPS + ICP show much better performance compared to the previous approach. Here, evenly distributed keypoints are perfectly matched using a 3D local descriptor that has enhanced the selection of correspondence points. It should also be noted that more keypoints are required for registration-based recognition. Also, the keypoints detected should be matched perfectly for alignment. At the same time, if the number of keypoints detected is less, even if perfectly matched, the registration may be affected. The output of the combinations [72] + RoPS + ICP; and [70] + RoPS + ICP have degraded and the fact is that it has fewer keypoints, and after applying descriptor-based matching, there is a probability of even less keypoints matching. This leads to poor initial and final alignment and degrades the recognition performance. Finally [68] + RoPS + ICP; and [73] + RoPS + ICP showed consistency performance using both approaches, where [68] + RoPS + ICP has improved and showed comparable performance on par with [73] + RoPS + ICP.

6 Experimental Results

The proposed technique has been validated on UND-J2 database. It is one of the biggest accessible ear database with 415 subjects with 1800 samples. Minolta vivid 910 3D scanner collects the database in two sessions with a 17-week time gap between them. The images in the database are influenced by pose changes, scaling, and occlusion due to earrings and hair. After eliminating all duplicates, we used subjects that has two or more samples.

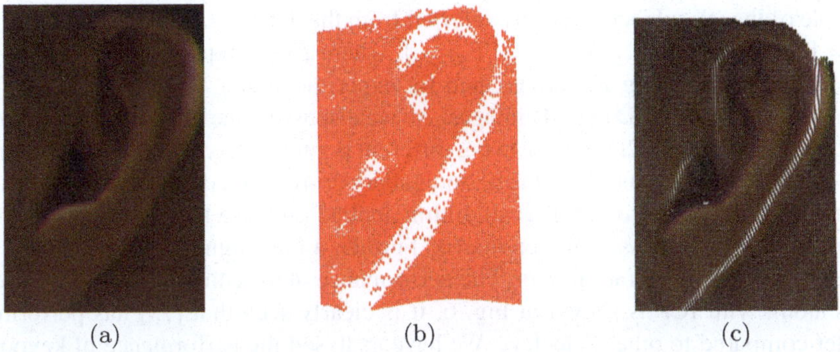

Fig. 5 A 2D-3D co-registered ear sample from UND-J2 database. **a** 2D ear image, **b** 3D ear image, **c** 2D-3D pre-processed co-registered ear image

6.1 Data Pre-processing

The co-registered images of the ear is first cropped and then pre-processed. The scale of 2D and 3D images is exactly the same, but due to the few invalid 3D points each pixel is not exactly mapped to 3D. For our experiment, we generated structured data with valid 3D points and corresponding 2D texture information.

Figure 5c shows an example of pre-processed ear along with its original 2D and 3D images. It can be easily noted that in Fig. 5a the area where the hair present (top right corner) can not be captured by the sensor and the corresponding area in 3D is reflected by the null points as shown in Fig. 5b. In fewer cases, in our experimentation, invalid keypoints mapped on 3D are carefully ignored.

6.2 Evaluation Procedure

This section presents experimental evaluations of the proposed technique. The following terms are defined to analyze the recognition performance. FAR (False Acceptance Rate) is the rate at which the recognition system accepts an unauthorized ear image and FRR (False Rejection Rate) is the rate at which the recognition system rejects an authorized ear image. EER (Equal Error Rate) is the measure of the likelihood at which the FAR and FRR are equal. ROC (Receiver Operating Characteristic) is another important measure used to evaluate the recognition performance. It plots FAR with respect to GAR (Genuine acceptance rate) (defined as 100–FRR) and shows the probability of a designed system in distinguishing the ear models. A gallery dataset $G = \{E_{11}, E_{21}, \ldots E_{i1} \ldots E_{n1}\}$ using n ear subjects, where E_{i1} is a sample of the ith subject randomly selected from the database, and the probe dataset $P = \{E_{12...q_1}, E_{22...q_2}, \ldots E_{i2...q_i} \ldots E_{n2...q_n}\}$, where $E_{i2...q_i}$ represents q_i samples of the subject i excluding E_{i1} are created for experimentation. The procedure is followed same for both 2D and 3D images. As we discussed, the first approach follows keypoint detection in 2D images using different popular detectors and mapped onto 3D images as keypoints. For a probe and gallery co-registered image, once the keypoints in 2D mapped to 3D, ICP algorithm applied on these keypoints for a coarse alignment to get a transformation matrix. Further, a fine alignment is performed on ear pairs to get a registration error. The performance of the combination of six detectors along with ICP is shown in Fig. 6. It is clearly seen that [73] has performed better compared to other detectors. We have analysed the performance of keypoint detector in the presence of noise. Analyses have been performed by adding Gaussian noise to the 3D ear images to verify how much it affects the recognition. For varying sigma, ($\sigma = 0.1, 0.3, 0.5$) we have added the noise to 3D images and verified the recognition using precision and recall curves. For each noise level, we obtain the PR curves for all six chosen keypoint detectors. The recognition performance varying sigma (0.1, 0.3 and 0.5) is shown in Fig. 7. It is clearly seen that the detector [73] which performed well without noise has degraded in the presence of noise. The other two detectors [71] and [68], has shown better performance in the presence of noise. Further, we experimented with the second approach, where the keypoints on 2D are first detected and then mapped to 3D. Followed by, feature descriptor of each keypoints are computed and stored. Consider, for example, a probe image with M keypoints and a gallery image with N keypoints, we get a matrix of size $M \times N$. The matrix is constructed by finding the distances between feature vectors, where each row is computed by calculating the distances between a keypoint in probe to all the keypoints in gallery. Likewise, M rows calculated for M keypoints in probe. The distances are sorted and a ratio of first smallest to the second smallest is found. If the ratio is greater than a pre-defined threshold, the two keypoints are considered as matching points. After the probe image has all these keypoints paired with the gallery image keypoints, a coarse alignment is performed between the ear pairs to obtain a transformation matrix. In addition, the ICP algorithm [37] is used to find the best alignment for the entire gallery and probe ear image. The error of registration is used as a matching score, indicating a lower error that matches better. The technique proposed is validated on UND-Collection J2 data and has obtained a encouraging recognition. Figure 8 shows the ROC curve for our proposed technique on UND-Collection J2 dataset. The performance of [70, 72] has significantly reduced, while the other four have shown better results when combined with descriptors.

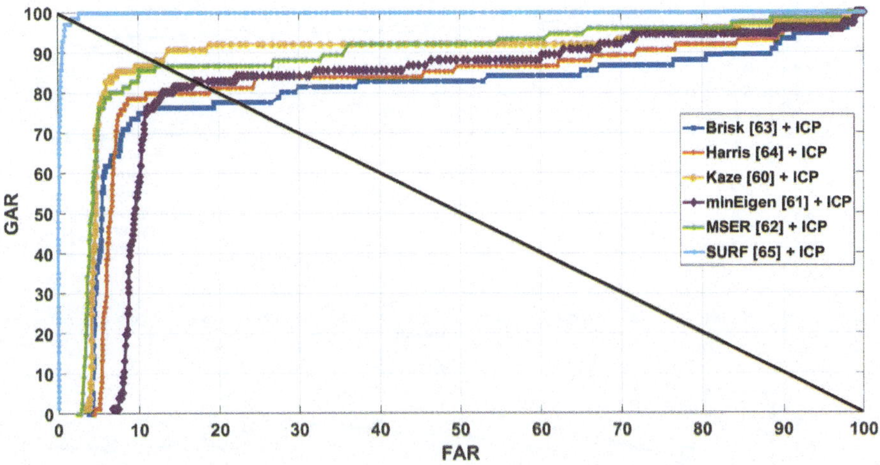

Fig. 6 FAR vs GAR for the proposed combinations of six keypoint detectors with ICP

6.3 Computational Efficiency

All experiments are performed using an Intel(R) Xeon(R) CPU E5-1620 @ 3.50 GHz CPU and 32 GB RAM with Windows 10 OS. Using the proposed method, we calculated the average time taken to fit two 3D ear samples to evaluate computational efficiency. It takes an average of 0.40 s with the first approach and 0.30 seconds with second approach to match a gallery image with a probe image.

7 Conclusion

This work presents a few methods of recognition of 3D Ear images with their co-registered 2D images. A detailed analysis is performed using several keypoint detectors combined with an ICP and a 3D descriptor. However, direct mapping of keypoints from 2D to 3D, along with ICP, provides better performance, while using a descriptor combination to find better matching keypoints has shown encouraging recognition performance. Inclusion of the descriptor in finding correspondence points allows use of 2D and 3D information, and we have achieved enhanced performance. The first step is to locate the keypoints in the 3D ear images of the probe and gallery using the proposed technique in order to find very close match points between the ear pairs. These points are aligned to find a transformation matrix. Fine alignment is done in the second stage with complete ear pairs, where the error of registration is used as matching score. The key contribution of the paper is to propose the detection and description technique for 3D data with 2D images and 3D image information. The proposed biometric trait can be supplemented with other biometric traits for secure

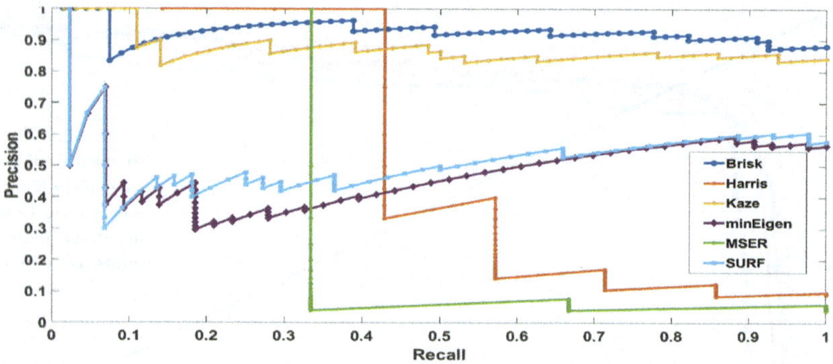

(a)

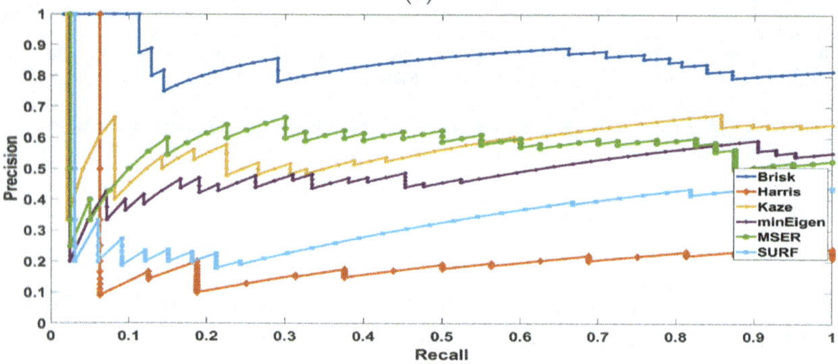

(b)

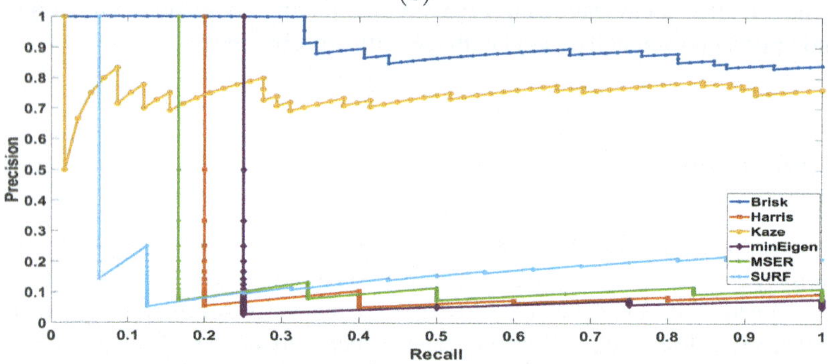

(c)

Fig. 7 Noise analysis of the proposed technique for varying Gaussian noise. **a** $\sigma = 0.1$, **b** $\sigma = 0.3$, **c** $\sigma = 0.5$

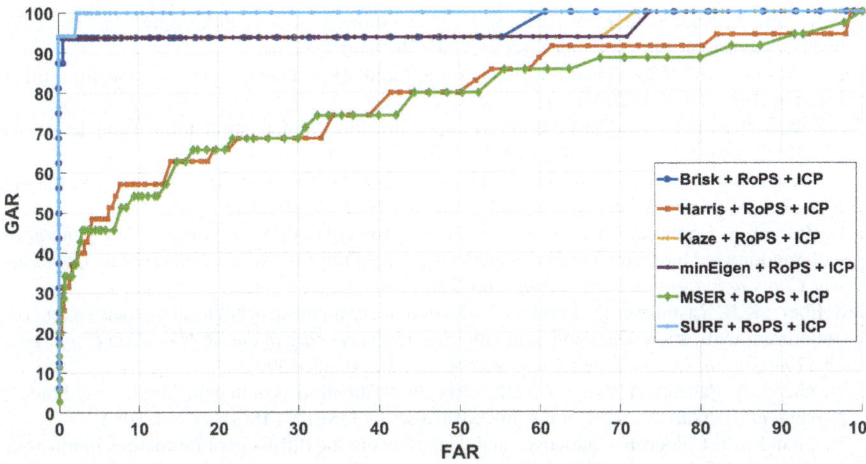

Fig. 8 FAR versus GAR for the proposed combinations of six keypoint detectors with a 3D descriptor and ICP

access and it can also be cascaded for two-step authentication process. Even though the proposed approach has only been tested on 3D ear data, it can be used in many other applications to match 2D-3D co-registered data.

Acknowledgements This research is supported by *SB/FTP/ETA* – 0074/2014 from the Department of Science and Technology, Government of India, from the Science and Engineering Research Board (SERB). Ph.D. Fellowship of Iyyakutti Iyappan Ganapathi is funded by Digital India Corporation Visvesvaraya Ph.D. Scheme of the Ministry of Electronics and Information Technology, Government of India.

Compliance with Ethical Standards

Conflict of Interest Aradhya Saini declares she has no conflict of interest. Sandeep Daniel declares he has no conflict of interest. Satyam Saini declares he has no conflict of interest. Ankush Mittal declares he has no conflict of interest.

References

1. S.S. Ali, I.I. Ganapathi, S. Prakash, P. Consul, S. Mahyo, Securing biometric user template using modified minutiae attributes. Pattern Recogn. Lett. **129**, 263–270 (2020)
2. S.S. Ali, I.I. Ganapathi, S. Mahyo, S. Prakash, *Polynomial vault: a secure and robust fingerprint based authentication* (IEEE Trans. Emer, Topics Comput, 2019)
3. S.S. Ali, I.I. Ganapathi, S. Prakash, Robust technique for fingerprint template protection. IET Biometr. **7**(6), 536–549 (2018)
4. R.D. Labati, A. Genovese, E. Muñoz, V. Piuri, F. Scotti, G. Sforza, Biometric recognition in automated border control: a survey. ACM Comput. Surv. (CSUR) **49**(2), 1–39 (2016)
5. A. Jain, L. Hong, S. Pankanti, Biometric identification. Commun. ACM **43**(2), 90–98 (2000)

6. A.K. Jain, K. Nandakumar, A. Ross, 50 years of biometric research: Accomplishments, challenges, and opportunities. Pattern Recog. Lett. **79**, 80–105 (2016)
7. R.V. Yampolskiy, V. Govindaraju, Behavioural biometrics: a survey and classification. Int. J. Biometr. **1**(1), 81–113 (2008)
8. S.P. Banerjee, D.L. Woodard, Biometric authentication and identification using keystroke dynamics: a survey. J. Pattern Recogn. Res. **7**(1), 116–139 (2012)
9. A. Serwadda, V.V. Phoha, Examining a large keystroke biometrics dataset for statistical-attack openings. ACM Trans. Inf. Syst. Secur. (TISSEC) **16**(2), 1–30 (2013)
10. L. Ballard, S. Kamara, F. Monrose, M.K. Reiter, Towards practical biometric key generation with randomized biometric templates, in *Proceeding of the 15th ACM Conference on Computer and Communications Security* (2008), pp. 235–244
11. S. Eberz, K.B. Rasmussen, V. Lenders, I. Martinovic, Evaluating behavioral biometrics for continuous authentication: Challenges and metrics, in *Proceeding of the ACM on Asia Conference on Computer and Communications Security* (2017), pp. 386–399
12. N. Zheng, A. Paloski, H. Wang, An efficient user verification system using angle-based mouse movement biometrics. ACM Trans. Inf. Syst. Secur. (TISSEC) **18**(3), 1–27 (2016)
13. A. Chandra, T. Calderon, Challenges and constraints to the diffusion of biometrics in information systems. Commun. ACM **48**(12), 101–106 (2005)
14. S. Eberz, K.B. Rasmussen, V. Lenders, I. Martinovic, Looks like eve: exposing insider threats using eye movement biometrics. ACM Trans Privacy Secur. (TOPS) **19**(1), 1–31 (2016)
15. O. Hamdy, I. Traoré, Homogeneous physio-behavioral visual and mouse-based biometric. ACM Trans. Comput-Human Inter. (TOCHI) **18**(3), 1–30 (2011)
16. G. Jaswal, A. Kaul, R. Nath, Knuckle print biometrics and fusion schemes-overview, challenges, and solutions. ACM Comput. Surveys (CSUR) **49**(2), 1–46 (2016)
17. J.A. Markowitz, Voice biometrics. Commun. ACM **43**(9), 66–73 (2000)
18. A.K. Jain, P. Flynn, A.A. Ross, *Handbook of biometrics*. Springer Science and Business Media (2007)
19. S. Prakash, P. Gupta, *Ear Biometrics in 2D and 3D: Localization and Recognition*, vol. 10 (Springer, 2015)
20. I.I. Ganapathi, S. Prakash, 3d ear based human recognition using gauss map clustering, in *Proceedings of the 10th Annual ACM India Compute Conference* (2017), pp. 83–89
21. I.R. Dave, I.I. Ganapathi, S. Prakash, S.S. Ali, A.M. Srivastava, 3d ear biometrics: acquisition and recognition, in *15th IEEE India Council International Conference (INDICON)* (IEEE, 2018), pp. 1–6
22. L. Yuan, Z. Chun Mu, Ear recognition based on local information fusion. Pattern Recogn. Lett. **33**(2), 182–190 (2012)
23. L. Nanni, A. Lumini, A multi-matcher for ear authentication. Pattern Recognit. Lett. **28**(16), 2219–2226 (2007)
24. Y. Liu, B. Zhang, D. Zhang, Ear-parotic face angle: a unique feature for 3d ear recognition. Pattern Recognit. Lett. **53**, 9–15 (2015)
25. D.J. Hurley, B. Arbab-Zavar, M.S. Nixon, The ear as a biometric, in *Handbook of biometrics* (2008), pp. 131–150
26. A. Abaza, A. Ross, C. Hebert, M.A.F. Harrison, M.S. Nixon, A survey on ear biometrics. ACM Comput. Surv. (CSUR) **45**(2), 22 (2013)
27. Ž. Emeršič, V. Štruc, P. Peer, Ear recognition: more than a survey. Neurocomputing **255**, 26–39 (2017)
28. D.B. Gore, Comparative study on feature extractions for ear recognition: comparative study on feature extractions for ear recognition. Int. J. Appl. Evol. Comput. **10**(2), 8–18 (2019)
29. S.M.S. Islam, M. Bennamoun, R. Owens, R. Davies, Biometric approaches of 2D-3D ear and face: a survey, in *Advances in Computer and Information Sciences and Engineering*, ed. by T. Sobh (2008), pp. 509–514
30. C. Middendorff, K.W. Bowyer, P. Yan, Multi-modal biometrics involving the human ear, in *Proceeding of IEEE Conference on Computer Vision and Pattern Recognition (CVPR 2007)* (2007) pp. 1–2

31. K. Pun Y. Moon, Recent advances in ear biometrics, in *Proceeding of IEEE International Conference on Automatic Face and Gesture Recognition (FG 2004)* (2004), pp. 164–169
32. A. Pflug, C. Busch, Ear biometrics: a survey of detection, feature extraction and recognition methods. IET Biometr. **1**(2), 114–129 (2012)
33. D. Singh S. Singh, A survey on human ear recognition system based on 2D and 3D ear images. Open J. Inf. Secur. Appl. **2014**, 21–30 (2014)
34. L. Yuan, Z.-C. Mu, F. Yang, A review of recent advances in ear recognition, in *Proceeding of Chinese Conference on Biometric Recognition (CCBR 2011)*, pp. 252–259
35. P. Srivastava, D. Agrawal, A. Bansal, Ear detection and recognition techniques: a comparative review. Adv. Data Inf. Sci. 533–543 (2020)
36. Y. Guo, F. Sohel, M. Bennamoun, M. Lu, J. Wan, Rotational projection statistics for 3d local surface description and object recognition. Int. J. Comput. Vis. **105**(1), 63–86 (2013)
37. P.J. Besl, N.D. McKay, Method for registration of 3-d shapes. Sensor Fusion IV: control Paradigms and Data Structures **1611**, 586–607 (1992)
38. S. Prakash, P. Gupta, Human recognition using 3D ear images. Neurocomputing **140**, 317–325 (2014)
39. H. Chen, B. Bhanu, Efficient recognition of highly similar 3D objects in range images. IEEE Trans. Pattern Anal. Mach. Intell. **31**(1), 172–179 (2009)
40. H. Chen, B. Bhanu, 3D free-form object recognition in range images using local surface patches. Pattern Recogn. Lett. **28**(10), 1252–1262 (2007)
41. H. Chen, B. Bhanu, Human ear recognition in 3D. IEEE Trans. Pattern Anal. Mach. Intell. **29**(4), 718–737 (2007)
42. H. Chen, B. Bhanu, Contour matching for 3D ear recognition, in *Proceeding of Seventh IEEE Workshops on Applications of Computer Vision (WACV/MOTION 2005)*, vol. 1. (2005), pp. 123–128
43. G. Passalis, I.A. Kakadiaris, T. Theoharis, G. Toderici, T. Papaioannou, Towards fast 3d ear recognition for real-life biometric applications, in *2007 IEEE Conference on Advanced Video and Signal Based Surveillance* (IEEE, 2007), pp. 39–44
44. S. Cadavid, M. Abdel-Mottaleb, 3-d ear modeling and recognition from video sequences using shape from shading. IEEE Trans. Inf. Forensics Secur. **3**(4), 709–718 (2008)
45. Z. Ding, L. Zhang, H. Li, A novel 3d ear identification approach based on sparse representation, in *2013 IEEE International Conference on Image Processing* (IEEE, 2013), pp. 4166–4170
46. J. Zhou, S. Cadavid, M. Abdel-Mottaleb, A computationally efficient approach to 3D ear recognition employing local and holistic features, in *Proceeding of IEEE Computer Society Conference on Computer Vision and Pattern Recognition Workshop (CVPRW 2011)* (2011), pp. 98–105
47. S.M. Islam, M. Bennamoun, R. Davies, Fast and fully automatic ear detection using cascaded adaboost, in *Proceeding of IEEE Workshop on Applications of Computer Vision (WACV 2008)* (2008), pp. 1–6
48. H. Zeng, J.-Y. Dong, Z.-C. Mu, Y. Guo, Ear recognition based on 3d keypoint matching, in IEEE 10th International Conference on Signal Processing Proceedings (IEEE, 2010), pp. 1694–1697
49. I.I. Ganapathi, S. Prakash, 3D ear recognition using global and local features. IET Biometr. **7**(3), 232–241 (2018)
50. I.I. Ganapathi, S.S. Ali, S. Prakash, Geometric statistics-based descriptor for 3D ear recognition. Vis. Comput. pp. 1–13 (2018)
51. X. Sun, G. Wang, L. Wang, H. Sun, X. Wei, 3D ear recognition using local salience and principal manifold. Graph. Mod. **76**(5), 402–412 (2014)
52. X. Dong, Y. Guo et al., 3d ear recognition using sift keypoint matching. Energy Procedia **11**, 1103–1109 (2011)
53. D.G. Lowe, Object recognition from local scale-invariant features, in *Proceedings of the Seventh IEEE International Conference on Computer Vision*, vol. 2 (IEEE, 1999), pp. 1150–1157
54. X.-P. Sun, S.-H. Li, F. Han, X.-P. Wei, 3D ear shape matching using joint α-entropy. J. Comput. Sci. Technol. **30**(3), 565–577 (2015)

55. S.M. Islam, R. Davies, M. Bennamoun, R.A. Owens, A.S. Mian, Multibiometric human recognition using 3D ear and face features. Pattern Recogn. **46**(3), 613–627 (2013)
56. S.M. Islam, M. Bennamoun, A.S. Mian, R. Davies, Score level fusion of ear and face local 3D features for fast and expression-invariant human recognition, in *Proceeding of International Conference Image Analysis and Recognition (ICIAR 2009)* (2009), pp. 387–396
57. J.D. Bustard, M.S. Nixon, 3D morphable model construction for robust ear and face recognition, in *Proceeding of IEEE Computer Society Conference on Computer Vision and Pattern Recognition (CVPR 2010)* (2010), pp. 2582–2589
58. I.I. Ganapathi, S. Prakash, I.R. Dave, P. Joshi, S.S. Ali, A.M. Shrivastava, Ear recognition in 3D using 2D curvilinear features. IET Biometr. **7**(6), 519–529 (2018)
59. P. Yan, K.W. Bowyer, Biometric recognition using 3D ear shape. IEEE Trans. Pattern Analys. Mach. Intell. **29**(8), 1297–1308 (2007)
60. S. Islam, M. Bennamoun, A. Mian, R. Davies, A fully automatic approach for human recognition from profile images using 2D and 3D ear data, in *Proceeding of International Symposium on 3D Data Processing Visualization and Transmission (3DPVT 2008)* (2008)
61. P. Yan, K.W. Bowyer, Multi-biometrics 2D and 3D ear recognition, in *Proceeding of International Conference on Audio-and Video-Based Biometric Person Authentication (AVBPA 2005)* (2005), pp. 503–512
62. P. Yan, K.W. Bowyer, Ear biometrics using 2D and 3D images, in *Proceeding of IEEE Conference on Computer Vision and Pattern Recognition (CVPR 2005)-Workshops* (2005), pp. 121–121
63. P. Yan, K. Bowyer, Empirical evaluation of advanced ear biometrics, in *Proceeding of IEEE Conference on Computer Vision and Pattern Recognition (CVPR 2005)-Workshops* (2005), pp. 41–41
64. D.L. Woodard, T.C. Faltemier, P. Yan, P.J. Flynn, K.W. Bowyer, A comparison of 3D biometric modalities, in *Proceeding of Conference on Computer Vision and Pattern Recognition Workshop (CVPRW 2006)* (2006), pp. 57–57
65. T. Theoharis, G. Passalis, G. Toderici, I.A. Kakadiaris, Unified 3D face and ear recognition using wavelets on geometry images. Pattern Recogn. **41**(3), 796–804 (2008)
66. G. Passalis, I. A. Kakadiaris, T. Theoharis, G. Toderici, T. Papaioannou, Towards fast 3D ear recognition for real-life biometric applications, in *Proceeding of IEEE Conference on Advanced Video and Signal Based Surveillance (AVSS 2007)* (2007), pp. 39–44
67. Y. Guo, M. Bennamoun, F. Sohel, M. Lu, J. Wan, 3d object recognition in cluttered scenes with local surface features: a survey. IEEE Trans. Pattern Anal. Mach. Intell. **36**(11), 2270–2287 (2014)
68. P.F. Alcantarilla, A. Bartoli, A.J. Davison, Kaze features, in *European Conference on Computer Vision* (Springer, 2012), pp. 214–227
69. J. Shi, et al., Good features to track, in *Proceedings of IEEE Conference on Computer Vision and Pattern Recognition*. IEEE **1994**, 593–600 (1994)
70. D. Nistér, H. Stewénius, Linear time maximally stable extremal regions, in *European Conference on Computer Vision*. (Springer, 2008) pp. 183–196
71. S. Leutenegger, M. Chli, R.Y. Siegwart, Brisk: binary robust invariant scalable keypoints. *International Conference on Computer Vision*. (IEEE, 2011), 2548–2555
72. C.G. Harris, M. Stephens et al., A combined corner and edge detector, In *Alvey Vision Conference*, vol. 15, no. 50. (Citeseer, 1988), pp. 10–5244
73. H. Bay, T. Tuytelaars, L. Van Gool, Surf: Speeded up robust features, in *European Conference on Computer Vision* (Springer, 2006), pp. 404–417

Efficient and Low Overhead Detection of Brain Diseases Using Deep Learning-Based Sparse MRI Image Classification

Avrajit Ghosh, Arnab Raha, and Amitava Mukherjee

Abstract Transfer Learning-based Deep Learning is the most popular technique adopted in biomedical image classification. In the first step of this approach, a pre-trained convolutional neural network (CNN) is used as a feature extractor. The second step involves fine-tuning the hyperparameters, namely learning rate, batch-size, and epoch-size. However, due to large number of input images and large image size, the CNN demands high computational cost and huge training time. This chapter on sparse image classification adds a compression layer before the input images are fed to the classifier. The compression layer sparsifies the images in the transform domain. In the cases of biomedical images, the wavelet domain provides a sparse basis for representation. Thus, images are compressed by a significant factor known as the compression ratio (CR) without significant loss of image quality determined by Percentage Root Deviation (PRD) according to the theory of approximate computing. In this paper, classification of images belonging to four different brain disease, *viz.*, Endema, Necrosis, Enhancing Tumor, Non-Enhancing Tumor, has been performed by sparse image classification of MRI images.

Keywords CNN · Brain disease · Diagnosis

A. Ghosh (✉)
Department of Electronics and Telecommunication Engineering,
Jadavpur University, Kolkata 700032, India
e-mail: avrajitghosh1001@gmail.com

A. Raha
Intel Corporation, Santa Clara, USA

A. Mukherjee
Center of Bio-computing, School of Engineering and Technology,
Adamas University, Kolkata, India

P. Kumar et al. (eds.), *Machine Learning for Intelligent Multimedia Analytics*,
Studies in Big Data 82,
https://doi.org/10.1007/978-981-15-9492-2_2

1 Introduction

During the training phase of a convolutional neural network (CNN), the maximum time is spent in the backpropagation method using the chain rule. The time required to update the weights and biases in this fully connected layer (FCL) depends on the number of neurons in the hidden layers and also the size of the input image. Note that this is the computational bottleneck of the whole training process as the FCLs contribute to 90% of all the parameters. The number of derivatives needed to calculate is equal to the number of weights and biases in the FCL. If the first convolution layer had 64 kernels, each kernel of dimension $(3 * 3 * 3)$ and each input image is $(512 * 512)$, and then, the total number of weight updatations to be performed is $64 * 512 * 512 * 3 * 3 * 3$ which is of the order of magnitude of 10^6.

As the convolution layer goes deep into the network, more number of features are added, so the last layer will have weights of the order of 10^9. To solve the issue of tackling with so much data and limited computational resources, we have utilized two methods :

1. Training a network from scratch and updating all the weights and biases would require enormous amount of computer memory and computational power which is not feasible when deployed using low-cost and low-power COTS platforms such as Raspberry Pi and Arduino platforms that can be easily made available to doctors for clinical analysis. So the method of transfer learning is implemented.
2. Reducing the dimension of the set of input images without degrading the quality of images beyond a level of perception. Lossless sparsifying transforms are known to reduce the size of image by a significant factor (CR=4 for our work) without any significant loss in information. This decrease in the dimension of the set of input images leads to reduction of the number of parameters in the fully connected layer resulting in fast training of the network.

Utilizing these two properties, it is possible to train the network and test the class of images within a brief amount of time that is essential for clinical purposes.

2 Classification Pipeline

Our proposed classification pipeline introduces a compression layer before feeding the images to the CNN. The original sizes of the images were $(512 * 512)$. After decomposing the 2D images by discrete wavelet transform (DWT), we obtain several versions of images, namely: LL (Approximate image), HH (Diagonal details of the image), LH (Vertical details of the image), HL (Horizontal details of the image) [1]. For the purpose of classification, only the LL image is taken which is of size $(256 * 256)$. So, for our implementation, we have a fixed value of compression ratio of $CR = 4$. Most of the information of the original image is preserved in the compressed image, but the size of the image is decreased by a factor of 4. Consequently, the

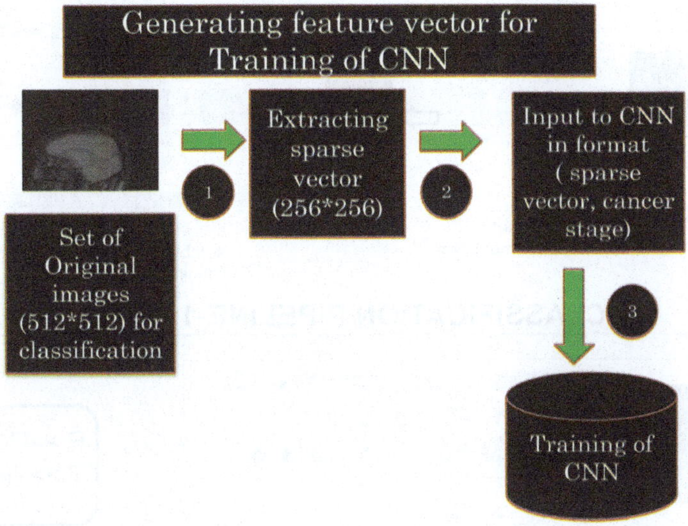

Fig. 1 Sparse image classification pipeline

number of parameters required for training the CNN is reduced drastically. This results in a significant decrease of training time and testing time. The compressed approximate images, that are obtained after applying DWT, are now given as an input to CNN (Fig. 1).

3 Contribution

The main contribution of this work lies in the reduction of the number of parameters and training time accuracy. Figure 2 shows the two classification pipelines. Classification pipeline-1 uses prior wavelet compression that reduces the size of the image by a factor of 4, if the first level approximate image is fed to the classifier. If the size of the input image is I, then after compression, the size is reduced to $\frac{I}{4}$. So, in Classification pipeline-1, the size of the input image is $\frac{I}{4}$, whereas in Classification pipeline-2, it is I. It is to be noted that approximate images can also be obtained from higher level decomposition. In this case, the size of image will decrease by a factor of 4 per decomposition level, but the quality of the image will degrade. This may lead to improved training time but at the cost of training accuracy. To observe the decrease in the number of parameters, we define the following terms:
In the convolutional layer:
Let, O_1 = Size (width) of output image
I = Size (width) of input image
K = Size (width) of kernels used in the Conv Layer

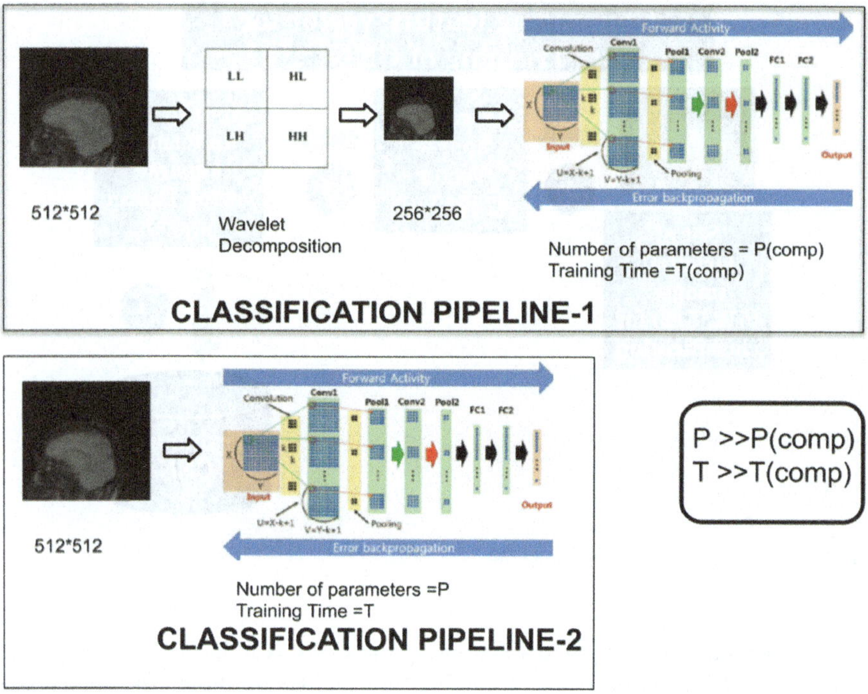

Fig. 2 Comparison of two classification pipelines (1) With prior DWT compression (2) Without prior compression

N = Number of kernels
S = Stride of the convolution operation
P = Padding
Then, the size of the output of the convolutional layer O_1 is given by

$$O_1 = \frac{I - K + 2P}{S} \tag{1}$$

The output of the convolutional layer will be a volume of size $O_1 \times N$. This volume will now be the input to the max pooling layer. In the max pooling layer
Let, O_2 = Size of the output image
S = Stride of the pooling
O_1 = Size of the input image
P_S = Pool Size

$$O_2 = \frac{O_1 - P_S}{S} \tag{2}$$

In the first layer of the fully connected layer, let W_1 = Number of weights of a FC Layer which is connected to a Conv Layer
B_1 = Number of biases of a FC Layer which is connected to a Conv Layer
O = Size (width) of the output image of the previous Conv Layer
N = Number of kernels in the previous Conv Layer
F = Number of neurons in the FC Layer

Then, $W_1 = O^2 \times N \times F$ and $B_1 = F$. The subsequent FCLs that are not connected to the convolutional-max pool layers have their number of weights and biases independent of the size of the input image.
So, it is evident that in Classification Pipeline-1, the output size of the max pooling layer is linearly proportionate to the size of the input layer. The corresponding convolutional layers have the input image size as the output size of the previous convolutional layer. Even the numbers of weights and biases are dependent on the size of the input image. So, definitely, $P(\text{comp})$ is significantly less than P. Thus, as training time is also a monotonically increasing function of the parameter size, $T(\text{comp})$ is less than T.

4 Image Compression

4.1 Discrete Wavelet Transform

To define orthogonal wavelet function, we define two basis functions known as the scaling function ($\psi_{j,k}(t)$) and the wavelet function ($\phi_{j,k}(t)$). Both the functions are defined by two parameters: translation and scaling parameter [1]. Different scaled versions of the original function represent different frequency resolutions.

$$\psi_{j,k}(t) = 2^{\frac{j}{2}} \psi(2^j t - k) \tag{3}$$

$$\phi_{j,k}(t) = 2^{\frac{j}{2}} \phi(2^j t - k) \tag{4}$$

In Eqs. 3 and 4, the parameter j denotes the dilation and k is the parameter that denotes the position of the oscillating function.
It is possible to decompose any function in L^2(R) using the scaling and the wavelet functions.

$$f[n] = \frac{1}{\sqrt{M}} \sum_k W_\psi[j_0, k]\phi_{j_0,k}[n] + \frac{1}{\sqrt{M}} \sum_{j=j_0}^{\infty} \sum_k W_\psi[j, k]\psi_{j,k}[n] \tag{5}$$

Equation 5 expresses a discrete signal as a linear combination of the two orthogonal bases, i.e., scaled version of the all the scaling functions and the scaled versions of all the wavelet functions. This is possible because all the versions of the scaling functions and the wavelet functions are orthogonal to each other [2]. So, the set of all the scaling and wavelet functions acts as a complete basis set to represent a discrete signal [3].

$$W^i_\psi[j_0, k] = \frac{1}{\sqrt{M}} \sum_n f[n]\psi_{j_0,k}[n] \quad (6)$$

$$W_\phi[j_0, k] = \frac{1}{\sqrt{M}} \sum_n f[n]\phi_{j_0,k}[n] \quad (7)$$

The coefficients of the scaling and the wavelet functions in Eq. 5 could easily be found out by taking the inner product of the function with each of the scaling and wavelet functions shown in Eqs. 6 and 7. This is the procedure of expressing a 1D discrete signal in terms of wavelet and scaling functions.

4.2 *Image Compression Using Wavelets*

To extend the representation to 2D or an image, the scaling and wavelet functions are two variable functions $\psi(x, y)$ and $\phi(x, y)$. In the case of an image, the translation is determined by two parameters (m, n), whereas the dilation parameter j remains the same as 1D case.

Hence, the scaling and wavelet functions in 2D are represented as :

$$\psi_{j,m,n}(x, y) = 2^{\frac{j}{2}}\psi(2^j x - m, 2^j y - n) \quad (8)$$

$$\phi_{j,m,n}(x, y) = 2^{\frac{j}{2}}\phi(2^j x - m, 2^j y - n) \quad (9)$$

The scaling function in 2D is obtained by passing low-pass filters through both the rows and columns. So, $\phi(x, y)$ can be written as a product of two separable scaling functions. By applying the scaling functions (low-pass filter) and the wavelet functions (high-pass filter) alternately on rows and columns, we get three 2D wavelet functions [4]. Each of the three outputs of the wavelet functions enhances a particular direction of the image: Horizontal, Vertical and Diagonal.

$$\phi(x, y) = \phi(x)\phi(y) \quad (10)$$

$$\psi^H(x, y) = \psi(x)\phi(y) \quad (11)$$

$$\psi^V(x, y) = \phi(x)\psi(y) \quad (12)$$

$$\psi^D(x, y) = \psi(x)\psi(y) \tag{13}$$

The coefficients of the 2D scaling and wavelet functions can be obtained by taking the inner product of the image with the corresponding basis functions (Fig. 3).

$$W_\phi(j_0, m, n) = \frac{1}{\sqrt{MN}} \sum_{x=0}^{M-1} \sum_{y=0}^{N-1} f(x, y)\phi_{j_0,m,n}(x, y) \tag{14}$$

$$W^i_\psi(j, m, n) = \frac{1}{\sqrt{MN}} \sum_{x=0}^{M-1} \sum_{y=0}^{N-1} f(x, y)\phi_{j_0,m,n}(x, y), i = H, V, D \tag{15}$$

After having obtained the coefficients, we can express the 2D image as a linear combination of the basis functions.

$$\begin{aligned} f(x, y) = & \frac{1}{\sqrt{MN}} \sum_m \sum_n W_\phi(j_0, m, n)\psi_{j_0,m,n}(x, y) \\ & + \frac{1}{\sqrt{MN}} \sum_{i=H,V,D} \sum_{j=j_0}^{\infty} \sum_m \sum_n W^i_\psi(j, m, n)\phi_{j,m,n}(x, y) \end{aligned} \tag{16}$$

4.3 Choosing Mother Wavelets

The compression factor of the image depends on the shape of then mother wavelet. If the shape of the signal or the image resembles closely with the mother wavelet, then for that signal, image can have a sparse representation in that wavelet basis. To get an estimate about the similarity of the shape, we perform correlation of the image with all the shifted and dilated versions of the mother wavelet [5] That mother wavelet which gives the maximum correlation would be the ideal basis. In this case, the desired mother wavelet is Coiflets-3 that is given in Table 1 (Fig. 4).

5 Image Classification Using Deep Learning

We have used transfer learning and trained the set of biomedical images in a pre-trained CNN. Transfer learning is a machine learning method, where a model developed for a task is reused as the starting point for a model on a second task. A pre-trained model in transfer learning includes three parts:

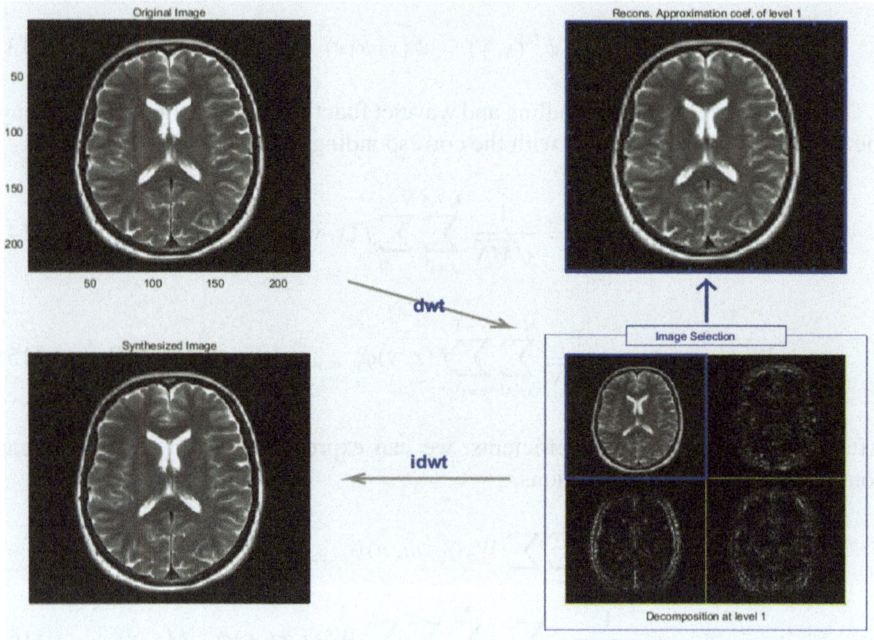

Fig. 3 Wavelet decomposition of the brain image returns LL, LH, HL and HH images. Approximate image (LL) is used as the input image to the CNN

Table 1 Maximum correlation value for mother wavelets

Mother wavelet	Correlation value
Haar	0.69
Daubachies	0.79
Coiflets-3	0.85
Biorthognal	0.79
Mexican hat	0.72
Morlet wavelet	0.77

1. Selecting the source model—A pre-trained model is to be selected which has been trained on numerous classes before. For our case, we chose GoogleNet, the winner of ImageNet Large Scale Visual Recognition Competition (ILSVRC) which is developed by Google.
2. Reuse Model—The pre-trained model can then be used as the starting point for a model on the second task of interest. This may involve using all or parts of the model, depending on the modeling technique used. For our purpose, we have changed the necessary FCLs to update the weights and bias by applying back-propagation in those selected layers only.

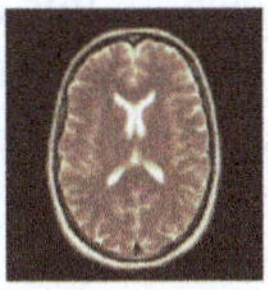

(a) Biorthogonal wavelet

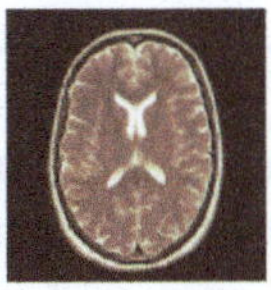

(b) Reverse Biorthogonal wavelet

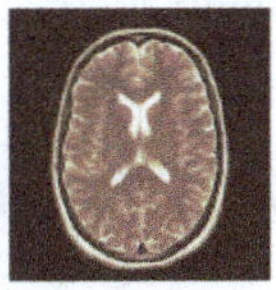

(c) Haar wavelet

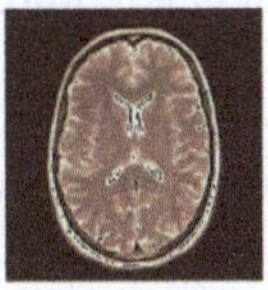

(d) Daubachies-4 wavelet

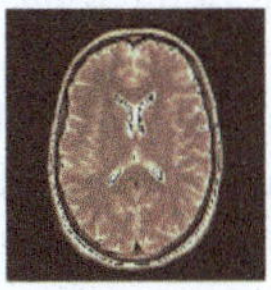

(e) Symlets wavelets

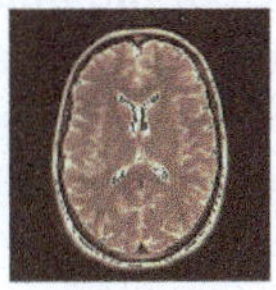

(f) Coiflets-3 wavelets

Fig. 4 Approximate images for DWT compression using different mother wavelets

3. Tune Model—The most important phase is the tuning of the hyperparameters concerned. We chose a set of optimum hyperparameters by trial and error method to get the maximum training accuracy.

Figures 5, 6, 7, 8, 9, 10, 11 and 12 are example images that have been included to describe the general working principle of a convolutional neural network. In our implementation of transfer learning, the sub-procedures are the same.

5.1 Structure of CNN

The CNN is made up of sequence of layers, and each layer transforms an input 3D volume to an output 3D volume with some functions that may or may not have parameters. We use three main types of layers to build this architecture: convolutional layer, pooling layer and fully connected layer (Fig. 5). We stack this sub-layers in order to form a full CNN [6]. A general convolutional net has the following layers-

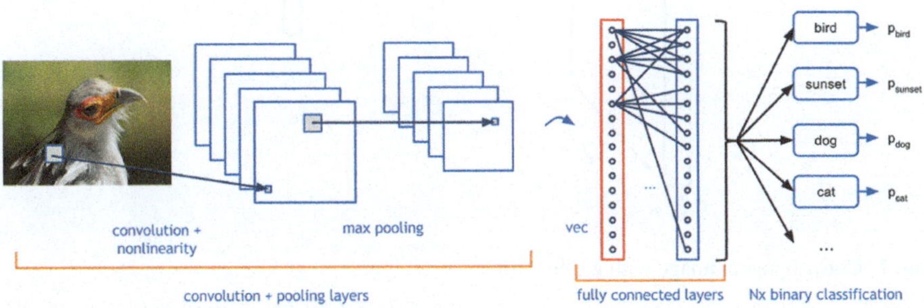

Fig. 5 Architecture of convolutional neural network

1. Input Layer—This layer holds the raw pixel value of the images. In this case, it is $256 * 256 * 3$ as GoogleNet takes an RGB image of this size as its input. So, we convert our grayscale image into RGB image by adding extra two dimensions to it. These added dimensions have the same pixel intensity distribution as the original grayscale image.
2. Convolutional Layer—This layer contains filter kernels that extract specific features of the image. After the filters have passed over the image, a feature map is generated for each filter.
3. Nonlinearity Layer (ReLU)—It applies an element-wise activation function such as the max(0, x) thresholding function. All the negative pixel values are forced to 0.

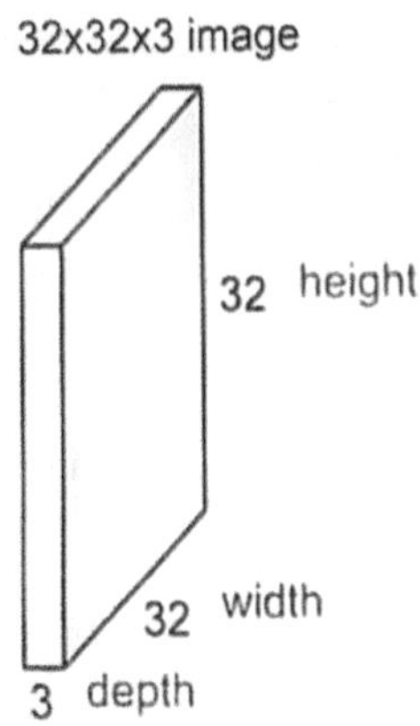

Fig. 6 Example of input image In our work it is $256 * 256 * 3$

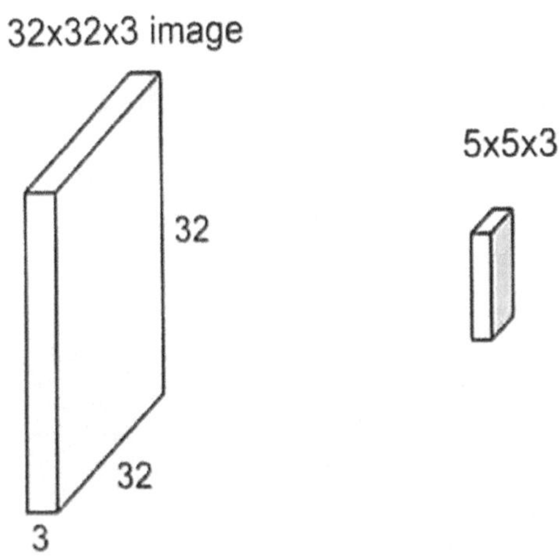

Fig. 7 Convolving an image with a single filter

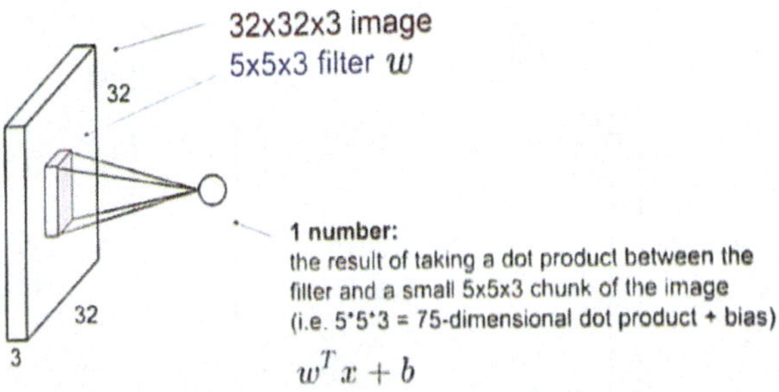

Fig. 8 Taking dot product with a single filter and a chunk of the input image

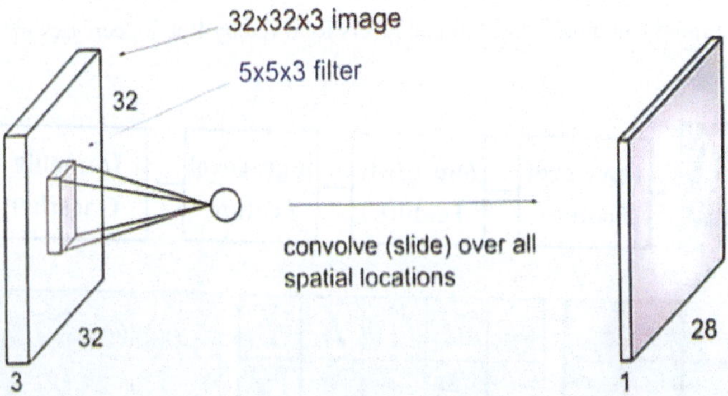

Fig. 9 Sliding the filter over the input image and taking the dot product results in a new image called the Feature Map

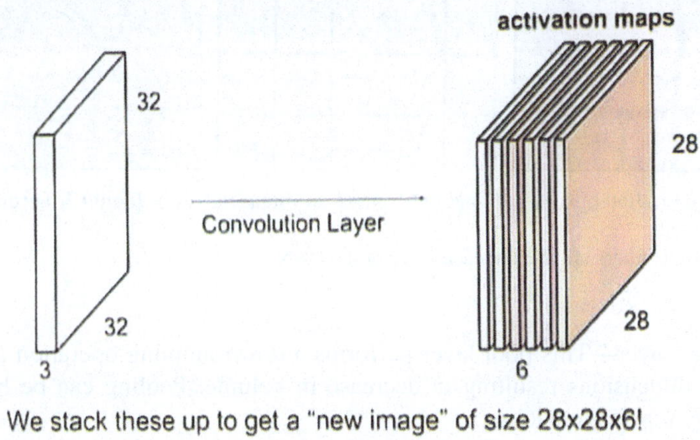

Fig. 10 In this demonstration if we use six filters on the input image, we get a new image of size $28 * 28 * 6$

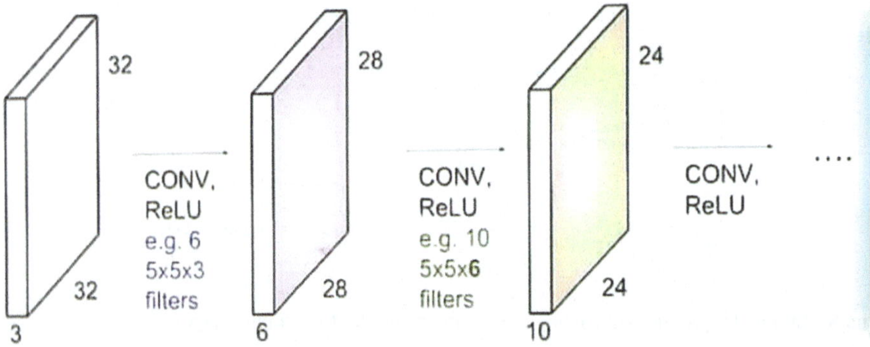

Fig. 11 Adding more and more convolutional layers followed by ReLU decreases the size of the feature map

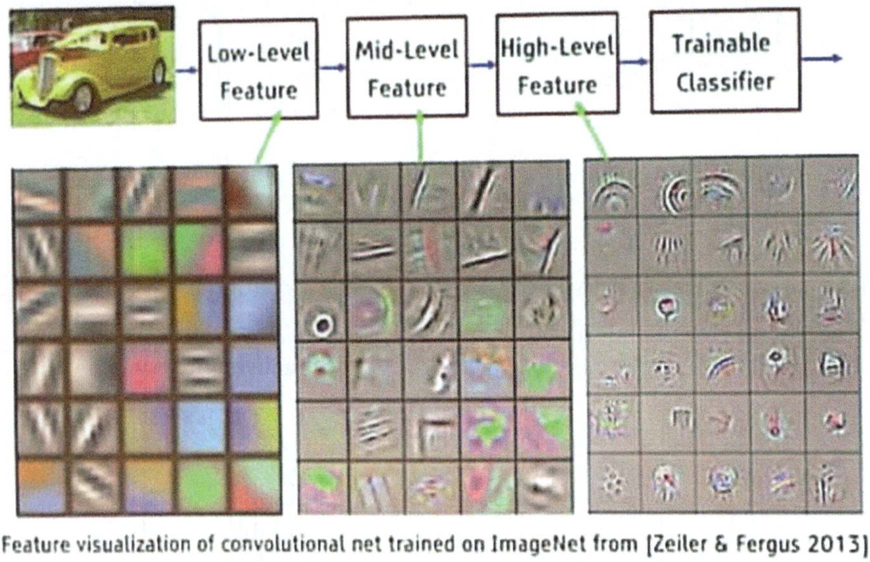

Fig. 12 Various filters are used in each layer of the CNN

4. Pooling Layer—This pool layer performs a downsampling operation along the spatial dimensions resulting in decrease in volume. Pooling can be both max pooling or average pooling.
5. Fully Connected Layer—This is a fully connected standard neural net for classification. The fully connected layer consists of:

    a. Input layer: The last feature map of the convolutional layer forms the input layer

b. Hidden layers: There are numerous hidden layers that are fully connected with weights
c. Output Layer: The last layer contains nodes corresponding to each of the variants of the image. The nodes in this layer show the probability that the classifier can determine it belonging to a particular class. The last layer of the fully connected layer consists of four nodes, each corresponding to the biomedical image variant.

Input Layer The input layer (Fig. 6) contains a three channeled input or an RGB image of size ($256 * 256 * 3$). In this case, all 280 images available in the database used consisting of both training and test images are fed to convolutional neural net. In our work, all the brain images are grayscale images. To convert them into RGB, we introduce two extra channels that have the same pixel intensity distribution as the original grayscale one. However, for training the classifier and updating the weights, only the training set is fed. The whole image database is divided in 7:3 ratio and separated into 196 training and 84 test images sets. We have performed the whole process of classification with different sets of test images and training images. It means that the images in test set used in the first round are not included in the test set used in the second round. Like this, all the images in our dataset have been used only once in the test set. Like this, we eradicate any bias (this 'bias' is different from the other parameter also known as 'bias' that is being updated in backpropagation) that has been shown to any particular set of images.

Convolution: In the visual demonstration, we take $5 * 3 * 3$ filter and slide it over the image and take the dot product along the way between the filter and chunks of the input image.

The first part of convolution is taken the dot product with the chunk of the input images with the filter. This results in a single number as shown in the figure as shown in Fig. 6.

Then, the single filter is slided over all the portions of the input image such that it covers the whole image, and on its way, it goes along and takes the dot product (Fig. 6). This combined process of taking the dot product and sliding through the whole of the input image is called image convolution. The output image of the convolution is known as the 'Feature Map' of the input image. It is to be noted that the size of the feature map is different from the size of the input image. This new size is dependent on the size of both input image and the filter. In our work, if we have input image of size $32 * 32 * 3$ and filter of size $5 * 5 * 3$, then the generated feature map is of size $28 * 28 * 1$.

Each feature map demonstrates a unique feature of the input image like edge, orientation toward a specific direction, etc. This feature is extracted through the filter. So, we apply various filters to the input images to extract various features, and hence, numerous feature maps are obtained. If we apply six filters in the first convolutional layer, the resultant image dimension is $28 * 28 * 6$. In the next convolutional layer, we apply a filter of dimension of $5 * 5 * 6$ which results in a feature map of dimension $24 * 24 * 1$. Using ten such $5 * 5 * 6$ filters will give an output dimension of $24 * 24 * 10$ (Fig. 10). It is to be noted that the dot product occurs between the feature map of dimension $28 * 28 * 6$ and the filter of dimension $5 * 5 * 6$ to result in a feature map of dimension $24 * 24 * 1$. So, the depth of the volume of image depends only on the number of filters we use.

Non Linearity Layer: After the feature map is generated, a ReLU function is applied on the image. Generally, the function is of the form of max (0, x). It leaves the positive pixel values of the feature map unchanged, but the negative pixel values are forced to zero. Now, if we have convolutional layers in sequence followed by ReLU, the sizes of feature maps decrease as we go deep into the network (Fig. 11).

In CNN, we use multiple layers of convolutional layers; in each layer, a specific set of filters are applied. As we go deeper to other convolution layers, the filters are doing dot products to the input of the previous convolution layers (Fig. 12).

Pooling Layers: The pooling layer is used to reduce the spatial size of the feature maps after the convolutional layers. This pooling layer is used to reduce the amount of parameters and hence computations when we finally proceed to the fully connected neural network (Fig.13).

Max pooling is the most common approach and GoogleNet uses max pooling (Fig.14).

Fully Connected Layer: The feature map in the last layer of the CNN is much reduced size than the original image because of subsequent convolutional and pooling operations. All the pixel values in the last layer of feature map are stretched (from 3D volume to a single array consisting of nodes) to form the input nodes of a fully connected neural network. The fully connected layers contain several hidden layers and an output layer consisting of the number of variants (Fig. 15).

Each node in a particular layer is connected by all the nodes in the previous layer by 'weights,' and a separate term called 'bias' is added to that node which is a term independent of the previous node. Here, we refer to a single value to node. It can also be said that the value of particular node in a layer is a linear combination of the values of the nodes in its previous layer plus an offset term known as bias. Here, $Z_{l,j+1}$ is the lth node in the $(j + 1)$th layer, which is a linear combination of all the nodes in the jth layer and is represented by $Z_{i,j}$, where $i = 1, 2 \ldots N$ (N is number of nodes in layer j) and plus an offset term 'b'. G is the nonlinear activation function (ReLU).

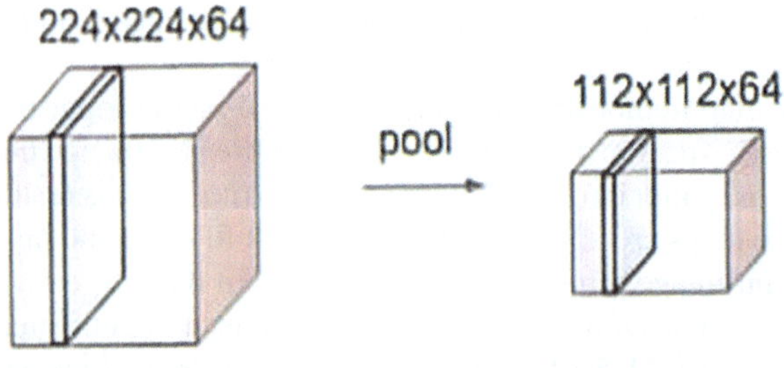

Fig. 13 Pooling reduces the spatial dimension of the feature map but the depth remains unchanged

$$Z_{l,j+1} = G\left(\sum_{i=1}^{N} W_i Z_{i,j} + b\right) \tag{17}$$

Calculating the values of the nodes in the next layer from the values of the nodes in the previous layer is known as 'Forward Propagation' [7]. It is a process of feeding input values to the neural network and getting an output which we call predicted value (Eq. 17). The second layer takes values from the first layer and applies multiplication, addition and activation operations (applying nonlinear functions) and passes these values to the next layer. Same process is repeated for subsequent layers, and finally, we get an output value from the last layer (Fig. 15).

After forward propagation, we get an output value which is the predicted value. To calculate error, we compare the predicted value with the actual output value. We use a loss function (mentioned below) to calculate the error value. Then, we calculate the derivative of the error value with respect to each and every weight in the neural network [7]. Backpropagation uses chain rule of differential calculus. In chain rule, first, we calculate the derivatives of error values with respect to the weight values of the last layer [8]. We call these derivatives gradients and use these gradient values to calculate the gradients of the second last layer. We repeat this process until we get gradients for each weight in our neural network. Then, we subtract this gradient value from the weight value to reduce the error value. In this way, we move closer (descent) to the local minima (means minimum loss). Suppose, the outputs of the fully connected layers are p1, p2, p3 and p4. It is mentioned later that the loss function is a function of all four variables p1, p2, p3 and p4. The probabilities of the four nodes are obtained by passing the linear combination of the previous node values $Z_{k,1}$ through an activation function G (ReLU function) (Fig. 16).

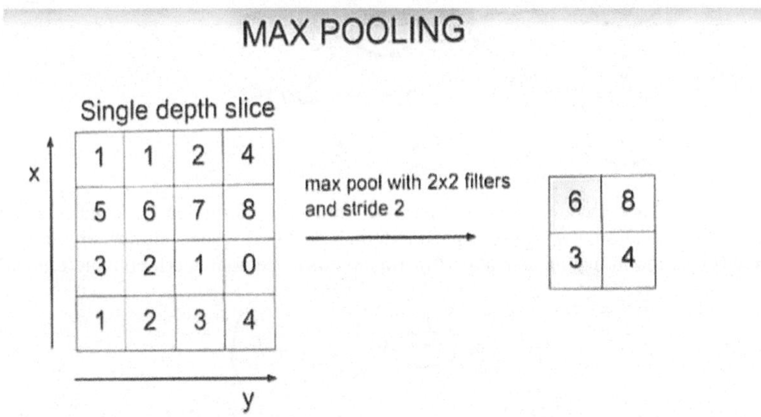

Fig. 14 Example of max pooling

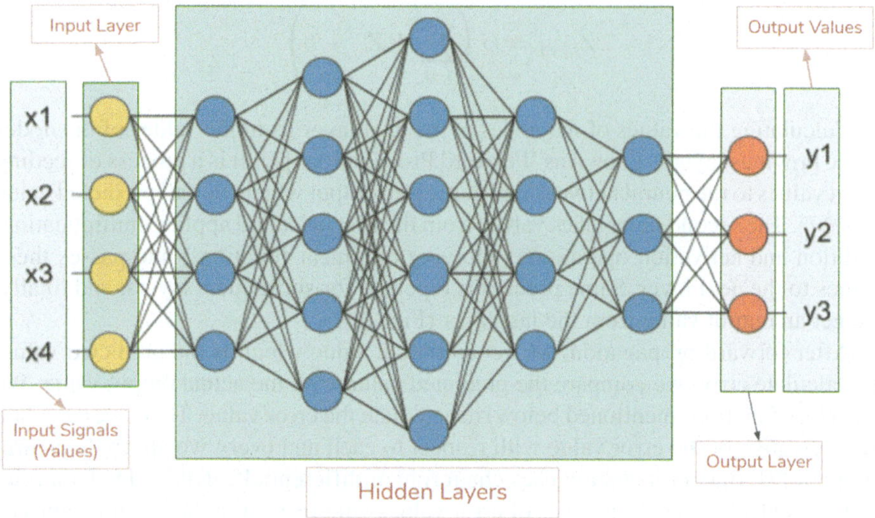

Fig. 15 In fully connected layers, each node in the hidden layer is connected to all the nodes in the previous layer

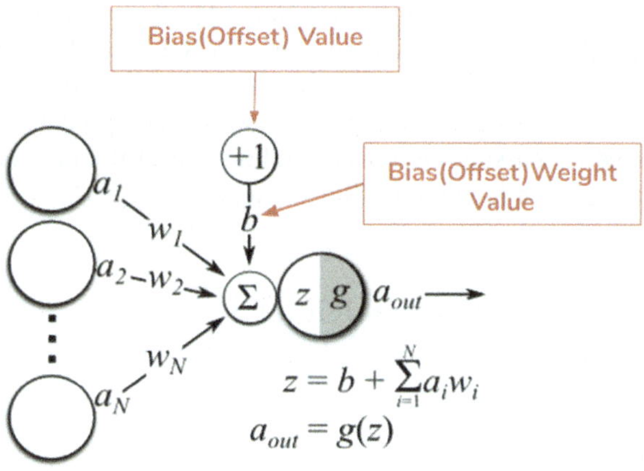

Fig. 16 The input node value is represented as a_i which are multiplied by weights w_i added to form a linear combination of a_i's, a separate bias term b is added to the resulting value, and then, the final value is passed through an activation function to introduce nonlinearity (sigmoid/tanh/ReLU)

$$p_i = G\left(\sum_{k=1}^{N} W_{k,1,i} Z_{k,1} + b_i\right) \tag{18}$$

$W_{k,1,i}$ determines the weight attached to the kth node of the first hidden layer just preceding the output layer and is connected to the ith output node. $Z_{k,1}$ is value of the kth node in the first hidden layer just preceding the output layers. The first hidden layer preceding the output layer has N nodes. Similarly, Eq. 18 describes how the values of the nodes in the first hidden layer are dependent on the values of the nodes in the second hidden layer preceding the output layer. The second hidden layer preceding the output layer has M nodes.

$$Z_{k,1} = G\left(\sum_{l=1}^{M} W_{l,2,k} Z_{l,2} + b_k\right) \tag{19}$$

Our goal with backpropagation is to update each of the weights in the network so that they cause the predicted output (p_{ij}) to be closer to the actual output (y_{ij}), thereby minimizing the loss function. To know how much change in $W_{k,1,i}$ affects the loss function w, calculate the partial derivative $\frac{\partial Loss}{\partial W_{k,1,i}}$. We evaluate this by applying the chain rule of differentiation.

$$\frac{\partial Loss}{\partial W_{k,1,i}} = \frac{\partial Loss}{\partial p_i} . \frac{\partial p_i}{\partial W_{k,1,i}} \text{for } k = 1, 2...N \text{ and } i = 1, 2, 3, 4 \tag{20}$$

With the help of Eq. 20, we get the derivatives of all the weights that connect the first hidden layer to the final output layer.

Gradient descent is an iterative optimization algorithm for finding the minimum of a function. To find a local minimum of a function using gradient descent, one takes steps proportional to the negative of the gradient (or approximate gradient) of the function at the current point. To update the weights in the direction of the minimum loss function, we use Eq. 21 where η is the learning rate which is discussed in Sect. 4.3.1.

$$W_{k,1,i}(\text{updated}) = W_{k,1,i} - \eta \frac{\partial \text{Loss}}{\partial W_{k,1,i}} \tag{21}$$

This set of updated weights is then used for forward feedback for the next epoch. Like this, more the number of epochs we use, we keep on updating the weights in a direction that minimizes the loss function and gives better classification accuracy. Here, we have only shown how the weights between the first hidden layer and the output layers are updated; similarly, the weights of the previous layers can also be updated through the same method of long chain differentiation.

5.2 Architecture of GoogleNet

GoogleNet achieved a top-5 error rate of 6.67%. The key innovation on the inception models is called the inception module (Fig. 17). This is a block of parallel convolutional layers with different sized filters (e.g., $1 \times 1, 3 \times 3, 5 \times 5$) and a 3×3 max pooling layer, the results of which are then concatenated. The network used a CNN inspired by LeNet but implemented a novel element which is dubbed an inception module. This module is based on several very small convolutions in order to drastically reduce the number of parameters. Their architecture consisted of a 22-layer

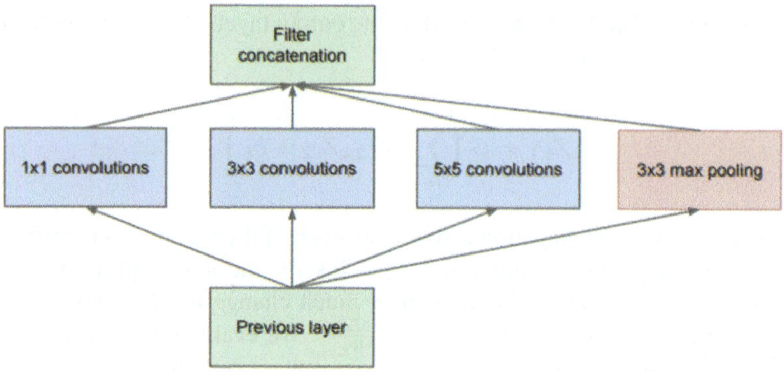

Fig. 17 Example of inception module

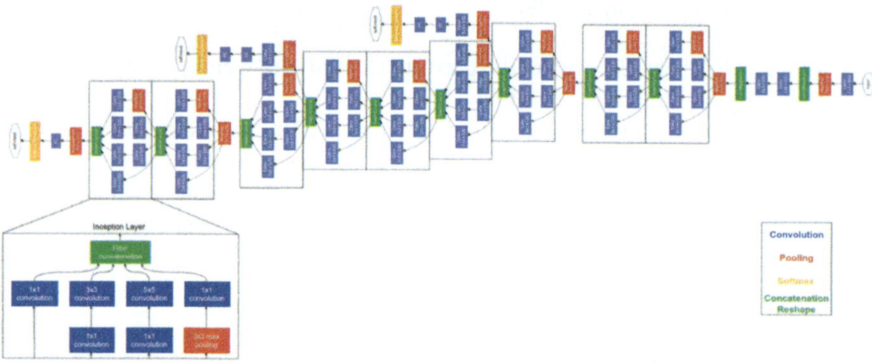

Fig. 18 GoogleNet architecture

deep CNN but reduced the number of parameters from 60 million (AlexNet) to 4 million (Fig. 18).

5.3 *Tuning the Hyperparameter of the Models*

Deep learning models are full of hyperparameters and finding the best configuration for these parameters in such a way that the high-dimensional space is not a trivial challenge. These hyperparameters act as knobs which can be tweaked during the training of the model. For our model to provide best result, we need to find the optimal value of these hyperparameters. Our main motive is to update the weights in the layers of the fully connected layer. Loss function in multi-class classification problem is called the cross-entropy loss or the log loss.

$$\text{Loss} = -\sum_{i=1}^{N}\sum_{j=1}^{M} y_{ij} \log p_{ij} \tag{22}$$

Suppose, there are M output classes (here four) and we test N images (here 84). We also assign labels to the variants, for example, for variants A, B, C and D the values of j are 1, 2, 3 and 4, respectively. In testing of the ith image, the jth output node shows the probability p_{ij}. This is the probability that image i belongs to the jth class where i = 1, 2, 3,…84 and j = 1, 2, 3, 4. Let, y_{ij} be actual binary indicator that the ith test image belongs to jth class or not. For example, if the 20th test image actually belongs to class *A-B-*, then $y_{20,1} = 0, y_{20,2} = 1, y_{20,3} = 0$ and $y_{20,4} = 0$. With the above notations, the loss function is a good measure of how the prediction done by the classifier deviates from the original label. Say, for the 20th image mentioned above, $p_{20,j}$ values are 0.6, 0.1, 0.2, 0.1, respectively. It means that it shows a probability 0.6 for variant A, 0.1 for variant B, 0.2 for variant C and 0.1 for variant *D*. Then, the loss value is log(0.1) or 10 which is higher. This shows that the probability estimating for the classes is poor because of the high loss value. However, if $p_{20,j}$ values would have been 0, 0.9, 0.1 and 0 for j=1, 2, 3 and 4, respectively, then, the loss value would be 0.045 which is very low and is an indication that the predicted value (0.9) and the true value (1) for output node of *B* are closer.

The three important parameters that have been tuned in our work to achieve best classification accuracy are-

(1) Learning Rate: In the gradient descent algorithm, we start with random model parameters and calculate the error for each learning iteration, keep updating the model parameters to move closer to the values that results in minimum cost. Gradient descent algorithms multiply the gradient (slope) by a scalar known as the learning rate (or step size) to determine the next point. This parameter indicates that how far the weights are moved in the direction of the gradient (Fig. 19). In other words, learning rate controls the rate or speed at which the model learns. If the learning rate is small, then training is more reliable, but it will take longer time because steps are tiny toward the minimum of the loss function. If the learning rate is high, then training may not converge or even diverge. Weight changes can be so higher that the optimizer overshoots the minimum and makes the loss worse. At extremes, a learning rate that is too large will result in weight updates that will be too large and the performance of the model (such as its loss on the training dataset) will oscillate over training epochs. Oscillating performance is said to be caused by weights that diverge (are divergent). Thus, our aim is to find the optimal learning rate which can quickly find the minimum loss.

(2) Number of Epoch: One epoch is that an entire dataset is passed through the neural network only ONCE. Updating the weights with a single pass or one epoch is not enough and may lead to underfitting. As the number of epochs increases, more number of times the weights are changed in the neural network and the curve goes from underfitting to optimal to overfitting curve. So, by fine-tuning, it is important to set a specific value of epoch to get an optimum classification accuracy (Fig. 20).

(3) Batch Size: One cannot pass the entire dataset into the classifier because of limited memory. So, we divide the entire dataset into batches. When each batch of training set is fed to the classifier, it is called one iteration. After several batches (which make up the entire dataset) have been fed to the classifier, it is called one epoch. Thus, one epoch is composed of several iterations.

6 Classification Results

In deep learning-based classification method, the whole database containing 280 images has been split 196 images into the training set and 84 images into the test set. Here, we have performed a threefold cross-validation which means we have performed the classification three times, and in each of those three experiments, we have used separate test set of images. In each experiment (out of 3), the numbers of epochs have been set. In each epoch, the whole of the training set (196) is used for classification. However, as this size is large, we divide 196 training sets into batches. Passing of each batch of training images into classifier is called one iteration. To sum it up, each experiment consists of a number of epochs and each epoch consists number

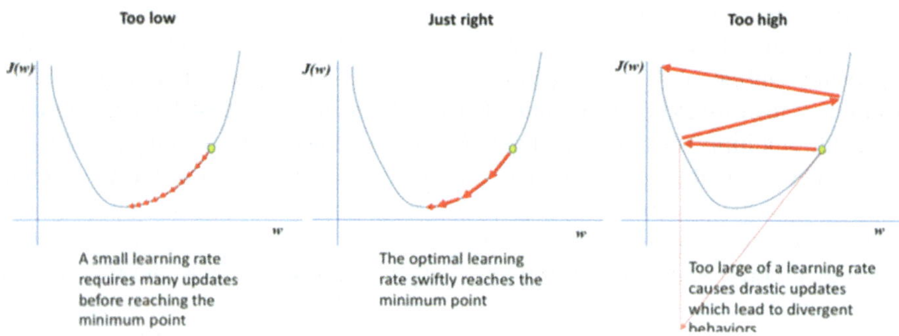

Fig. 19 Effect of various values of learning rate

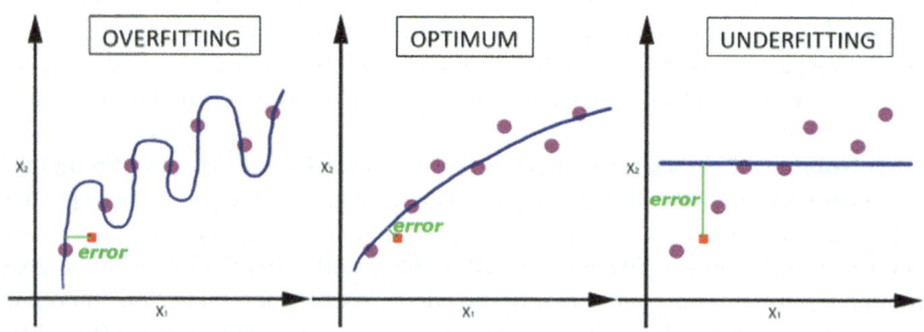

Fig. 20 Effect of number of epochs

Table 2 Classification using CNN with parameters Batch-size = 50 Number of Epochs = 20 Learning rate = 0.0009

Actual/predicted	A	B	C	D
A	0.8564	0.0476	0.0476	0.0476
B	0.0952	0.8088	0	0.0952
C	0.1428	0	0.7544	0.0952
D	0.1428	0	0.1428	0.7068

Mean classification accuracy= 76.89 %

Table 3 Classification using CNN with parameters Batch-size= 50 Number of Epochs=20 Learning rate=0.0015

Actual/predicted	A	B	C	D
A	0.904	0	0.0476	0.0476
B	0	0.883	0	0.117
C	0	0	0.756	0.244
D	0	0	0.244	0.756

Mean classification accuracy= 78.57%

Table 4 Classification using CNN with parameters Batch-size = 50 Number of Epochs = 20 Learning rate=0.0024

Actual/predicted	A	B	C	D
A	0.826	0	0.087	0.087
B	0	0.756	0	0.244
C	0.087	0.087	0.714	0.112
D	0.159	0	0.159	0.682

Mean classification accuracy= 74.37%

of iterations. The final classification accuracy is the mean of the three individual classification accuracy. Size of the images in deep learning-based feature extraction is $256 * 256 * 3$. Our sparse image classification pipeline has been implemented on the BRATS-2013 dataset that has four labels.

The labels are defined as:

1. Label-A: necrosis
2. Label-B: edema
3. Label-C: non-enhancing tumor
4. Label-D: enhancing tumor

By tuning the three hyperparamaters, the training accuracy has been noted. It has been noted that the maximum accuracy has been achieved for a specific value of batch size and epoch number. Keeping the first two hyperparamters fixed, the learning rate has been tuned. For a particular value of the learning rate, maximum training accuracy of 78.57% is achieved (Tables 2, 3 and 4; Fig. 21).

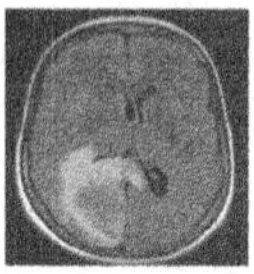

(a) Edema: Test Accuracy = 89.63%

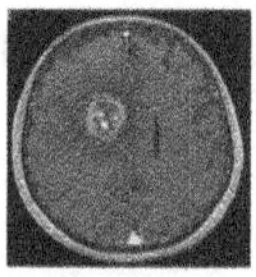

(b) Necrosis: Test Accuracy = 84.36%

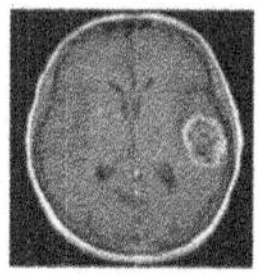

(c) Enhancing Tumor: Test Accuracy= 76.98%

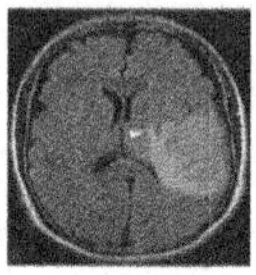

(d) Non-enhancing tumor: Test Accuracy= 74.65%

Fig. 21 Test Accuracies for four class labels of brain images

7 Conclusion and Future Work

The main contribution of the paper lies in utilizing the sparsity of the set of images that reduces the training and inference time of the CNN. But the detailed information of the image is partially lost which is the reason for slight decrease in training accuracy. Future directions of this work could focus on performing an optimization selecting the best compression ratio that produces desirable training accuracy.

In case of a general purpose computing system model, the proposed efficient neural network could be deployed in popular resource-constrained COTS systems such as RaspPi/Arduino containing image-acquiring sensors along with WiFi module. In these systems, the image is first acquired, then sparsified by the underlying microcontroller, and subsequently, the neural network processing can happen either locally within the system or the vectorized image can be transmitted to a server from where we fetch back the final inference result.

References

1. I. Daubechies, The wavelet transform, time-frequency localization and signal analysis. IEEE Trans. Inf. Theor. **36**(5), 961–1005 (1990). https://doi.org/10.1109/18.57199
2. A. Mojsilovic, M.V. Popovic, A.N. Neskovic, A.D. Popovic, Wavelet image extension for analysis and classification of infarcted myocardial tissue. IEEE Trans. Biomed. Eng. **44**(9), 856–866 (1997). https://doi.org/10.1109/10.623055
3. O. Rioul, P. Duhamel, Fast algorithms for discrete and continuous wavelet transforms. IEEE Trans. Inf. Theor. **38**(2), 569–586 (1992). https://doi.org/10.1109/18.119724
4. C.-T. Huang, P.-C. Tseng, L.-G. Chen, Analysis and VLSI architecture for 1-D and 2-D discrete wavelet transform. IEEE Trans. Sig. Proc. **53**(4), 1575–1586 (2005). https://doi.org/10.1109/TSP.2005.843704

5. Z. Xiong, K. Ramchandran, M.T. Orchard, Y.-Q. Zhang, A comparative study of DCT- and wavelet-based image coding. IEEE Trans. Circ. Syste. Video Technol. **95** (1999)
6. Y. Sun, B. Xue, M. Zhang, G.G. Yen, Evolving deep convolutional neural networks for image classification. IEEE Trans. Evol. Comput. **24**(2), 394–407 (2020). https://doi.org/10.1109/TEVC.2019.2916183
7. M.U. Yaseen, A. Anjum, O. Rana, N. Antonopoulos, Deep learning hyper-parameter optimization for video analytics in clouds. IEEE Trans. Syst. Man Cybern. Syst. **49**(1), 253–264 (2019). https://doi.org/10.1109/TSMC.2018.2840341
8. X. Zhang, S. Huang, X. Zhang, W. Wang, Q. Wang, D. Yang, Residual inception: a new module combining modified residual with inception to improve network performance, in 2018 25th IEEE International Conference on Image Processing (ICIP), Athens, 2018, pp. 3039–3043, https://doi.org/10.1109/ICIP.2018.8451515
9. L. Chen, P. Bentley, K. Mori, K. Misawa, M. Fujiwara, D. Rueckert, DRINet for medical image segmentation. IEEE Trans. Med. Imag. **37**(11), 2453–2462 (2018). https://doi.org/10.1109/TMI.2018.2835303
10. Q. Zhu, P. Zhang, Z. Wang, X. Ye, A new loss function for CNN classifier based on predefined evenly-distributed class centroids. IEEE Access **8**, 10888–10895 (2020). https://doi.org/10.1109/ACCESS.2019.2960065
11. S.J. Pan, Q. Yang, A survey on transfer learning, in *IEEE Transactions on Knowledge and Data Engineering 2010*

Continual Deep Learning Framework for Medical Media Screening and Archival

Pallavi Saha and Apurba Das

Abstract Health care is a domain of essential services and requires highest accuracy in diagnosis. AI can just aid the doctors achieving the aforementioned target but not replace them. AI always requires doctor's validation to improve itself. Application of deep learning in health informatics enables us to recognize the target disease at early stage of maturity. The primary section of the proposed piece of work helps us to automatically diagnose tuberculosis (TB) from chest X- Rays in minimal execution time and memory footprint. Next medical media screening and analysis would drive the process of automated archival strategy, tagged to self-generated meta-data. The classical deep learning method needs to be migrated to continuous deep learning framework before the final archival based on (a) collection or acquisition of periodic medical media and (b) feedback from human experts. Gradual self-improvement of the AI engine narrowing down the fuzzy zone between confident positive and confident negative of diagnosis would be the key achievement of proposed continuous deep learning framework.

Keywords Deep learning · Medical media archival · Tuberculosis · Chest X-ray images · Cross-validation · Continual learning

1 Introduction

During the last couple of decades, Artificial Intelligence (AI) has been synonymously pronounced with Machine Learning (ML) even though Machine Learning is just a subset of AI. With the advancement of deep neural network architectures, the word

P. Saha (✉) · A. Das
Computer Vision Lab (IoT), Tata Consultancy Services, Bangalore, India
e-mail: saha.pallavi@tcs.com

A. Das
e-mail: das.apurba@tcs.com

A. Das
PES University, Bangalore, India

P. Kumar et al. (eds.), *Machine Learning for Intelligent Multimedia Analytics*,
Studies in Big Data 82,
https://doi.org/10.1007/978-981-15-9492-2_3

Deep Learning (DL) again replaced or aliased the word AI even though DL is just a subset of ML. All other methods of AI including non-neural-network based ML have been losing the importance day by day. Although, DL based architectures promises to achieve close-to-human accuracy only with large number of labelled data, the research community invested their efforts in automated augmentation of training data or cross-modal associative labelling [1] of training data. There are also attempts to develop data labeling tools in unsupervised or semi-supervised manner. Hence, fully automated deep learning has become an active area of research including but not restricted to medical image processing, medical image analytics and diagnosis. It is especially important in clinical domain because of rarity of labelled data due to unavailability of domain experts or medical practitioners. Budd et al. [2] have done one extensive survey on active learning and human-in-loop learning for medical images. They have categorized and evaluated major four aspects of continual learning:

1. **Active Learning** to choose the best data to annotate for optimal model performance.
2. **Interpretation and Refinement** to steer models to optima for a given prediction iteratively.
3. **Practical considerations** to develop scalable applications.
4. **Related Areas** to benefit human-in-the-loop computing as they evolve.

Similarly, Baweja et al. [3] has attempted to address the issue of *catastrophic forgetting* through reinforcement learning based continual learning for segmentation of white matter lesions in human brain. It is not only restricted to medical image analytics research but also in novel neural network architecture for continual learning in the video analytics domain like activity recognition [4–6].

In this book chapter, we have proposed an end-to-end continuous deep learning framework which evolves with the aid of human-in-loop principle to start with. In order to design the continual deep learning framework (CDLF), it is utmost important to pose a light weight model which can be used as a solution accelerator in the larger CDLF. We have here considered the medical image analytics application of diagnosis of tuberculosis from chest X-ray.

Tuberculosis is a contagious malady, and is one of the well known death cause globally standing ahead of HIV or AIDS [7]. It is a life threatening disease and challenges approximately 1.8 million lives every year [8]. Early detection and proper treatment of tuberculosis can help to combat with this ailment [9]. As it is one of the leading death cause and difficult to cure, World Health Organization (WHO) suggests organized and wide application of screening to destroy the disease. More than 80% of the affected population, TB is basically prevalent in poor countries, such as South-East Asia, Western Pacific, and African. The affected populace mostly belong from the nation which has weak medical infrastructure.

Mycobaterium tuberculosis is the key microbe in tuberculosis [10]. In this kind of chronic disease, the bacteria usually enter into the lungs while breathing and then it is transmitted to the entire body through blood circulatory system [11]. Since it's a contagious disease, a person who is in contact to the infected person also get infected and it reproduces by increasing the number of affected populations. Curtailing the time required for the diagnosis of TB can help us to mitigate the spread of the disease and minimize the propagation rate. The main aim of this research is to cut down the patient's waiting time with the help of a solution to detect TB disease in premature state. In the aforementioned of situation, an algorithm which can automatically identify the tuberculosis bacteria in chest X-Rays will be inexpensive, fast and effective. Several Researchers and Scientist have addressed this problem in range of publications [12, 13] by employing various machine learning and deep learning techniques. If a person is affected by tuberculosis, then the tuberculosis bacteria multiplies in the body and attacks the lung or other body parts. When an active TB affected person coughs or sneezes and some other person inhales the discharged droplets, the disease spreads. In the past few years, it has been observed that convolutional neural networks (CNNs) have shown significant improvement in results on images, videos as compared to the traditional image processing and machine learning methods. It has also been widely applied on biomedical segment for various kind of analysis such as cancer detection and classification [14], tumor segmentation [15, 16], extraction of cerebrospinal fluid (CSF) [17, 18] from CT Scan and many more. Various research works has been carried out in the field of deep learning for the diagnosis of tuberculosis bacteria on chest X-Rays due to inexpensive and availability of data [19]. Satapathy et al. [20] designed an algorithm based on Siamese neural network for the automatic diagnosis of Pneumonia from Chest Radio-graphs. Hasan et al. [4] proposed a continuous learning framework for Activity Recognition by employing deep hybrid feature models.

The cited publications adopts deep learning models such as AlexNet, GoogleNet and ResNet to perform detection and classification [21, 22]. Although the mentioned networks are very effective, these networks have huge number of parameters, trained on large number of images and more than thousand classes (labels). As a result, this kind of network is very computationally expensive and requires high memory consumption. In this current work, we designed an algorithm with less computation and memory consumption that can diagnose the TB disease from chest X-Rays. The rest of this chapter is articulated as follows.

Our contributions have been presented in Sect. 2 with description of novel features/ algorithms. The proposed architecture of continual deep learning framework (CDLF) has been discussed in Section 3. The proposed solution accelerator of the automatic detection of TB from chest X-Rays has been described in Sect. 4. In the Sect. 4.2 of Sect. 4, the experimental results are discussed in detail with the achieved accuracy. In Sect. 5, few more potential applications of proposed continuous deep learning framework (CDLF) has been discussed. Finally, we have concluded our findings in Sect. 6 and lists out the future scope of the work.

2 Novelty and Contributions

Our contributory work in this chapter can be segregated into two parts.

I. The first novel contribution is the proposal of continuous deep learning framework. The architecture proposed is generic across domains. The necessity of having active learning framework is well-accepted as described in the Sect. 1. The novel architecture proposed in the following section ensures:

   (a) Improvement of accuracy with regular data fetching.
   (b) Approval of data to the training pipeline according to statistical divergence analysis.
   (c) Considering a light weight deep learning model as initiator (to be discussed in the following sections).
   (d) Inclusion of new class to the model.
   (e) Analysis of inter and intra class distance to decide inception of additional model dynamically.
   (f) Cross-validation.
   (g) Intelligent archival.

II. The second novel contribution is the inception of a light weight deep learning network with optimized computational complexity and still ensuring promising accuracy.

The aforementioned novelties have been discussed in Sects. 3 and 4, respectively.

3 Architecture: Continual Deep Learning Framework (CDLF)

During the expected journey from research to solution deployment, it is important to honor and understand the need of scalable architecture which evolves with time and data on a regular basis. Here, the problem of tuberculosis is essentially a binary classification problem. But the framework proposed here is a generic model which can be employed for multi-class classification requirements and even in video analytics like multi-class object classification, action recognition, action sequencing and so on. The problem of any learning framework improvement can be practically well-defined as (a) Improvement of classification accuracy for the already trained system trained for existing X number of classes, (b) Enabling the model to incorporate new class ensuring similar classification accuracy for the existing classes.

The aforementioned is proposed to be achieved by the technical architecture and solution flow described in Fig. 1.

- Primarily, the basic inference engine is trained with the available data of acceptable volume which can achieve just acceptable accuracy of the multi-class classification problem. All the hyper-parameters are tuned, network is designed keeping the

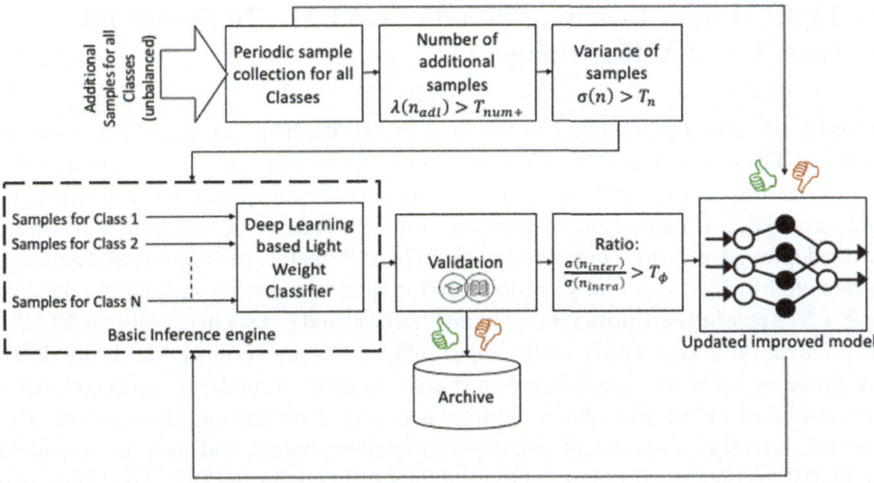

Fig. 1 Architecture of continual deep learning framework

scale in mind. One similar network for tuberculosis diagnosis has been discussed in Sect. 4 which is binary classification in this target problem senario.

- After the deployment, regularly addition data (in this case, image data) is being collected by the engine for all existing classes and new classes.
- A watch keeper program (WKP#1) keeps an eye on the number of additional samples collected during live running of the deployed system on a regular basis/periodic basis. When the number of additional samples, $\lambda(n_{addl})$ crosses the preconfigured threshold, T_{num+}, the variance of samples are ensured ($\sigma(n) > T_n$) by eliminating singularity.
- With the aforementioned 2-factor confirmation, the inference gets audited by experts (in our target solution, the experts are the doctors). The cross validation not only gets confirmed by the experts but also provides flexibility to correct the labels in case the basic inference engine misclassifies the samples. The right samples with right labels, hence, gets archived for future consumption, reporting and evolution of the engine.
- The ratio of variance of interclass ($\sigma(n_{inter})$) and intraclass ($\sigma(n_{intra})$) feature vector is being calculated through another watch keeper (WKP#2). If the same is higher than another predefined configuration, T_n, it signifies, the additional data samples are ready for next session of training to achieve better model accuracy. This new training engine is initialized by the basic inference engine having flexibility in unfreezing layers and modifications on shape of output vectors.
- The new training gets triggered in enterprise server. After the training is completed and converged, the validation watch keeper program (WKP#3) confirms improved model. Finally, the new model replaces the basic inference engine. The said model snapshot gets the nomenclature of basic model and the complete iteration gets executed ensuring evaluation of the continual deep learning framework guided by WKP#1, #2 and #3.

4 Light Weight Learning Module (LWLM): To Detect TB from Chest X-Ray Images

In order to serve the mankind in the field of bio-medical, an automated solution to detect TB disease from chest X-Rays accurately is only one of the parameter for acceptance. Apart from this, the solution should also be computationally inexpensive and have reduced memory consumption, so that it execute in real time and can be easily deployed in various portable devices. The architecture of our proposed network is depicted in Fig. 4. Initially, all the input images are normalized and re-sized to 512×512 resolution. Further it is split into train set and test set in the ratio of $80:20$. Normalization is performed on the input images to remove noise and bring down the range of intensity values between 0 and1 in order to make it computationally effective. During training phase, the train set images are fed into the network. The network broadly comprises of five convolutional segments, and then accompanied by global average pooling layer (GAP), then a fully connected (FC) layer with two output units and $softmax$ classifier is employed for classification. GAP helps us to eliminate large number of trainable parameters from the model. In this model, GAP is applied before FC layer to reduce the model complexity and overfitting problems. The GAP has been depicted in Fig. 2.

Softmax is a function that used for multi-class classification and it provides probability of an object belonging to certain class. Softmax is often used in neural networks, to map the non-normalized output of a network to a probability distribution over predicted output classes. The standard (unit) softmax function $\sigma : \mathbb{R}^K \rightarrow \mathbb{R}^K$ is defined by the formula as depicted in Eq. 1.

$$\sigma(\mathbf{z})_i = \frac{e^{z_i}}{\sum_{j=1}^{K} e^{z_j}} \text{ for } i = 1, \ldots, K \text{ and } \mathbf{z} = (z_1, \ldots, z_K) \in \mathbb{R}^K \quad (1)$$

Here, the standard exponential function is applied onto each element z_i of the input vector $\mathbf{z}$ and normalize these values by dividing by the sum of all these exponentials; this normalization ensures that the sum of the components of the output vector $\sigma(\mathbf{z})$ is 1.

Each of the convolutional segment includes two 3 convolution, one 1×1 convolution blocks with Rectifier Linear Unit (ReLU) as activation function followed by max-pooling (abbreviates each of the feature maps to the maximum value) after

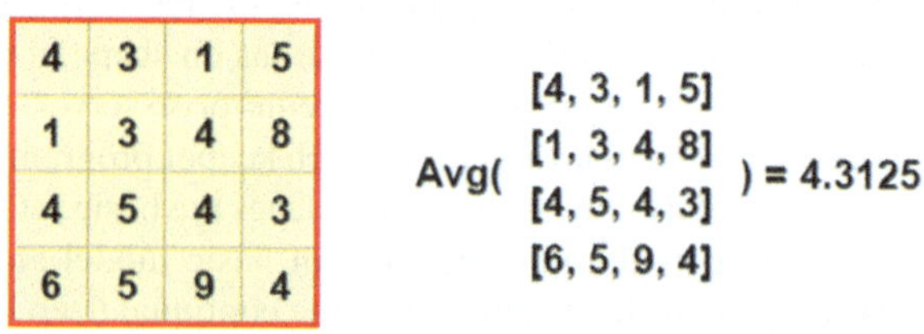

Fig. 2 Illustration of Global Average Pooling (GAP)

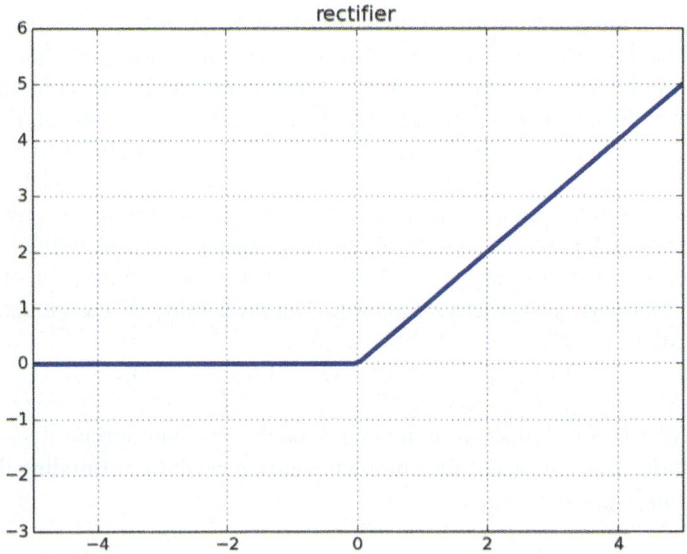

Fig. 3 Rectified linear unit

each convolution. ReLU is a non-linear activation function that will output the input directly if it's positive, otherwise it will be zero as depicted in Eq. 2 and Fig. 3. It is applied to increase non linearity in the images.

$$f(x) = \max(0, x) \tag{2}$$

The kernel size used in the max-pooling layer is 3 with stride 2 similar to the AlexNet [23]. During convolution phase, in order to maintain the resolution of the input images zero padding is done. In each of the convolutional blocks, first input images is fed into one 3 convolution and then the output is fed into another 3 convolution. Simultaneously, another parallel connection is established where the input images is fed into 1×1 convolution with stride 4. The stride 4 is chosen for this operation only in the first convolutional block to make equal resolutions with other convolutions. Then the output of the former and later connection is added, and then further fed into the next convolutional block. The next 4 convolutional blocks performs the operation with stride 1. In a neural network, the first few layers are responsible for extracting low level features of the images (shape, color, edge etc.) and gradually over the layers it starts learning high level features specific to those images/ patterns/ objects only [24]. So, in the initial layers of the network less filters are used and then over the next layers number of filters are increased as those layers will be involved in the extraction of complex features. The same approach has been undertaken in the proposed work too. In the first convolutional block 16 filters were used, then it got further incremented in the multiplicative of 16 so on for the next 4 blocks (16, 32, 48, 64, 80 for five blocks). Empirical tests confirmed that the depth of 5 block leads to good classification performances. The network is compiled with RMS Prop optimizer (Eqs. 6 and 7) and the error function has been computed using binary cross entropy. In machine learning, the true probability p_i is the true label, and the given distribution q_i is the predicted value of the current model. For logistic regression, which (among other things) can be used to classify observations into two possible classes (often simply labelled 0 and 1. The output of the model for a given observation, given a vector of input features x, can be interpreted as a probability, which serves as the basis for classifying the observation. The probability is modeled using the logistic function $g(z) = 1/(1 + e^{-z})$ where z is some function of the input vector x, commonly just a linear function. The probability of the output $y = 1$ is given by Eq. 3.

$$q_{y=1} = \hat{y} \equiv g(\mathbf{w} \cdot \mathbf{x}) = 1/(1 + e^{-\mathbf{w} \cdot \mathbf{x}}), \tag{3}$$

where the vector of weights $\mathbf{w}$ is optimized through some appropriate algorithm such as gradient descent. Similarly, the complementary probability of finding the output $y = 0$ is simply given by Eq. 4.

$$q_{y=0} = 1 - \hat{y} \tag{4}$$

Having set up our notation, $p \in \{y, 1 - y\}$ and $q \in \{\hat{y}, 1 - \hat{y}\}$, we can use binary cross entropy to get a measure of dissimilarity between p and q is as described in Eq. 5.

$$H(p, q) = -\sum_i p_i \log q_i = -y \log \hat{y} - (1 - y) \log(1 - \hat{y}) \tag{5}$$

The train data-sets are shuffled on every epoch so that during every epoch so that data are selected randomly for training and this will reduce the chances of overfitting. As the number of images for both the classes (positive and negative) cases for the detection of TB from chest X-Rays received from the data set are not equal, few samples were augmented by few augmentation techniques (noise addition, rotation, scaling). The model is trained for 200 epochs with 10 batch size and Random normal initializer is used for weight initialization. RMS Prop optimizer has been used with learning rate 0.001. The proposed model has been depicted in Fig. 4.

4.1 RMSProp

RMSProp (Root Mean Square Propagation) is a method in which the learning rate is adapted for each of the weights and biases. The principle algorithm is to divide the learning rate for a weight by a running average of the magnitudes of recent gradients for that weight [25]. So, first the running average is calculated in terms of means square,

$$v(w, t) := \gamma v(w, t - 1) + (1 - \gamma)(\nabla Q_i(w))^2 \tag{6}$$

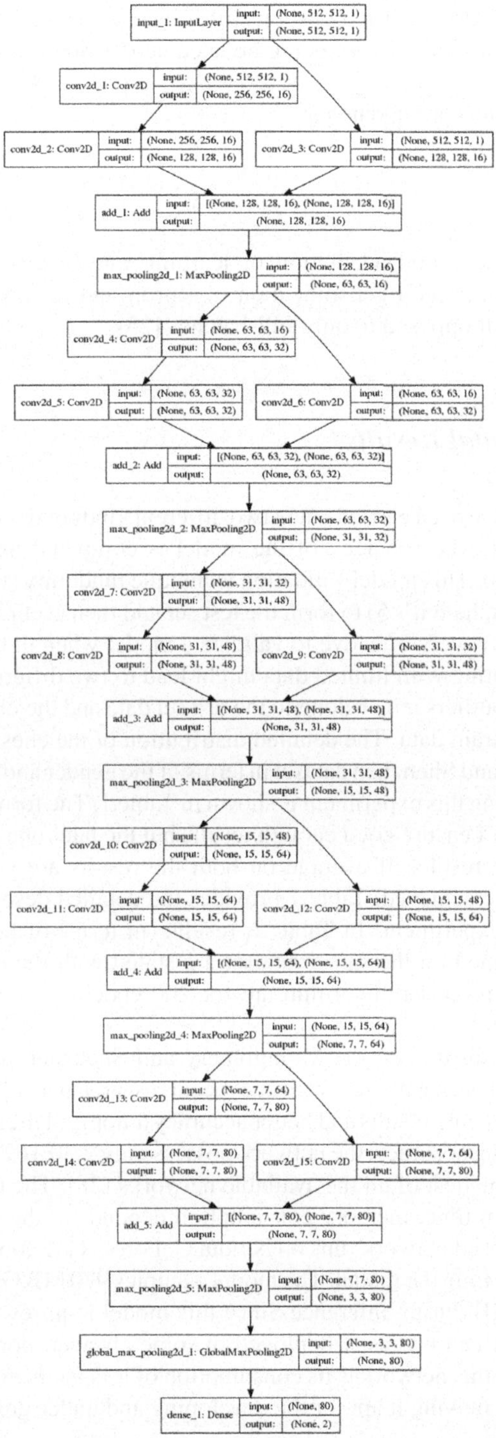

Fig. 4 Proposed light weight deep neural network architecture to detect TB from chest X-ray images

where, γ is the forgetting factor, Q is objective function that needs to be minimized, η is learning rate, weights w's are being updated across time t and v is the running average.

And the parameters are updated as,

$$w := w - \frac{\eta}{\sqrt{v(w,t)}} \nabla Q_i(w) \tag{7}$$

RMSProp has shown good adaptation of learning rate in different applications. RMSProp can be seen as a generalization of Rprop and is capable to work with mini-batches as well opposed to only full-batches [25].

4.2 *Experimental Results*

The experiments have been carried out in two different kinds of data sets Montgomery and Shenzhen [26]. The accuracy of the model is computed using K-fold cross-validation technique. This model validation technique randomly select samples from the dataset (k-times, here it's 5) to form the test set and then accuracy is evaluated on them. This method is very effective to validate a model when the amount of data is limited. Experimenting with limited data might lead to two different situations: one where most of the outliers may belong to the test set data and the other where outliers may belong to the train data. The detailed distribution of the chest X-Rays received from Montgomery and Shenzhen dataset in terms of the gender and active TB patients utilized in performing this experiment is shown in Table 1. The former one produces a model which doesn't ensure good enough results but the later one generates a model assuring promising results. If average on both the results are performed, reliable and good results are achieved. Table 1 provides the detailed description of the data sets used for this experiment. In Table 2, results (in terms of percentage) for the comparison performed on the same data set are stated with the cited publications. The results of our proposed algorithm are for 200 epochs consuming 231 ms per step.

The comparison results on the Montgomery and Shenzhen datasets of various work performed has been gathered from one recent research paper [28]. The proposed model showed promising results and better accuracy among all the compared models. Apart from achieving accuracy, the network possess only 2,30,162 parameters which is either equal or smallest of all the available networks [28]. The training, inference speed and execution time, memory consumption depends in the complexity of the network. The proposed network runs on Nvidia GeForce RTX 206, and requires less than an hour time to train the model. The model occupies 970 MB GPU memory while training and 200 MB during inference.Since this model requires very less memory during inference, it can be easily deployed in some cheaper, portable devices.The prime advantage of this network is its consumption of less memory which allows use of cheap hardware, making it apt for the developing and under-developed countries.

Table 1 Overview of the datasets used for training and evaluation of tuberculosis classification

	Montgomery	Shenzhen
Patients	138	662
Without TBC	80	326
With TBC	58	336
Male	63	442
Female	74	213
Others/ Unknown	1	7
Cross-validation	27	28

Table 2 Comparison of accuracies between state-of-art and proposed algorithm on the standard datasets

Algorithm/ Architecture	No. of parameters	Montgomery dataset (%)	Shenzhen dataset (%)
Hwang et al. [21]	Millions	67.4	83.7
Jaeger et al. [27]	Millions	78.3	84.1
Islam et al. [22]	Millions	77.4	88.0
Pasa et al. [28]	230,000	79.1	84.4
Proposed light network	230,162	84.21	90.48
Proposed lighter network	176,332	78.9	83.2

The aforementioned Table 2 depicts the effectiveness of the proposed network. It is observed that, we could achieve very comparable (almost same) accuracy with the lighter weight network as shown in the last row or Table 2. Interestingly, we could achieve similar accuracy to that of the Pasa et al. [28] with 20% reduced number of parameters. The second last row of the same table shows that the light weight network proposed by us could achieve much better accuracy with respect to Pasa et al. [28] using similar number of parameters.

The proposed light weight network architecture (Fig. 4) number of kernels grown in integral multiple of 16 (i.e., 16, 32, 48 and so on) across depth of the network, whereas in the proposed lighter weight network the number of kernels chosen are in integral multiple of 14 (i.e., 14, 28, 42 and so on). All the other rows are comparable as far as the accuracy and number of parameters are concerned as the second smallest network (GoogLeNet) has 7 million [29]. Other architectures used in the studies cited above have up to 60 million parameters [23]. Sample outputs with True positive (TP), False Positive (FP), True Negative (TN) and False Negative (FN) each have been depicted in Fig. 5. The definitions of the said performance metrics are presented in Table 3.

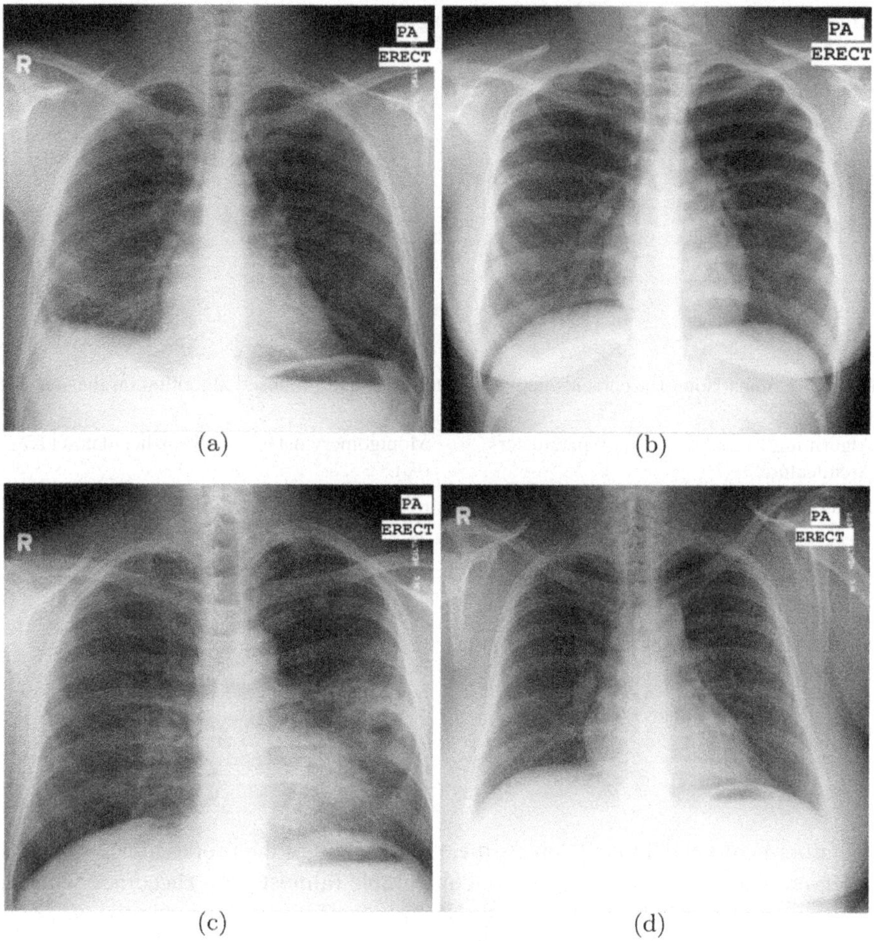

Fig. 5 Sample outputs of the proposed light weight neural network based diagnosis of tuberculosis from Chest X-ray images (Dataset: [26]): **a** True positive: TB has been correctly detected as positive by the network; **b** True negative: Normal patient's x-ray image has been correctly detected as TB negative; **b** False negative: TB patient's X-ray image has been diagnosed with TB negative erroneously; **d** False positive: Normal patient's X-ray image has been diagnosed as TB positive erroneously

Table 3 Performance metrics for TB prediction from chest X-ray images

Performance Metric	Ground Truth	Network prediction
TP	Unhealthy	Unhealthy
TN	Healthy	Healthy
FN	Unhealthy	Healthy
FP	Healthy	Unhealthy

This proves that the proposed light weight and lighter weight networks have highest parameter efficiency addressing overfitting.

5 Potential Applications of CDLF on Biomedical Imaging

5.1 *Tele-Ophthalmology: Diabetic Retinopathy*

Diabetic retinopathy is a medical condition in which damage occurs to the retina due to diabetes mellitus. It is one of the major cause of permanent blindness if kept untreated [30]. Hence, early detection of this disease is of utmost importance. Interestingly, like other medical imagery, availability of fundus image samples representing healthy and different stages of diabetes is a challenge. On the other hand, there is a large scale of difference of opinion is evident. Hence, it is the need of the hour to get one light weight deep learning model in place with an ability to improve the same by additional transfer learning with trusted experts' cross-validation. The evolution of the deep learning model to better and better accuracy is only achievable if the framework proposed in this chapter is employed. The CDLF honors the trust factor of the organization and the patients on specific experts or group of experts. The said framework along with a light weight deep learning classifier and severity analysis model can save blindness of millions of patients by earliest diagnosis (Fig. 6).

5.2 *Skin Cancer*

Another such severe disease which is being attempted to be diagnosed through deep learning on medical images is skin cancer [31]. It has been always again a challenge to get one light weight primary system in hand for automated diagnosis and keep

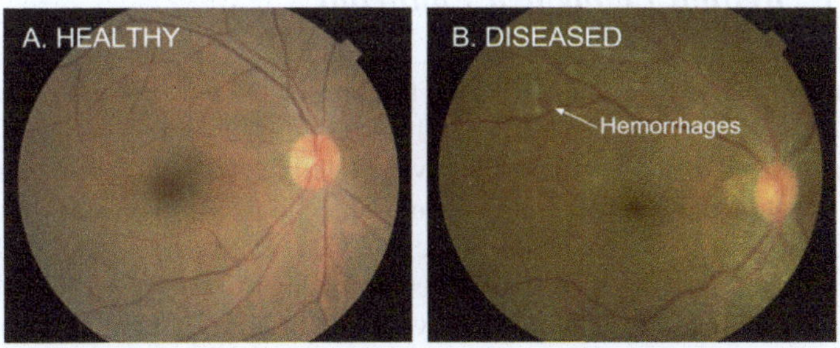

Fig. 6 Samples of retinal fundus images: The left one is of a healthy retina (**a**), whereas the image on the right is a retina with referable diabetic retinopathy (**b**) due to a number of hemorrhages (red spots) present [32]

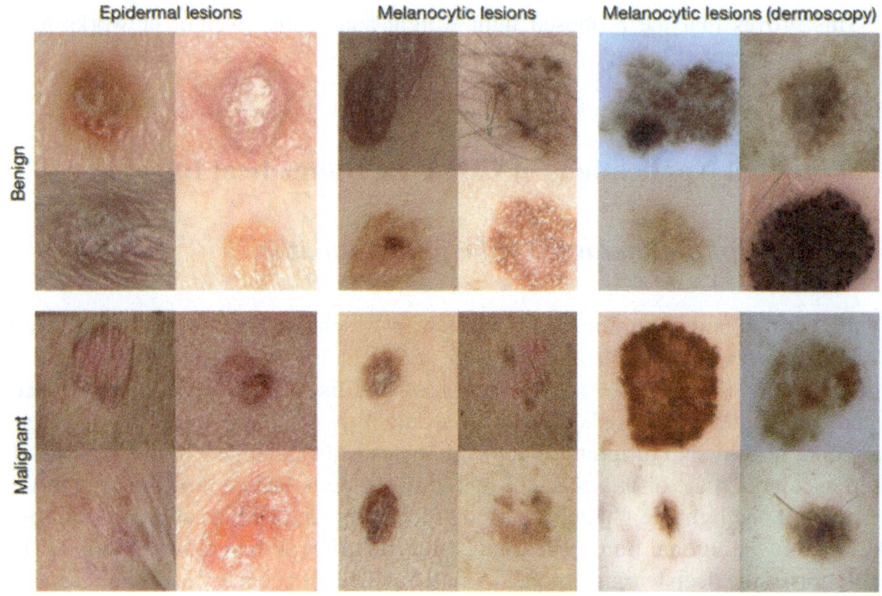

Fig. 7 Different kinds and stages of skin cancer [33]

the said system improving by adding more data samples, labels and cross-validation through expert medical practitioners. Although the publication in 2020 [32] claimed to achieve >99% accuracy in detection of skin cancer on standard dataset, reality is different when it comes to unseen practical data. Hence, the requirement of CDLF is always there in order to provide a flexible system which honors expert's opinion while evolution towards better accuracy with additional data training and labeling periodically. Different stages and kinds of skin cancer images have been depicted in Fig. 7.

5.3 *RPE: Retinal Pigmented Epithelial*

The retinal pigment epithelium (RPE) is the pigmented cell layer just outside the neurosensory retina that nourishes retinal visual cells. It is firmly attached to the underlying choroid and overlying retinal visual cells [34]. With the increase in cell therapy in preclinical phase of treatment and stem cell research of biomanufacturing it is important to validate transplant function in a noninvasive method. Recently, researchers [35] have developed characterization methodology with available set of image data samples through deep convolutional neural network. Deep learning is operated over quantitative bright-field absorbance microscopy (QBAM) captured data to predict tissue function and cellular donor identity Due to unavailability of really large data samples, the said work had to post process the inferences further by classical machine learning method through hand crafted features. The employment of our proposed CDLF can actually help this domain of research to even achieve improved accuracy of result by continuous addition of labeled data samples.

6 Conclusion

In this age of Industry 4.0, IoT and cyber-physical systems are taking over almost all the domains of importance. Healthcare, being a part of essential services, is not an exception. It is, on the contrary, a specialized field to look at from the AI and deep learning perspective, too. It is, in one hand, important to achieve even better diagnosis accuracy; on the other hand availability of labeled data and even labeling experts are prominent challenge. This current work, hence, proposed a detailed architecture of Continual Deep Learning Framework (CDLF) which is generic as far as the application domain is concerned. Also, the necessity of light weight yet high accurate model requirement of CDLF in medical diagnosis is honored. The current work not only attempts to create a generic self-learning framework but also targets to achieve highly accurate medical diagnosis system to ensure earliest detection of disease, powered by AI.

The proposed framework is not only restricted to the application of medical diagnosis but also inspires other industrial verticals like retail and manufacturing. For object detection or verification based planogram compliance, retail shrinkage detection, manufacturing part defect classification and severity mapping are just few application of the proposed continuous deep learning framework (CDLF).

For the researchers in this field, this proposed generic framework can be the starting point to employ. The framework is flexible to be modified based on specific target solutions in cross-domain. The "human in loop" idea for cross-validation is to be honored for first few iteration of active learning. Future research might look for parameters to make the requirement of cross-validation lean. There might be further detailed mathematical model of the proposed continuous deep learning framework (CDLF) to not only drive multi-threaded/ multi-processor implementation but also partitioning the processing requirement between local and enterprise level servers.

Permission of dataset usage The required permission has been obtained from Jaeger [26] for using the dataset in our experiments. The approval letter has been attached as supplementary material with the manuscript.

Author Contributions Pallavi Saha and Apurba Das conceived the study. Pallavi Saha developed the model, carried out the numerical experiments, analyzed the data. Pallavi Saha wrote the first draft of the manuscript and Apurba Das reviewed, edited and finalized the final manuscript.

References

1. M. Zhu, A. Badii, Cross-modal semantic-associative labelling, indexing and retrieval of multimodal data, in *Multiple Sensorial Media Advances and Applications: New Developments in MulSeMedia*, pp. 234–257, 01 2011
2. S. Budd, E. Robinson, B. Kainz, A Survey on Active Learning and Human-in-the-Loop Deep Learning for Medical Image Analysis, 10 2019
3. C. Baweja, B. Glocker, K. Kamnitsas, Towards continual learning in medical imaging, in *CoRR*, abs/1811.02496 (2018)
4. M. Hasan, A.K. Roy-Chowdhury, A continuous learning framework for activity recognition using deep hybrid feature models. IEEE Trans. Multimedia **17**(11), 1909–1922 (2015)
5. M. Hasan, A.K. Roy-Chowdhury, Continuous learning of human activity models using deep nets, in *Computer Vision – ECCV 2014*, ed. by D. Fleet, T. Pajdla, B. Schiele, T. Tuytelaars (Springer International Publishing, Cham, 2014) , pp. 705–720
6. R. Minhas, A.A. Mohammed, Q.M.J. Wu, Incremental learning in human action recognition based on snippets. IEEE Trans. Circ. Syst. Video Technol. **22**(11), 1529–1541 (2012)
7. Y. Yang, K.E. Niehaus, T.M. Walker, Z. Iqbal, A. Sarah Walker, D.J. Wilson, T.E.A. Peto, D.W. Crook, E. Grace Smith, T. Zhu, D.A. Clifton, Machine learning for classifying tuberculosis drug-resistance from DNA sequencing data. Bioinformatics **34**(10), 1666–1671 (2017)
8. M. Beccaria, T. Mellors, J. Petion, C. Rees, M. Nasir, H. Systrom, J. Sairistil, M.-A. Jean-Juste, V. Rivera, K. Lavoile, P. Severe, J. Pape, P. Wright, J. Hill, Preliminary investigation of human exhaled breath for tuberculosis diagnosis by multidimensional gas chromatography - time of flight mass spectrometry and machine learning. J. Chromatography B **1074–1075**, 01 (2018)
9. A. Mwinga, P. Bernard Fourie, Prospects for new tuberculosis treatment in africa. Tropical Med. Int. Health : TM IH **9**, 827–832 (2004)
10. L. Bird, The tuberculosis signature. Nature Rev. Immunol. **10**(10), 677 (2010)
11. C.-Y. Chiang, C. Weezenbeek, T. Mori, D. Enarson, Challenges to the global control of tuberculosis, in *Respirology (Carlton, Vic.)*, 18, 04 2013
12. J. Melendez, C. Sánchez, R. Philipsen, P. Maduskar, R. Dawson, G. Theron, K. Dheda, B. Ginneken, An automated tuberculosis screening strategy combining x-ray-based computer-aided detection and clinical information. Sci. Rep. **6**, 25265 (2016)
13. A. Karargyris, J. Siegelman, D. Tzortzis, S. Jaeger, S. Candemir, Z. Xue, K.C. Santosh, S. Vajda, S. Antani, L. Folio, G. Thoma, Combination of texture and shape features to detect pulmonary abnormalities in digital chest x-rays. Int. J. Comput Assisted Radiol. Surg. **11**, 06 2015
14. K. Sau, P. Saha, *Classification of Skin Cancer: ANN Trained with Scaled Conjugate Gradient Algorithm*, pp. 134–143, 06 2019
15. F. Milletari, N. Navab, S.-A. Ahmadi, *V-net: Fully Convolutional Neural Networks for Volumetric Medical Image Segmentation*, pp. 565–571, 10 2016
16. O. Ronneberger, P. Fischer, T. Brox, *U-net: Convolutional Networks for Biomedical Image Segmentation*, vol. 9351, pp. 234–241, 10 2015
17. K. Sau, P. Saha, A. Ghosh, A. Chanda, *Segmentation of Ventricle and Feature Extraction*
18. P. Saha, S. Chatterjee, S. Roy, S. Sen, *Normal Pressure Hydrocephalus Detection Using Active Contour Coupled Ensemble Based Classifier*, pp. 543–553, 01 2020
19. X. Wang, Y. Peng, L. Lu, Z. Lu, M. Bagheri, R.M. Summers, Chestx-ray8: Hospital-scale chest x-ray database and benchmarks on weakly-supervised classification and localization of common thorax diseases, in *2017 IEEE Conference on Computer Vision and Pattern Recognition (CVPR)*, pp. 3462–3471 (2017)
20. A. Acharya, R. Satapathy, A deep learning based approach towards the automatic diagnosis of pneumonia from chest radio-graphs. Biomed. Pharmacol. J. **13**, 449–455 (2020)
21. S. Hwang, H.-E. Kim, J. Jeong, H.-J. Kim, *A Novel Approach for Tuberculosis Screening Based on Deep Convolutional Neural Networks*, p. 97852W, 03 2016
22. M. Islam, M. Abdul Aowal, A. Tahseen Minhaz, K. Ashraf, *Abnormality Detection and Localization in Chest x-rays Using Deep Convolutional Neural Networks*, 05 2017

23. A. Krizhevsky, I. Sutskever, G. Hinton, Imagenet classification with deep convolutional neural networks. Neural Inf. Process. Syst. **25**, 01 (2012)
24. K. Simonyan, A. Vedaldi, A. Zisserman, *Deep Inside Convolutional Networks: Visualising Image Classification Models and Saliency Maps*, preprint, 12 2013
25. T. Tieleman, G. Hinton, Lecture 6.5—RmsProp: Divide the gradient by a running average of its recent magnitude. COURSERA: Neural Netw. Mach. Learn. (2012)
26. S. Jaeger, S. Candemir, S. Antani, Yì-Xiáng Wáng, P.-X. Lu, G. Thoma, Two public chest x-ray datasets for computer-aided screening of pulmonary diseases. Quantitative Imaging Med. Surg. **4**, 475–477 (2014)
27. S. Jaeger, A. Karargyris, S. Candemir, L. Folio, J. Siegelman, F. Callaghan, Z. Xue, K. Palaniappan, R. Singh,S. Antani, Automatic tuberculosis screening using chest radiographs. IEEE Trans. Med. Imaging **33**, 233–245 (2014)
28. F. Pasa, V. Golkov, F. Pfeiffer, D. Cremers, D. Pfeiffer, Efficient deep network architectures for fast chest x-ray tuberculosis screening and visualization. Sci. Rep. **9**(1), 6268 (2019)
29. C. Szegedy, W. Liu, Y. Jia, P. Sermanet, S.E. Reed, D. Anguelov, D. Erhan, V. Vanhoucke, A. Rabinovich, Going deeper with convolutions, in *CoRR*, abs/1409.4842 (2014)
30. B. Tymchenko, P. Marchenko, D. Spodarets, Deep learningapproach to diabetic retinopathy detection, in *Proceedings of the 9th International Conference on Pattern Recognition Applications and Methods*, (2020)
31. M.A. Kadampur, S. Al Riyaee, Skin cancer detection: Applying a deep learning based model driven architecture in the cloud for classifying dermal cell images. Inform. Med. Unlocked. **18**, (2020)
32. L. Peng, V. Golshan, Deep learning for detection of diabetic eye disease. Google AI Blog, (2019)
33. A. Esteva, B. Kuprel, R.A. Novoa, J. Ko, S.M. Swetter, H.M. Blau, S. Thrun, Dermatologist-level classication of skin cancer with deep neural networks. Nature. **542**(7639), 115–118 (2017)
34. M.M. Boyer, G.L. Poulsen, T.M. Nork, Relative contributions of the neurosensory retina and retinal pigment epithelium to macular hypoflurocence. Arch. Ophthalmol. **118**, (2000)
35. N.J. Schaub, N.A. Hotaling, P. Manescu, S. Padi, Q. Wan, R. Sharma, A. George, M. Ouladi, J. Chalfoun, M. Simon, C.G. Simon Jr., P. Bajcsy, K. Bharti, Deep learning predicts function of live retinal pigment epithelium from quantitative microscopy. J. Clinic. Invest. **130**(2), 1010–1023 (2000)

KannadaRes-NeXt: A Deep Residual Network for Kannada Numeral Recognition

Aradhya Saini, Sandeep Daniel, Satyam Saini, and Ankush Mittal

Abstract The Kannada Numeral system is unique to Kannada language. In our knowledge very little work has been previously attempted towards the recognition of Kannada handwritten numerals. Therefore, this chapter proposes a framework to correctly recognise Kannada handwritten numerals. The proposed KannadaRes-NeXt framework is based on deep residual network ResNeXt to correctly classify the images of handwritten Kannada numerals. The *Kannada-MNIST* dataset comprises of 60,000 images for training with an equal number of images for each of the ten glyphs in the Kannada numeral system drawn by native speakers of the Kannada language. This dataset also comprises of *main test set* with 10,000 images for testing purpose drawn by native speakers. *Dig-MNIST* dataset serves as an out-of-domain test dataset and comprises of 10,240 images of glyphs drawn by people unfamiliar with the Kannada language. In order to further evaluate the robustness of our Kannada Numeral Recognition system we have introduced two other datasets of 10,000 *AugKannada-MNIST* test set images and 10,240 *AugDig-MNIST* dataset images to their respective test sets. KannadaRes-NeXt is observed to have achieved an accuracy of 97.36% on the *Kannada-MNIST* test set, an accuracy of 79.06% on *Dig-MNIST* dataset, accuracies of 96.32 and 77.31% on augmented *main test set* and augmented *Dig-MNIST* test set respectively. These results also indicate efficiency of our proposed framework as KannadaRes-NeXt performs better than other deep learning techniques for Kannada Numeral Recognition.

A. Saini (✉)
Indian Institute of Technology, Roorkee, India
e-mail: aradhya.saini91@gmail.com

S. Daniel
VIT Chennai, Chennai, India
e-mail: sandeepjoshuadaniel@gmail.com

S. Saini
Jaypee Institute of Information Technology, Noida, India
e-mail: satyamsaini97@gmail.com

A. Mittal
Raman Classes, Roorkee, India
e-mail: dr.ankush.mittal@gmail.com

P. Kumar et al. (eds.), *Machine Learning for Intelligent Multimedia Analytics*,
Studies in Big Data 82,
https://doi.org/10.1007/978-981-15-9492-2_4

Keywords Deep residual network · Recognition · Kannada numerals · ResNeXt · CNN · VGG16 · Handcrafted

1 Introduction

In recent times there has been rapid progress in the technology which has increased the popularity of handheld devices. In the recent scenario, mobiles phones and tablet personal computers (PCs) have turned out to be an integral part of our day by day lives. This has led to the development of a system termed as handwriting recognition which serves as an alternative to various input devices and keyboard. Various approaches have been invented in the past for the recognition of the printed characters of Indian languages. However, fewer attempts have been made for handwritten character recognition. Consequentially, a high degree of recognition accuracy for the handwritten characters is yet to be achieved [20]. Handwritten numeral recognition is a subset of handwritten character recognition. This method defines the ability to identify the numbers correctly which are provided as an input image by the machine [43]. Handwritten numeral recognition introduces more complexity when compared to printed numeral recognition due to its variations observed in the writing style. The handwritten numeral recognition system is segregated into two streams which are online and offline [24]. Offline numeral recognition is considered much harder than online recognition due to unavailability of any dynamic information pertaining to strokes [24]. Optical Character Recognition (OCR) is termed as offline character recognition system [25]. The motivation for our work towards handwritten Kannada numeral recognition is drawn from :

- The advancements in various fields have been communicated in the English language due to its global usage hence the local Indian languages are seen more as barriers rather than assets, in the path of knowledge dissemination for the localities. In order to bridge this gap, recognition and translation of local scripts is essential.
- The ultimate aim is to build a multilingual OCR program to translate and process the multilingual documents automatically and efficiently.
- Recognition of the handwritten characters is a popular and interesting research area. Nowadays it has various potential applications in fields such as automatic bank cheque processing [2], share certificate sorting [2], form filling [33], word processing, postal automation [24], signature verification and automatic zip code reading from the postal addresses present on envelopes [2].
- A number of works have already been carried out for Roman, Arabic and Japanese scripts [25, 40]. Most of the research in the Indian scripts is carried out for Bangla and Devnagari [20, 25, 33, 40]. There is still deficiency in efficient Indic scripts handwritten character recognition systems [25]. As for the Kannada script very few works have been carried out towards recognition of Kannada characters [5, 9, 20, 30]. Moreover, lesser number of works have been reported exclusively for Kannada numeral recognition [34, 40].

The aforementioned reasons have been the inspirations driving our work for development of Kannada Numeral Recognition system.

Kannada is a Dravidian language derived from the Brahmi script and is commonly spoken among the people of the south-western state of Karnataka in India, among an exceptional number of linguistic minorities in states of Maharashtra, Tamil Nadu, Telangana, Andhra Pradesh, and Kerala in India as well as by people abroad. Karnataka, a state in India, has recognised Kannada as the official and administrative language. Kannada is also considered as the transactional language in Karnataka. Kannada is the major south Indian character which is verbal by about 50 million people [18]. Due to its widespread usage Kannada is also classified into one of the groups among the official languages of India. Handwritten character recognition faces a number of composition style problems such as in terms of shape, thickness of characters etc. [24, 25, 43]. Kannada text recognition faces few major challenges [25] and the curious case of Kannada glyphs for 3,7,6 poses as a challenge during numeral recognition [32]. The motivation stems from a limited amount of research being carried out in the direction of Kannada Numeral Recognition. Few important reasons supporting our work are listed as follows:

- Consider the study of handwritten Kannada numerals and build a recognition system as the initial work to meet the objective of processing bi-lingual documents. This acts as a strong motivation.
- Apart from the perspective of philanthropy, the existing technological approaches for recognition of handwritten Kannada numerals are prone to limitations as elaborated in the previous sections. Thus the importance of the solution to this problem statement and the absence of a reliable, robust, apt as well as updated technique to solve this task serves as the motivation for us.
- This work is considered as an attempt to take a step towards provision of stability from the aforementioned existing situations for Kannada Numeral Recognition.

There are 4 vowels along with 34 consonants in Kannada script. Distinct glyphs are used to represent numbers 0–9 which are different from the modern Hindu-Arabic numerals used today. Unlike other historical numeral systems these are used extensively in navigating through day to day lives of more than 50 million people. We can see the glyphs present in the *Kannada-MNIST* dataset [32] used in Fig. 1. These images give us the shape of the numerals used in modern society in Karnataka by native speakers of the language. As seen in the figure all images, even of the same number aren't exactly similar. Some of the glyphs may seem similar to the arabic numerals commonly used like 0 seems to be the same in both scripts and even the 7 in the Kannada script seems similar to 2 in the arabic script.

The rest of the chapter is discussed as follows: Sect. 2 describes Related Work and main contributions of our work. Proposed algorithm is discussed in Sect. 3 along with elaborated description of various feature extractors and classifiers and their respective real-life applications. In Sect. 4 experimental setup along with dataset description is discussed followed by discussion of results in Sect. 5. The analysis of these results is mentioned in Sect. 6 and the conclusion along with future work is discussed in Sect. 7.

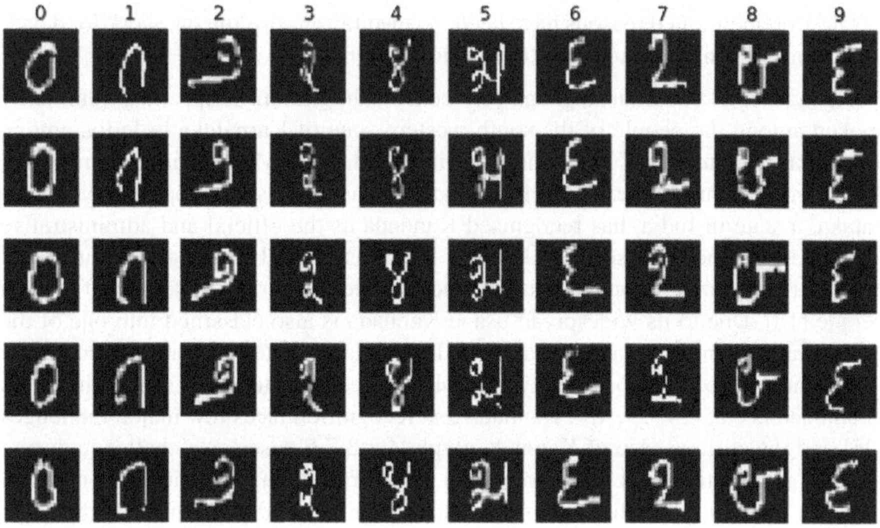

Fig. 1 Handwritten Kannada-MNIST numerals

2 Related Work

In the present scenario, computer vision and machine learning are powerful tools for object recognition [7, 36, 38, 41]. In [11] an accuracy of 98% was achieved on a dataset of 2300-digit images using a chain histogram which is a handcrafted feature extraction method. In [14] an accuracy of 91% is achieved on a test set of 50 numerals using K-nearest neighbors. In [16] SVMs were used to achieve 98% accuracy on a dataset of 5000 40 $\times$ 40 images. Even various optimization techniques have been introduced for achieving higher detection accuracy [35, 37].

Until recently there have been only small steps towards the problem of recognizing handwritten digits written in the Kannada script using image processing and computer vision techniques. The work in [28] is based on energy features and nearest neighbour classifier for Kannada handwritten digit recognition. An average recognition accuracy of 94.80% has been achieved on own Kannada handwritten digits dataset as ground truth datasets are unavailable. In [34] linear interpolation, subspace algorithm OLDA and nearest neighbour classifier techniques are computed on 1400 numeral data samples to achieve a recognition accuracy of 99.3%. However, in this work the number of samples is less and handcrafted methods are used for computation. A dataset of about 50,000 numeral string samples has been used for Kannada Numeral string recognition. Even though efficient techniques such as CNN (Convolutional Neural Network) and CTC (Connectionist Temporal Classification) based architecture has been implemented for numeral string recognition still the performance has been evaluated on comparatively lesser number of images in the dataset [48]. DCNN classifier along with distance and zone matrix is used for recognition of handwritten Kannada numeral script [18]. A testing accuracy of 93.13% has been achieved using CNN on a dataset of 20,200 samples of handwritten Kannada digits for handwritten digit classification system [45]. In [32] there has been creation of an MNIST type dataset for the Kannada script termed as *Kannada-MNIST* with a train set of 60,000 28 × 28 images and *main test set* of 10,000 28 × 28 images. Additionally, they created *Dig-MNIST* dataset with 10,000 images of digits drawn by non-native speakers of the language with limited familiarity with the script. They achieved an accuracy of 97.1% using a CNN using Adadelta as the optimizer.This up-to-date dataset is very beneficial for proposing and evaluating new models [19].

Even though powerful deep learning model such as CNN has already been variedly discussed above for Kannada Numeral Recognition still we aim to improve the accuracy for numeral recognition. Also, in the aforementioned literature we can observe the scarcity of images in Kannada Numeral Recognition dataset. This condition is catered to in [32] by introduction of *Kannada-MNIST* dataset with the hope to address the rarity of MNIST-sized dataset for Kannada language. For this same dataset, through our proposed deep residual network termed as KannadaRes-NeXt we have aimed to achieve an accuracy higher than in [32]. The description of the dataset is given in Sect. 4.1

2.1 Main Contributions of Our Work

The main contributions of our work are summarized as below:

1. We have proposed KannadaRes-NeXt as an efficient deep residual network for Kannada Numeral Recognition system. *Kannada-MNIST* and *Dig-MNIST* pose a sterner challenge with regards to generalizability than the MNIST or KMNIST datasets [32]. Through implementation of KannadaRes-NeXt architecture on both the datasets we have achieved commendable results.
2. Despite stringent challenges faced with regards to both the aforementioned datasets our KannadaRes-NeXt architecture is observed to have achieved higher Kannada Numeral recognition accuracies as compared to accuracies achieved on these datasets in [32].
3. We have introduced novelty in our work by testing our KannadaRes-NeXt architecture for robustness on 20,000 *AugKannada-MNIST+Kannada-MNIST* test set images (augmented *main test set*)and 20,480 *AugDig-MNIST +Dig-MNIST* dataset images (augmented *Dig-MNIST* test set).We have contributed 10,000 *Aug Kannada-MNIST* test set images and 10,240 *AugDig-MNIST* dataset images to their respective test sets for evaluation of our proposed architecture.
4. Our architecture has achieved high recognition rates on both these newly introduced datasets also.
5. It is observed that proposed KannadaRes-NeXt architecture has achieved marginally higher accuracies on all the four datasets: *Kannada-MNIST* test set, *Dig-MNIST* test set,*AugKannada-MNIST+ Kannada-MNIST* test set,*AugDig-*

MNIST +Dig-MNIST dataset when compared with other efficient deep learning models also such as Convolutional Neural Networks (3 convolutional layers), VGG16.

3 Proposed Algorithm

ResNeXt uses more parameters than its predecessors due to its architecture which gives it the advantage in learning from images as compared to other models. Only the *Kannada-MNIST* train dataset has been used for training the model. The *Kannada-MNIST* test and *Dig-MNIST* datasets along with respective augmented datasets *AugKannada-MNIST+Kannada-MNIST* test set and *AugDig-MNIST+Dig-MNIST* set have been used for testing the performance of the various models along with the proposed model. The proposed KannadaRes-NeXt model takes in a single image and classifies it into one of ten classes. For the Convolutional Neural Networks (CNN) model the image is passed through multiple convolution layers as seen in Fig. 3 from where the features of the image are extracted. After this step the images are classified into one of the ten classes. Apart from Res-NeXt other models have also been tested to compare the results as seen in the flowchart of the proposed system in Fig. 2. For handcrafted features Local Binary Patterns (LBP), Histogram of Oriented Gradients (HOG) and Discrete Cosine Transform (DCT) & Discrete Wavelet Transform (DWT) feature extraction techniques have been used. Support Vector Machine (SVM) and K-Nearest Neighbor (k-NN) classifiers have been trained on the features extracted by the aforementioned techniques and the results have been recorded. Our approach unlike previous attempts uses the most recent successful CNN used on the Imagenet dataset which is the ResNesXt model.

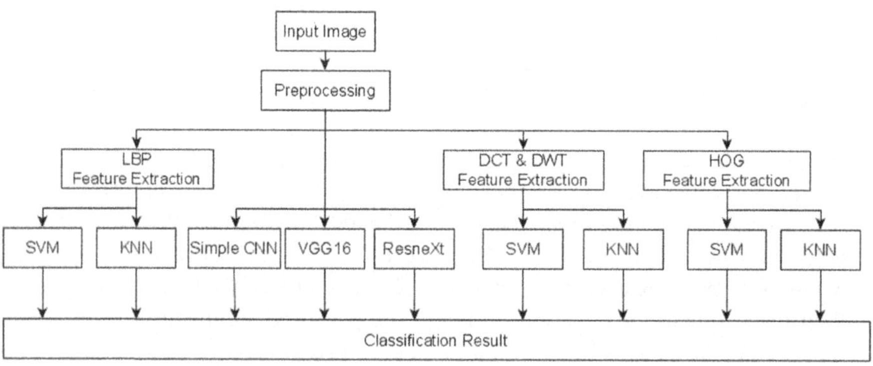

Fig. 2 Flowchart of the proposed system

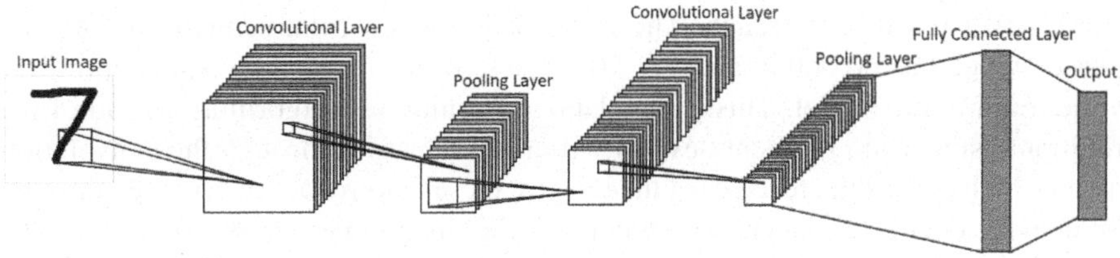

Fig. 3 Representation of a CNN model

3.1 Convolutional Neural Networks (CNN)

CNNs are a branch of Deep Neural Networks which are commonly used for analysing data based on images. CNNs use Convolutional layers as the hidden layers, which are the layers located between the input and output layers. The convolution layers convolve the image/feature map it receives and passes it to the next layer.

The input to a CNN is a tensor which has the shape (number of images) × (image height) × (image width) × (image depth), as the input. As it is passed through convolutional layers the image gets converted to a feature map which progressively reduces the dimensions of the width and height while increasing the depth.

Occasionally in place of a convolutional layer a pooling layer may be used. The pooling layer combines multiple pixels into one in a particular neighbourhood in a feature map. This is done to reduce the computational efforts needed. The size of the neighbourhood used for pooling is normally small, typically 2 × 2. Additionally, pooling can be of two types, *max* and *avg*, Max pooling takes the maximum value of all the values in the neighbourhood while average pooling takes the mean value of all the values in the neighbourhood.

Each neuron in the layer computes the value to be outputted by applying a function on the input values which arrives as the output from the previous layer. The function is decided upon by a vector of weights and biases. The vector containing the weights and biases is called the filter. During the learning process filter is modified so as to reduce the value of the loss function and increase the performance of the model [1, 23]. The architecture of a CNN model is shown in Fig. 3.

For comparison with the Res-NeXt model we have used a Simple CNN with three convolutional layers [15] as well as the established VGG16 CNN model [42].

3.2 Proposed ResNeXt Model

In recent years Convolution Networks and Neural Networks have cemented their superiority over traditional image feature extraction techniques like HOG [11], LBP [29], DWT and DCT [14, 16]. Since we have 6,000 images per class for training the model therefore, training the deep neural networks without using pre-trained weights is a feasible approach. The network is trained from scratch on the train dataset which has 60,000 images for 10 classes. We have used the Adam optimizer while training the model. This is an adaptive learning rate algorithm instead of the traditional stochastic gradient descent. After being passed through the convolution layers which extract the features of the images, these features are then passed through the densely connected layers which classify the image into one of 10 classes. The ResNeXt model [47] is based on the Resnet CNN with an additional dimension called cardinality. This means that in each block instead of a linear function, which was used for traditional neurons, nonlinear function is used for each path. The cardinality increases the number of paths and thus increases the number of functions improving on the performance of previous models as per the following equation:

$$y = \sum_{i=1}^{C} T_i(x) \tag{1}$$

where y is considered to be the output and $T_i(x)$ is the function transforming x analogous to a neuron. As seen in Fig. 4 the architecture of the proposed model is shown. Also, shown in this figure is the model comprising of multiple blocks and each block consists of numerous paths. The number of paths is decided upon by the

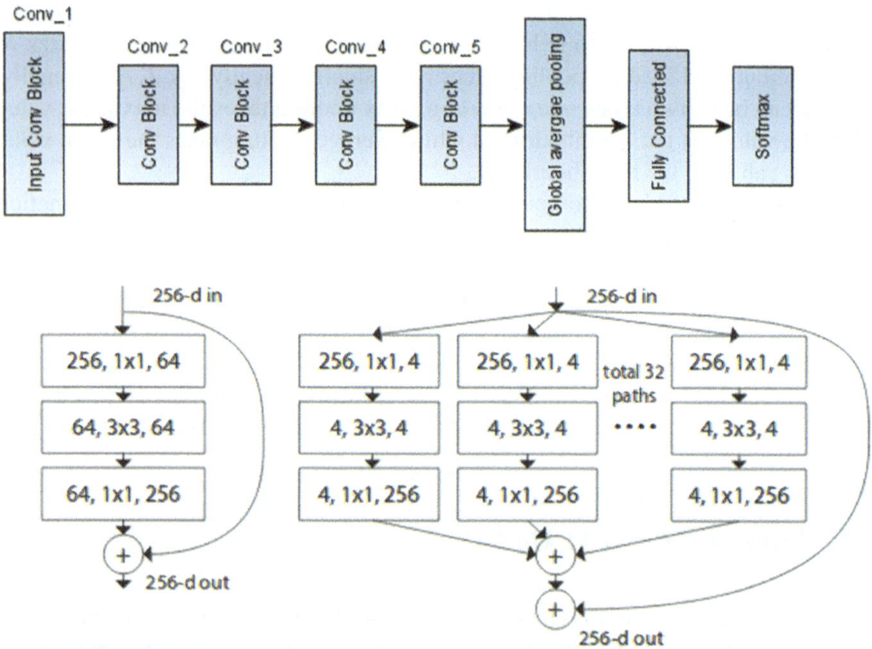

Fig. 4 Overall architecture of the ResNeXt model consisting of multiple Conv Blocks (top), Architecture of a Conv Block with cardinality = 32 (bottom)

cardinality. Each path consists of multiple convolution layers. All the paths in a block contain the same set of layers as each other.

The images need to be padded with null values to increase the size from a 28 × 28 image to a 32 × 32 image. This means that two columns of null values have been added to the left as well as to the right side of the image and two rows of null values are added to the bottom and top of the image. This is done due to the nature of CNNs where the length and width of the input images are reduced while the depth increases when being processed by the hidden convolution layers. The ResNeXt-50 model has been used without any pre-training on the datasets with the cardinality set on 32.

3.3 LBP Feature Extraction

Local Binary Patterns (LBPs) are texture descriptors. LBPs compute the local representation of a texture by comparison of each of the pixels with its surrounding neighbourhood of pixels. First the centre pixel is compared against its neighbouring eight pixels. In case the intensity of the pixel in the centre is larger or equal to the neighbouring value, then its value is made 1 otherwise this value is reset to 0. To calculate the LBP value for the center pixel, starting from any neighbouring pixel and moving clockwise or counter clockwise with a consistent ordering over all the pixels in the image, the values are stored as an 8-bit binary number which is later converted to an integer when required as shown in Fig. 5. This process is computed continuously over all the pixels which exist in the image to arrive at an LBP value for each pixel. The final step is creating a histogram for these values.

The following notation is used for the LBP operator: $\mathrm{LBP}^{u2}_{P,R}$. The subscript represents using the operator in a (P, R) neighbourhood which means the sampling of P points in a circle of radius R. The superscript u2 means using only uniform patterns and labelling all the remaining patterns with a single label [22]. After the image $f_l(x, y)$ labelled by LBP has been obtained Eq. 2 is used to define the LBP histogram.

$$H_i = \sum_{x,y} I\left\{f_l(x, y) = i\right\}, i = 0, \ldots, n-1, \quad (2)$$

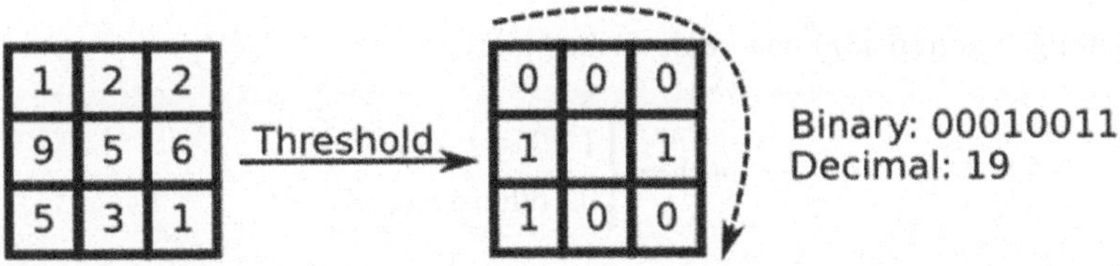

Fig. 5 Computation of LBP algorithm

where n is the number of different labels which are obtained from the LBP operator, and $I\{A\}$, if A is true is 1 if false is 0 [6, 31].

When the image patches are of different sizes (image patches whose histograms are going to be compared), the histograms must be normalized to get a comprehensible description. They are normalized using Eq. 3 where N_i is the normalized histogram and H_i is the non-normalized histogram.

$$N_i = \frac{H_i}{\sum_{j=0}^{n-1} H_j} \tag{3}$$

3.4 DCT and DWT Feature Extraction

DWT or Discrete Wavelet Transformation is used for dimensionality reduction. The information is encoded in different sub bands. It facilitates with corresponding frequency (of this signal) and the time associated with this signal enabling analysis in both the spatial as well as the frequency domain. When DWT is applied over a 2D image the image is processed using a 2D filter in each dimension. Low and High pass filters are applied to every row and column in every decomposition as can be seen in Fig. 6. The result of the decomposition achieved with DWT is made up of a low frequency sub band (cA/LL) and three high frequency sub bands (cH, cV, cD or LH, HL, HH respectively). In this case, cA/LL is the approximation coefficient, cH/LH is the Horizontal (detail) coefficient, the cV/HL component is the Vertical (detail) coefficient and the cD/HH component is diagonal (detail) coefficient. In our chapter the approximation coefficients matrix has been used since it is closer to the original image which comprises of reduced image dimensions. Although there are many forms of DWTs like Haar wavelets [26], Daubechies wavelets and dual-tree complex wavelet, we have used Haar wavelets for our experiment. The Haar wavelet is a sequence of rescaled square shaped functions. The Haar wavelets mother function can be described as:

$$\psi(t) = \begin{cases} 1 & 0 \leq t < \frac{1}{2}, \\ -1 & \frac{1}{2} \leq t < 1, \\ 0 & \text{otherwise.} \end{cases} \tag{4}$$

The scaling function $\varphi(t)$ can be described as:

$$\varphi(t) = \begin{cases} 1 & 0 \leq t < 1, \\ 0 & \text{otherwise.} \end{cases} \tag{5}$$

The Discrete Cosine Transform (DCT) is necessary in order to extract features. It converts the data within the image into an elementary frequency component. Here the sequence of the data points is shown as the sum of the cosine function which vibrates at different frequencies. In case of DCT higher value coefficients are stored

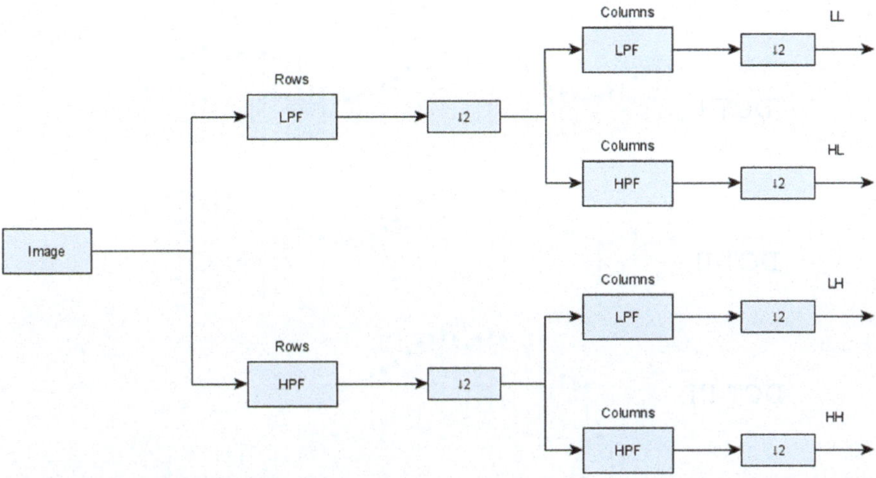

Fig. 6 Wavelet decomposition of an image

on upper left of the corner while the lower value coefficients are stored on bottom right of this matrix. The coefficients are saved in a zig-zag pattern as seen in Fig. 8.

Formally, DCT is a function which is both invertible linear $f : R^N \rightarrow R^N$, where R is the set of all real numbers. There are a number of different types of DCT functions [8].

$$X_k = \frac{1}{2}(x_0 + (-1)^k x_{N-1}) + \sum_{n=1}^{N-2} x_n \cos\left[\frac{\pi}{N-1}nk\right] \qquad k = 0, \ldots, N-1 \tag{6}$$

Equation 6 is known as DCT-I, it is equivalent (upto a scale factor of 2) to a Discrete Fourier Transform of $2N - 2$ real numbers with even symmetry.

$$X_k = \sum_{n=0}^{N-1} x_n \cos\left[\frac{\pi}{N}\left(n + \frac{1}{2}\right)k\right] \qquad k = 0, \ldots, N-1. \tag{7}$$

Equation 7 is known as DCT-II and is the most commonly used form of DCT. This transform is equivalent (upto a scale factor of 2) to a Discrete Fourier Transform of $4N$ real inputs of even symmetry where even indexed elements are zero.

$$X_k = \frac{1}{2}x_0 + \sum_{n=1}^{N-1} x_n \cos\left[\frac{\pi}{N}n\left(k + \frac{1}{2}\right)\right] \qquad k = 0, \ldots, N-1 \tag{8}$$

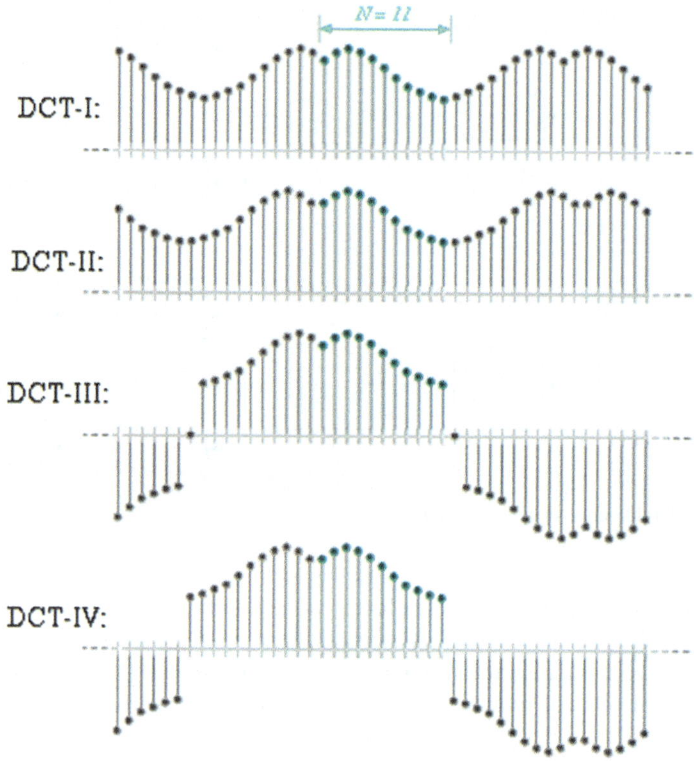

Fig. 7 The even/odd extensions of DCT input data, for $N = 11$ data points

Equation 8 is known as DCT-III, because it is the inverse of the DCT-II function. It is sometimes also called the inverse DCT (IDCT).

$$X_k = \sum_{n=0}^{N-1} x_n \cos\left[\frac{\pi}{N}\left(n + \frac{1}{2}\right)\left(k + \frac{1}{2}\right)\right] \qquad k = 0, \ldots, N-1 \tag{9}$$

Equation 9 is called DCT-IV. The DCT-IV matrix turns into an orthogonal matrix and its own inverse (since it is symmetric) if it is further multiplied by a scale factor of $\sqrt{\frac{2}{N}}$.

There are more DCT types ranging from DCT-V to DCT-VIII. These are different from DCT-I to DCT-IV in a way because types I-IV treat both boundaries consistently around the point of symmetry. They are either even/odd around a data point for both the boundaries or even halfway between the two data points for both boundaries as seen in Fig. 7. DCTs of types V-VII alternatively imply boundaries which are even/odd around a data point for one of the boundary and halfway between two of the data points for another boundary.

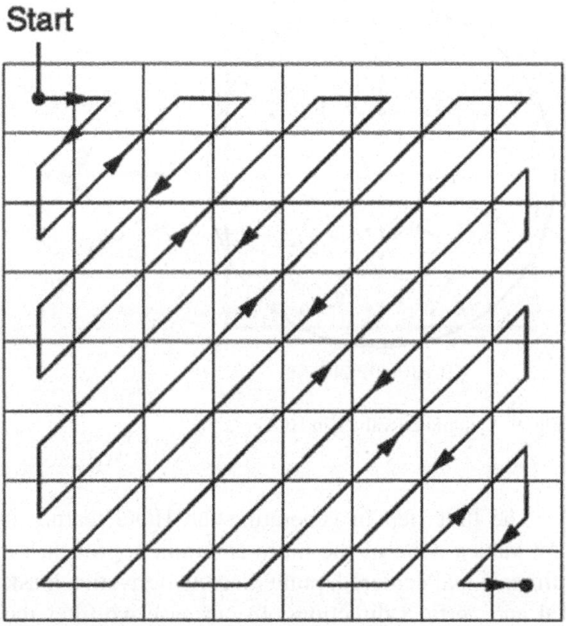

Fig. 8 Zig-Zag pattern in which coefficients are stored in DCT

In our case, first the pre-processed images from 28 × 28 pixels are processed by DWT. A single level two dimensional DWT of this image is calculated using the haar wavelet. DWT breaks down the image in the approximation coefficient, vertical coefficient, horizontal coefficient and also the diagonal coefficient but only the approximation coefficient is used. After performing the first transformation, DCT is applied upon the approximation coefficient (cA) segment of the corresponding DWT output. The DCT is known to output 2D discrete cosine transform of approximation coefficient.

3.5 HOG Feature Extraction

The HOG descriptor is used for identifying the structure (shape) of the object. HOG feature is able to provide information based on edge detection as well as the edge direction. This is achieved via extraction of the gradient as well as the orientation (may say magnitude as well as the direction) of the edges. Also, the orientations can be seen to be calculated in the 'localized' regions. This implies that the entire image is divided into certain number of smaller regions and corresponding to each of these regions, the gradients and orientations are calculated as seen in Fig. 9. After this the HOG feature generates histogram for each of these segmented regions separately. The generated histograms are sourced using the gradients along with the orientations of pixel values.

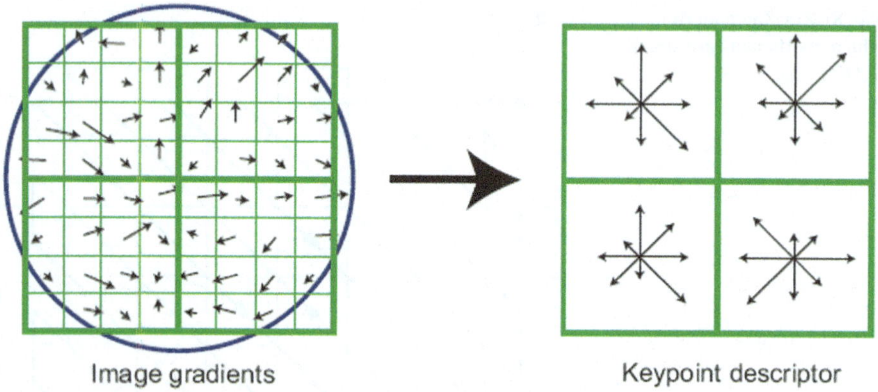

Fig. 9 Calculated values in HOG

The first step in generating the HOG feature is the calculation of the gradient values. One of the more common approaches to this involves applying a one dimensional centered, point discrete derivative mask on one or both of the horizontal and vertical directions. In our case we filter the image for the intensity using $[-1, 0, 1]$ and $[-1, 0, 1]^{\top}$, as the filter kernels.

In second step the cell histograms are created. Each pixel which is present in a cell casts a weighted vote for a histogram channel on the basis of its orientation which in turn is based on the values calculated in the first step. The cells may be rectangular or radial and depending on whether the gradient is signed or unsigned the histogram channel may also be spread evenly over 0° to 180° or 0° to 360°.

To maintain for changes in contrast and illumination, the gradients are normalized by grouping the cells together into the larger blocks. The HOG descriptor is considered as the vector which results from the concatenation of all of the normalized cell histograms obtained from all the regions within the blocks. These blocks frequently overlap, so it is probable that each cell contributed more than once. There are 2 main block shapes used for this purpose which are the R-HOG and C-HOG. The R-HOG are square grids which are represented by 3 values: the number of cells contained per block, the number of pixels available per cell and the number of channels present per histogram. The C-HOG are circular and normally have 2 variants: the ones which have a single central cell and the ones which have an angularly divided central cell. Regardless of the type of C-HOG these blocks are defined by these 4 values: the number of angular bins, the figure of the sum of radial bins, the radius of the central bin and the expansion factor for the radius of the additional radial bins.

There exist four methods for normalizing blocks. Let v be a non-normalized vector which contains all the histograms present in a block, $\|v\|_k$ is its k-norm for $k = 1, 2$ and e is a small constant.

$$\text{L2-norm:} \quad f = \frac{v}{\sqrt{\|v\|_2^2 + e^2}} \tag{10}$$

$$\text{L1-norm:} \quad f = \frac{v}{(\|v\|_1 + e)} \tag{11}$$

$$\text{L1-sqrt:} \quad f = \sqrt{\frac{v}{(\|v\|_1 + e)}} \tag{12}$$

In addition to Eqs. 10, 11 and 12 another method can be used called L2-hys which involves using L2-norm and then clipping v such that it cannot exceed the value of 0.2 and then normalizing the result.

3.6 Classifiers

For classification purpose KNN (K-Nearest Neighbors) and SVM (Support Vector Machine) have been used.

3.6.1 KNN

The K-Nearest Neighbors model [4] is a supervised machine learning classification technique. The algorithm assumes that similar data points are in close range of each other. The data points are vectors which exist in a multidimensional feature space. During the training phase the feature vectors as well as the class labels are stored. In the classification phase, the unlabelled vector is classified by assigning the label which is the most frequent among k (k is a user defined natural number) training samples that are nearest to these data points as seen in Fig. 10. The distance between data points is calculated by using the Euclidean distance metric.

KNN provides us with highly consistent results. As the amount of data available approaches infinity a 2-class is guaranteed an error rate less that twice the Bayes

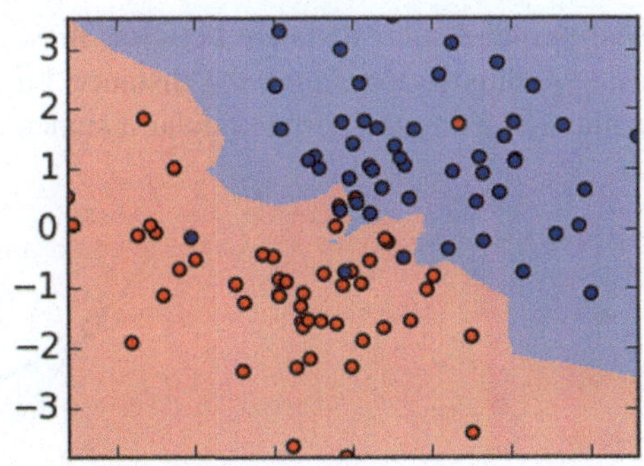

Fig. 10 Classification performed using KNN

error rate [44]. For a Multi-Class KNN classification the upper bound on the error rate is given as Eq. 13 [10].

$$R^* \le R_{kNN} \le R^* \left(2 - \frac{MR^*}{M-1}\right) \tag{13}$$

3.6.2 SVM

The Support Vector Machine (SVM) is also a supervised classification algorithm in machine learning. Here every data point is plotted in the form of an n-D(dimensional) space (here, n is number of computed features) in which value of each of these features is equivalent to the value of the coordinate. After this, classification is then performed by identifying the hyper-plane that optimally helps in differentiating between the classes as seen in Fig. 11.

Given a training set of n points where the points are in the form (x_i, y_i) where y_i is either positive 1 or negative 1, which indicates the class the data point belongs to and $\vec{x}_i$ is a real vector of z dimensions which belongs to the feature space. A hyperplane can be drawn using the data points $\vec{x}_i$ which satisfies Eq. 14, where $\vec{w}$ is a normal vector to the hyperplane.

$$\vec{w} \cdot \vec{x} - b = 0 \tag{14}$$

If the data is linearly separable, then it is possible to create two parallel hyperplanes which separate the two classes of data such that the distance between the two classes is as large as possible. The maximum margin hyperplane is the hyperplane between them. The margin is the perpendicular distance between the two hyperplanes. These equations can be represented by Eqs. 15 and 16 (applicable for normalized data).

$$\vec{w} \cdot \vec{x} - b = +1 \tag{15}$$

$$\vec{w} \cdot \vec{x} - b = -1 \tag{16}$$

The perpendicular distance between these two parallel hyperplanes can be given as $\frac{2}{\|\vec{w}\|}$, thus to maximise the distance $\|\vec{w}\|$ must be minimized. This distance is computed using the perpendicular distance from a point to a plane.

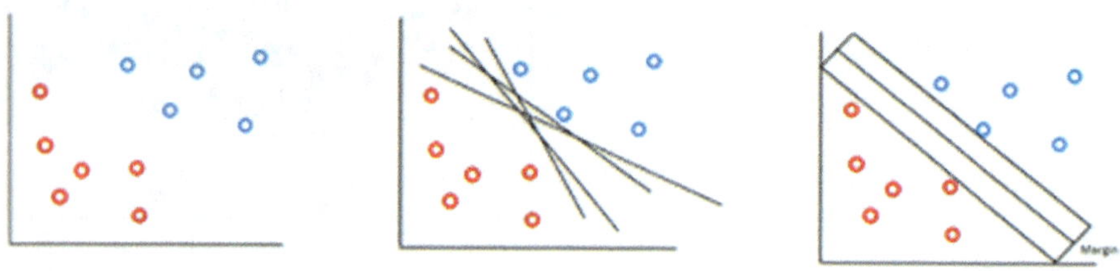

Fig. 11 SVM as used for classification

In some cases where data is not linearly separable hinge loss is used, as seen in Eq. 17

$$\max(0, 1 - y_i(\vec{w} \cdot \vec{x_i} - b)) \tag{17}$$

Where y_i is the i-th target (either positive 1 or negative 1) and $\vec{w} \cdot \vec{x_i} - b$ is the value of the output at present. This function is zero if $\vec{x_i}$ satisfies Eq. 18. i.e if it lies on the correct side of the margin, if it is on the incorrect side of the hyperplane, then the functions value is directly proportional to the perpendicular distance of the point to the hyperplane.

$$y_i(\vec{w} \cdot \vec{x_i} - b) \geq 1, \qquad \text{for all } 1 \leq i \leq n \tag{18}$$

In that case we need to minimize Eq. 19,

$$\left[\frac{1}{n}\sum_{i=1}^{n} \max\left(0, 1 - y_i(\vec{w} \cdot \vec{x_i} - b)\right)\right] + \lambda\|\vec{w}\|^2 \tag{19}$$

where λ is the determining factor in the trade-off between increasing the margin size and making sure that $\vec{x_i}$ lies on the correct side of the margin. So, for small enough values of λ the second term seen in the loss function tends to become so small it will be practically negligible, which will result in it behaving like an SVM with linearly separable data.

In cases where multiple classes exist there are traditionally two methods used. The One vs All approach or the One vs One approach. In the One vs All approach, suppose we have n classes, we create n classifiers such that for each classifier we have one positive class and the rest are treated as a single negative class. In smaller dataset this method is usable, however as the number of classes increase consequentially the size of the negative class also increases which may lead to a biased dataset. In the One vs One we create $\frac{n^2}{2}$ classifiers such that each classifier can classify between two classes. During the testing phase a scorecard is maintained in the form of a square matrix, such that each column and row represents one class, for each data point as it is being classified by each classifier. The row or column with the highest score is the class to which the data point is classified. This method avoids creating a bias but becomes computationally expensive with higher number of classes.

3.7 *Real-Life Applications: An Insight*

3.7.1 Convolutional Neural Networks (CNNs)

- CNNs are expected to be the future with their incorporation in driver less vehicles, in robotics that can imitate human actions, while aiding human genome mapping ventures, in forecasting earthquakes and natural hazards, and maybe even during self-diagnosis of medical conditions.

- In facial recognition CNN network is useful in decomposition of a face into its major components. Identification of every face in the picture is followed by focusing on each face in the presence of external factors such as light, angle, pose, etc. The unique features are then identified and compared with already existing information in the database in order to match a face with a name.
- CNNs can potentially provide powerful tools for classifying, identifying, and predicting patterns in climate and environmental data.

3.7.2 ResNeXt

- ResNeXt uses RestNeXt-101-FPN as a backbone for Mask R-CNN. This has given unprecedented results on COCO dataset and has been able to outperform when ResNet model works as a backbone.
- The ResNeXt-29 model has been used to compare and explore the impact of contextual information for semantic segmentation. Feature representation of shallow networks has been explored for image classification.
- In [21] a universal CNN has been trained which is capable of handling low-,mid- and high-level vision tasks which is then extended to exploit relatively recent models such as ResNet and ResNeXt. Also, a comparative analysis follows.

3.7.3 LBP

- The aim is to develop a real-time multi view face detector which is capable of real world environments which include mobile devices, video surveillance and content based video retrieval. The latest solution utilizes LBP functionality in a coarse to fine identification technique (pyramid architecture) which is embedded in a fast AdaBoost learning-based classification scheme [17].
- Visual speech recognition aims at determination of what a person is saying straight through the movements of the lip. This indeed plays an essential role in speech recognition under various noisy conditions or for those listeners who have hearing impairment. The Volume LBP approach is then applied to the task which is considered as visual speech recognition [49].
- In this work the images are first filtered using the LBP operator and then extraction of Haar-like features is performed. Consequentially these are then used with AdaBoost for developing a cascade of classifiers, thus providing a robust approach for eye detection [46].
- The goal is determination of the emotional state of the face. An approach to identification of facial expression from images which are static has been developed with the help of LBP histograms. These are computed over non-overlapping blocks for the face description such as happiness,surprise,sadness,anger,neutral,fear, and disgust regardless [13].

3.7.4 SVM

- SVM-Fold is a Binary one-vs-the-rest SVM model trained to learn relative weights between different one-vs-all classifiers. This is helpful to solve the incomparable yield prediction scores for individual structural classes [27].
- Real facial recognition is another application where SVM has been applied heavily. The use cases for this technology include specific surveillance applications in government security systems, attendance marking in educational institutes and user identification in shops and malls for automatic bill generation.
- Handwriting recognition is another domain for SVM model application. SVM based models are able to outperform ANN based models on the Letter Recognition data set which was taken from the UCI Machine Learning Repository.

3.7.5 K-Nearest Neighbor (K-NN)

- K-Nearest Neighbor (KNN) is a very basic, very common, extremely powerful and effective pattern recognition algorithm. KNN is a straight forward classifier, where samples are categorized based on their nearest neighbor's class. KNN is used in a variety of applications such as economic forecasting, data compression and genetics. K-nearest neighboring algorithm and non-linear regression method is used to forecast market prices for a sample of six major Jordanian stock exchange companies to help investors make the right investment decisions [3].
- A variant of KNN algorithm when applied on categorical data helps in compressing the data in order to decrease the memory requirements at the time of execution [39].
- Combining KNN with genetic algorithm helps perform global search in medical databases (which are high volume in nature, highly redundant and full of irrelevant attributes) and hence provide optimal solution. This method enhances the accuracy in diagnosis of heart disease [12].

4 Experimental Setup

A description of the dataset used along with the environment in which the model has been trained and tested is listed below in Sects. 4.1 and 4.2.

4.1 Dataset Description

The dataset used has been created by Vinay Uday Prabhu in [32]. The dataset is split into 3 categories: *Kannada-MNIST* train, *Kannada-MNIST* test(*main test set*) and *Dig-MNIST* test dataset. The train set contains 60,000 28 × 28 single channel images. The *main test set* and *Dig-MNIST* set contain 10,000 and 10,240 images of the same format each respectively. To compare the difference between these three

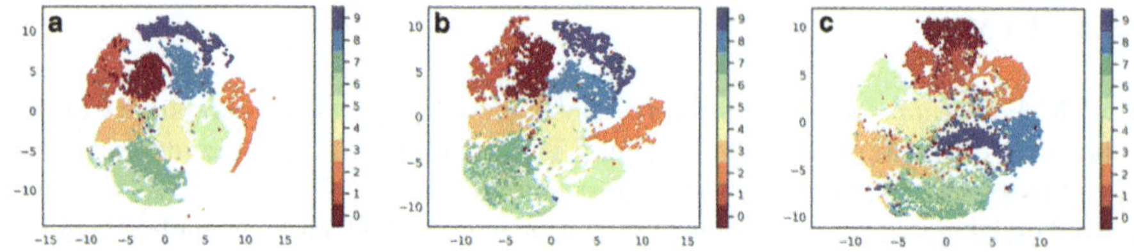

Fig. 12 **a** 2-D plot of the *Kannada-MNIST* train data; **b** 2-D plot of the *Kannada-MNIST* test data; **c** 2-D plot of the *Dig-MNIST* data

datasets we reduced the images within each set into a two-dimensional dataset using PCA and T-SNE for visualization purposes. As observed in Fig. 12 the *Kannada-MNIST* train and test data are largely similar with the relative location of each digit being the same. However the *Dig-MNIST* dataset is very different in the sense that the relative locations of some digits is completely different than what can be observed previously from the test and train dataset. This is where the complexity arises while testing. However, since we are reducing the image to just two dimensions from 784 dimensions (pixels), we can expect the digits to be largely similar since it is likely that data loss is taking place.

For further testing the robustness of our model we have introduced 20,000 *Aug Kannada-MNIST+Kannada-MNIST* test set images and 20,480 *AugDig-MNIST+Dig-MNIST* dataset images.These datasets include an additional 10,000 *AugKannada-MNIST* test set images and 10,240 *AugDig-MNIST* dataset images for testing the evaluation of our proposed architecture which are obtained after shearing *Kannada-MNIST* test(*main test set*) and *Dig-MNIST* test dataset respectively with a shear angle of 0.5 rad.

4.2 Experimental Scenarios

In order to test our proposed model, different scenarios were created to evaluate its performance.

1. **Scenario 1**: In this scenario both the *Kannada-MNIST* test(*main test set*) and *Dig-MNIST* test dataset have been used. These datasets are used without any modifications in order to obtain benchmark results on the model.
2. **Scenario 2**: In this scenario the images from the *AugKannada-MNIST+Kannada-MNIST* test set images and *AugDig-MNIST+Dig-MNIST* dataset images have been used. The modified images were combined with the normal respective dataset images and are used for testing the model's performance and robustness.

In both scenarios the models were trained only on the 60,000 images within the train set.

5 Results

The results of the aforementioned two scenarios (described in the previous section) are discussed below. Results of Scenario 1 describe the performance of our proposed model on the *Kannada-MNIST* test(*main test set*) and *Dig-MNIST(Digi)* test dataset without any modification. The results of Scenario 2 describe the performance on augmented datasets thus testing the robustness on the model. Despite the testing already being performed on *Dig-MNIST* test set this evaluation gives a better measure of how the model would perform under real world circumstances.

5.1 Experimental Scenario 1

In this scenario, models were trained on 60,000 images from the *Kannada-MNIST* train set and tested on the *Kannada-MNIST* test set and *Dig-MNIST* set of images. All 10,000 *main test set* images and 10,240 *Dig-MNIST* set images have been used for the testing of the models. The results are as seen in Table 1.

5.2 Experimental Scenario 2

In this scenario the models were trained on the images from the *Kannada-MNIST* train set and tested on the augmented *main test set* and augmented *Dig-MNIST* test set images. The images were augmented by shearing with an angle of 0.5 rad. While training 1% of the train data was used for validation so 600 images out of 60,000 were used for validation and the remaining 99% images were used for training. All the 20,000 augmented *main test set* test images and 20,480 augmented *Dig-MNIST* images have been used for the testing purpose. The results are shown in Table 2.

Table 1 Result of scenario 1

Model used	*Kannada-MNIST* test results		*Dig-MNIST* results	
	Accuracy (%)	Precision (%)	Accuracy (%)	Precision (%)
KNN with LBP feature extraction	36.16	36.58	14.80	15.57
SVM with LBP feature extraction	36.53	36.41	14.77	16.47
KNN with DCT and DWT feature extraction	75.34	81.55	47.66	61.92
SVM with DCT and DWT feature extraction	86.81	87.21	52.07	62.37
SVM with HOG feature extraction	94.02	94.12	82.29	82.99
KNN with HOG feature extraction	94.16	94.24	81.68	83.05
Simple CNN (3 Convolution layers)	97.15	97.26	78.60	78.13
VGG16	97.32	96.46	78.63	79.43
KannadaRes-NeXt (Proposed)	97.36	97.39	79.06	80.72

Table 2 Result of scenario 2

Model used	Augmented *main test* results		Augmented *Dig-MNIST* results	
	Accuracy (%)	Precision (%)	Accuracy (%)	Precision (%)
KNN with LBP feature extraction	29.64	30.86	11.56	13.68
SVM with LBP feature extraction	31.97	32.31	13.98	15.12
KNN with DCT and DWT feature extraction	29.83	32.75	14.82	17.16
SVM with DCT and DWT feature extraction	30.59	33.14	14.26	16.81
SVM with HOG feature extraction	50.24	76.36	47.12	62.76
KNN with HOG feature extraction	51.34	76.39	46.63	59.08
Simple CNN (3 Convolution layers)	91.62	93.52	69.76	70.93
VGG16	94.83	95.16	70.81	72.57
KannadaRes-NeXt (Proposed)	96.32	97.42	77.31	83.78

6 Analysis

In scenario 1 our proposed model KannadaRes-NeXt has achieved the highest accuracy with the *main test* dataset. Although the models which used HOG feature extraction had lower accuracies in the *main test* dataset but slightly higher in the *Dig-MNIST* dataset as compared to proposed framework as seen in Fig. 13. This is probably due to better generalization on the overall data and not just for *Kannada-MNIST* train and test data. But since the *Dig-MNIST* dataset is drastically different from the train and test data there is probably a trade-off between performance in the *Dig-MNIST* dataset and performance in the *main test* dataset.

In scenario 2 we see a large drop off in the performance of all other models which rely on handcrafted feature extraction methodologies while the CNN models have a minimal drop in performances when compared to KannadaRes-NeXt model as seen in Fig. 14. This goes to exhibit the superiority of CNNs as compared to conventional methods of image classification.The results of VGG16 are comparable to the results of the proposed KannadaRes-NeXt model in both scenarios. However, the KannadaRes-NeXt model is more versatile since it uses shortcut connections to overcome the vanishing gradient problem. The lower accuracies in the augmented *Dig-MNIST* dataset could be attributed to the dissimilarity between the datasets which was explored in 4.1. The relatively low drop in performance between scenarios could be attributed to the robustness of CNNs when compared to alternate methods of image classification. Of all the deep neural network models KannadaRes-NeXt has the overall better performance in both the scenarios with all the four datasets. The advantage is due to additional features in the network like the use of cardinality which allows for the use of a higher number of complex transformations.

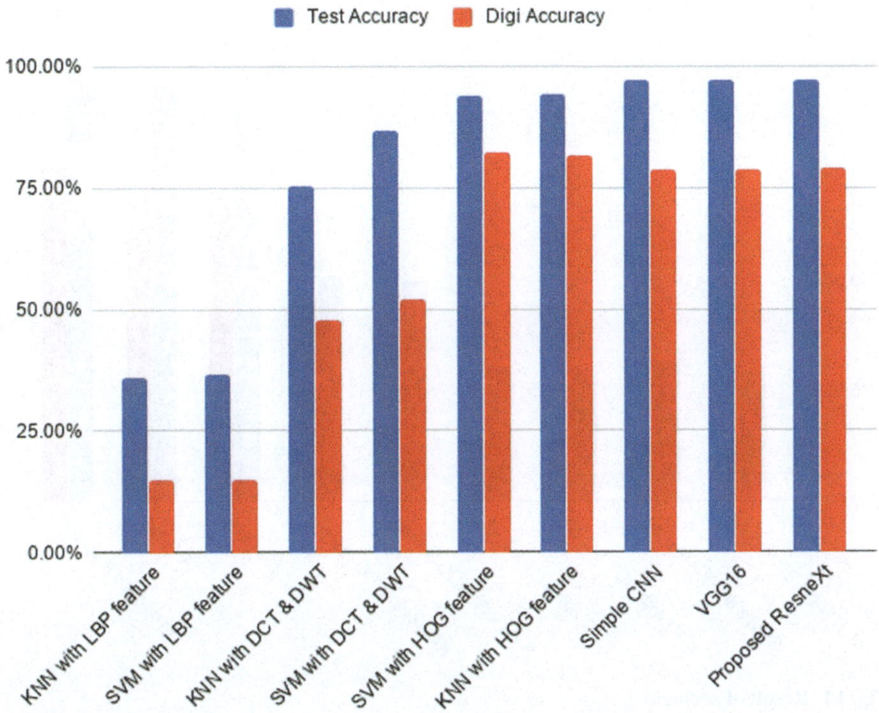

Fig. 13 Result of scenario 1

7 Conclusion and Future Work

The proposed KannadaRes-NeXt architecture is an efficient Kannada Numeral Recognition framework. The development of KannadaRes-NeXt empowers us to avoid challenges faced by traditional handcrafted features like feature extraction with minimal information loss. Also accuracies achieved using handcrafted features as well as deep learning models have been observed to be lower compared to those of KannadaRes-NeXt.

The results show the viability of KannadaRes-NeXt for Kannada numeral recognition based tasks. Some of the observations pertaining to the results are as follows:

- We have achieved efficient accuracies of 97.36% for the *Kannada-MNIST* test-set and 79.06% for the *Dig-MNIST* test-set for Kannada Numeral Recognition.
- The accuracy on *Kannada-MNIST* test set is 96.8% using CNN [32] while 97.36% using our proposed KannadaRes-NeXt model. For *Dig-MNIST* 76.1% accuracy is achieved using CNN [32] while an accuracy of 79.06% is achieved using KannadaRes-NeXt.This serves our purpose of achieving higher accuracies than in [32] on both the datasets.

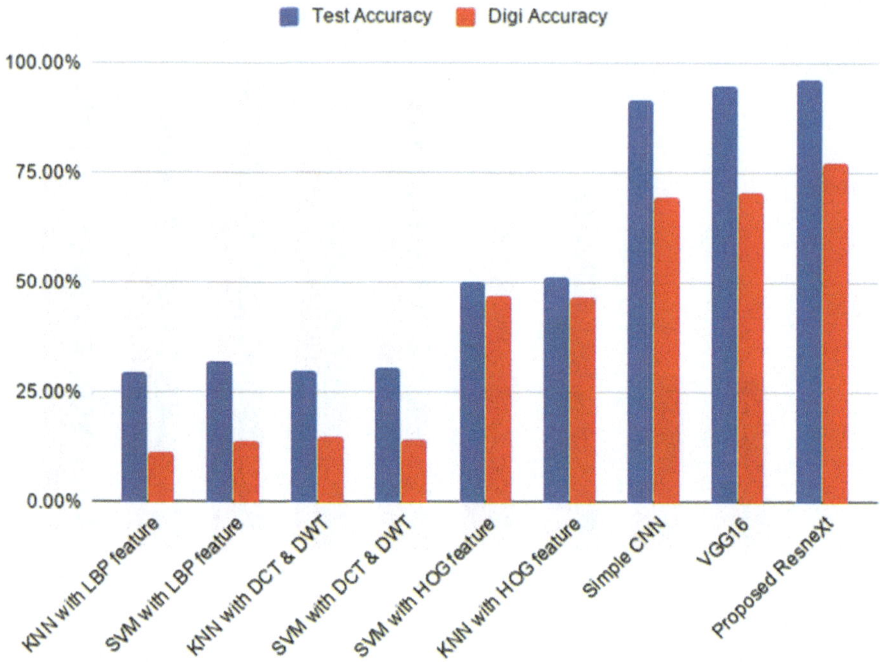

Fig. 14 Result of scenario 2

- Despite the irregularities in the new augmented datasets our proposed model has achieved remarkably high accuracies of 96.32% on *AugKannada-MNIST+ Kannada-MNIST* test set and 77.31% on *AugDig-MNIST +Dig-MNIST* dataset. This further establishes the robustness of our framework.
- It is observed that KannadaRes-NeXt architecture has achieved higher accuracies on all the four datasets when compared with other efficient and state-of-art deep learning models such as Convolutional Neural Networks (3 convolutional layers), VGG16.

  Suggestions for future direction are as follows:

1. There should be multiple standard handwritten character as well as numeral databases for different lingual scripts which are considerably large in size.
2. In order to address irregularities in strokes research should focus on representations based on various image transformations.
3. Degraded paper quality and degraded document quality should be undertaken during experiments. The work should be able to address paper quality along with noise by using corrective measures.
4. The methodology used to achieve aims in this work may be extended in the future to identify Kannada characters, also digits from different numeral systems or the alphabets of these languages i.e. for varied scripts recognition systems.

5. Also the feature selection, classification in OCR should be developed for different font styles.

Compliance with Ethical Standards

Conflict of Interest Aradhya Saini declares she has no conflict of interest. Sandeep Daniel declares he has no conflict of interest. Satyam Saini declares he has no conflict of interest. Ankush Mittal declares he has no conflict of interest.

References

1. H.H. Aghdam, E.J. Heravi, *Guide to Convolutional Neural Networks*, vol. 10 (Springer, New York, NY, 2017), pp. 978–3
2. S. Ahlawat, R. Rishi, Off-line handwritten numeral recognition using hybrid feature set-a comparative analysis. Procedia Comput. Sci. **122**, 1092–1099 (2017)
3. K. Alkhatib, H. Najadat, I. Hmeidi, M.K.A. Shatnawi, Stock price prediction using k-nearest neighbor (knn) algorithm. Int. J. Bus. Human. Technol **3**(3), 32–44 (2013)
4. N.S. Altman, An introduction to kernel and nearest-neighbor nonparametric regression. Am. Stat. **46**(3), 175–185 (1992)
5. Asha, K., Krishnappa, H.: Kannada handwritten document recognition using convolutional neural network, in *2018 3rd International Conference on Computational Systems and Information Technology for Sustainable Solutions (CSITSS)* (IEEE, 2018), pp. 299–301
6. O. Barkan, J. Weill, L. Wolf, H. Aronowitz, Fast high dimensional vector multiplication face recognition, 1960–1967 (2013). https://doi.org/10.1109/ICCV.2013.246
7. R. Bora, A. Bisht, A. Saini, T. Gupta, A. Mittal, Isl gesture recognition using multiple feature fusion, in *2017 International Conference on Wireless Communications, Signal Processing and Networking (WiSPNET)* (IEEE, 2017), pp. 196–199
8. V. Britanak, P.C. Yip, K.R. Rao, *Discrete Cosine and Sine Transforms: General Properties, Fast Algorithms and Integer Approximations* (Elsevier, 2010)
9. D. Chaithra, K. Indira, Handwritten online character recognition for single stroke kannada characters, in *2017 2nd IEEE International Conference on Recent Trends in Electronics, Information & Communication Technology (RTEICT)*. (IEEE, 2017), pp. 548–552
10. T. Cover, P. Hart, Nearest neighbor pattern classification. IEEE Trans. Inf. Theo. **13**(1), 21–27 (1967)
11. N. Dalal, B. Triggs, Histograms of oriented gradients for human detection (2005)
12. B. Deekshatulu, P. Chandra et al., Classification of heart disease using k-nearest neighbor and genetic algorithm. Procedia Technol. **10**, 85–94 (2013)
13. X. Feng, M. Pietikainen, A. Hadid, Facial expression recognition with local binary patterns and linear programming. Pattern Recog Image Anal C/C Raspoznavaniye Obraz. Analiz Izobrazhenii **15**(2), 546 (2005)
14. P. Ghadekar, S. Ingole, D. Sonone, Handwritten digit and letter recognition using hybrid dwt-dct with knn and svm classifier, in *2018 Fourth International Conference on Computing Communication Control and Automation (ICCUBEA)* (IEEE, 2018), pp. 1–6
15. T. Guo, J. Dong, H. Li, Y. Gao, Simple convolutional neural network on image classification in *2017 IEEE 2nd International Conference on Big Data Analysis (ICBDA)*. (IEEE, 2017), pp. 721–724
16. D. Gupta, S. Choubey, Discrete wavelet transform for image processing. Int. J. Emerg. Technol. Adv. Eng. **4**(3), 598–602 (2015)

17. A. Hadid, M. Pietikainen, T. Ahonen, A discriminative feature space for detecting and recognizing faces, in *Proceedings of the 2004 IEEE Computer Society Conference on Computer Vision and Pattern Recognition, 2004. CVPR 2004*, vol. 2 (IEEE, 2004), pp. II–II
18. V.C. Hallur, R.S. Hegadi, R.S. Hegadi, Handwritten kannada numerals recognition by using zone features and cnn classifier. Int. J. Technol. Human Interact. (IJTHI) **15**(4), 63–79 (2019)
19. W. Jiang, L. Zhang, Edge-siamnet and edge-triplenet: new deep learning models for handwritten numeral recognition. IEICE Trans. Inf. Syst. **103**(3), 720–723 (2020)
20. S. Karthik, K.S. Murthy, Deep belief network based approach to recognize handwritten kannada characters using distributed average of gradients. Cluster Comput. **22**(2), 4673–4681 (2019)
21. Kokkinos, I.: Ubernet: training a universal convolutional neural network for low-, mid-, and high-level vision using diverse datasets and limited memory, in *Proceedings of the IEEE Conference on Computer Vision and Pattern Recognition* (2017), pp. 6129–6138
22. P.P. Koltsov, Comparative study of texture detection and classification algorithms. Comput. Math. Math. Phys. **51**(8), 1460 (2011)
23. A. Krizhevsky, I. Sutskever, G.E. Hinton, Imagenet classification with deep convolutional neural networks, in *Advances in Neural Information Processing Systems*, pp. 1097–1105 (2012)
24. M. Kumar, M. Jindal, R. Sharma, S.R. Jindal, Offline handwritten numeral recognition using combination of different feature extraction techniques. Nat. Acad. Sci. Lett. **41**(1), 29–33 (2018)
25. M. Kumar, M.K. Jindal, R.K. Sharma, S.R. Jindal, Character and numeral recognition for non-indic and indic scripts: a survey. Artif. Intell. Rev. **52**(4), 2235–2261 (2019)
26. B. Lee, Y. Tarng et al., Application of the discrete wavelet transform to the monitoring of tool failure in end milling using the spindle motor current. Int. J. Adv. Manuf. Technol. **15**(4), 238–243 (1999)
27. I. Melvin, E. Ie, R. Kuang, J. Weston, W.S. Noble, C. Leslie, Svm-fold: a tool for discriminative multi-class protein fold and superfamily recognition, in *BMC Bioinformatics*, vol. 8. (Springer, 2007), p. S2
28. G. Mukarambi, B. Dhandra, Energy-based features for kannada handwritten digit recognition. Int. J. Comput. Vis. Rob. **10**(2), 156–166 (2020)
29. T. Ojala, M. Pietikainen, T. Maenpaa, Multiresolution gray-scale and rotation invariant texture classification with local binary patterns. IEEE Trans. patt. Anal. Mach. Intell. **24**(7), 971–987 (2002)
30. Pasha, S., Padma, M.: Handwritten kannada character recognition using wavelet transform and structural features, in *2015 International Conference on Emerging Research in Electronics, Computer Science and Technology (ICERECT)*, pp. 346–351. (IEEE, 2015)
31. M. Pietikäinen, Local binary patterns. Scholarpedia **5**(3), 9775 (2010). https://doi.org/10.4249/scholarpedia.9775. Revision #188481
32. Prabhu, V.U.: Kannada-mnist: a new handwritten digits dataset for the kannada language. arXiv preprint arXiv:1908.01242 (2019)
33. Pramanik, R., Dansena, P., Bag, S.: A study on the effect of cnn-based transfer learning on handwritten indic and mixed numeral recognition, in *Workshop on Document Analysis and Recognition*, pp. 41–51. (Springer, 2018)
34. Prasad, M.M.: Improvement in the online handwritten kannada numeral recognition with the difference feature (2019)
35. A. Saini, An optimized interconnection network based on genetic algorithm, in *2016 Fourth International Conference on Parallel, Distributed and Grid Computing (PDGC)*, pp. 468–473. (IEEE, 2016)
36. A. Saini, T. Gupta, R. Kumar, A.K. Gupta, M. Panwar, A. Mittal, Image based indian monument recognition using convoluted neural networks, in *2017 International Conference on Big Data, IoT and Data Science (BID)* (IEEE, 2017), pp. 138–142
37. A. Saini, et al., Designing an interconnection network using genetic algorithm (2015)
38. S. Saini, V. Tripathi, A. Verma, Employing night-time light images for wealth assessment in india: a machine learning perspective, in *Proceedings of ICETIT 2019*. (Springer, 2020), pp. 613–622

39. J. Salvador-Meneses, Z. Ruiz-Chavez, J. Garcia-Rodriguez, Compressed knn: K-nearest neighbors with data compression. Entropy **21**(3), 234 (2019)
40. N. Sharma, U. Pal, F. Kimura, Recognition of handwritten kannada numerals, in *9th International Conference on Information Technology (ICIT'06)*, (IEEE, 2006), pp. 133–136
41. P. Shukla, T. Gupta, A. Saini, P. Singh, R. Balasubramanian, A deep learning frame-work for recognizing developmental disorders, in *2017 IEEE Winter Conference on Applications of Computer Vision (WACV)* (IEEE, 2017), pp. 705–714
42. Simonyan, K., Zisserman, A.: Very deep convolutional networks for large-scale image recognition. arXiv preprint arXiv:1409.1556 (2014)
43. Thomas, S.: A study of representation learning for handwritten numeral recognition of multilingual data set, in *Information and Communication Technology for Sustainable Development* (Springer, 2018), pp. 475–482
44. G. Toussaint, Geometric proximity graphs for improving nearest neighbor methods in instance-based learning and data mining. Int. J. Comput. Geom. Appl. **15**(02), 101–150 (2005)
45. G.D. Upadhye, U. Kulkarni, Pattern classification of handwritten kannada digits using customized cnn, in *International Conference on Soft Computing and Signal Processing* (Springer, 2019), pp. 613–620
46. P. Viola, M. Jones, Rapid object detection using a boosted cascade of simple features, in *Proceedings of the 2001 IEEE Computer Society Conference on Computer Vision and Pattern Recognition. CVPR 2001*, vol. 1 (IEEE, 2001), pp. I–I
47. S. Xie, R. Girshick, P. Dollár, Z. Tu, K. He, Aggregated residual transformations for deep neural networks, in *Proceedings of the IEEE Conference on Computer Vision and Pattern Recognition* (2017), pp. 1492–1500
48. H. Zhan, P.N. Chowdhury, U. Pal, Y. Lu, Handwritten digit string recognition for indian scripts, in *Asian Conference on Pattern Recognition* (Springer, 2019), pp. 262–273
49. G. Zhao, M. Pietikäinen, A. Hadid, Local spatiotemporal descriptors for visual recognition of spoken phrases, in *Proceedings of the International Workshop on Human-Centered Multimedia* (2007), pp. 57–66

Secure Image Transmission in Wireless Network Using Conventional Neural Network and DOST

Manoj Diwakar and Pardeep Kumar

Abstract In digital science, organizations have to need secure transmission especially for image data. Images are encrypted to provide the secure transmission over the network. To encrypt, image processing concepts play a very important role. With this motivation, this paper is proposed image encryption in discrete orthonormal Stockwell transform (DOST) and CNN. For image encryption, pixel shuffling method and Arnold transform are used in DOST domain. The proposed methods are compared and analyzed the performance using mean error, PSNR and Entropy difference. The obtained results are better in compared to existing methods.

Keywords Arnold transforms · Block shuffling · Wavelet transform

1 Introduction

In digital science, images are widely used to transmit over the internet. Some of the organizations have to need the secure transmission as well as encrypted data. In case of images, the data are more need to secure, because image contains many objects, like line, circle, and so on. Every object has its own significance. So we have to take care of image encryption [1–5].

Wireless network communication is able to address various mobility issues and provides freedom to mobile users to roam around and still remaining connected to network, without worrying about their location. It has many advantages over traditional wired networks such as low cost of deployment, easy to install, offers much higher flexibility and addresses various mobility issues. It has started becoming the choice for various applications like Health Monitoring, Stock Trading, Banking, and Research. However, as wireless technology popularity increases and it starts to

---

M. Diwakar (✉)
CSE, Graphic Era Deemed to be University, Dehradun, India
e-mail: manoj.diwakar@gmail.com

P. Kumar
CSE and IT, Jaypee University of Information Technology, Solan, India

P. Kumar et al. (eds.), *Machine Learning for Intelligent Multimedia Analytics*, Studies in Big Data 82,
https://doi.org/10.1007/978-981-15-9492-2_5

support more and more applications, user demand the same level of availability and reliability as provided by wired counterparts. But research work is still undergoing in order to provide user with ubiquitous support. Now despite the gaining popularity of wireless networks, it has some serious problems which require immediate addressing that includes unreliability, low availability and utmost important issue is Security. Since the mode of communication used by wireless networks is Air, it makes them more vulnerable to attacks. Therefore protection of information being communicated over network is of paramount importance. A network is deemed as secure if it fulfills three security features: Data Privacy/Confidentiality, Integrity of Data, and finally the Access Control. If any network unsuccessful to address any of these security features, it is deemed as unsecure and leads to network traffic and various network services being compromised by the attacker. In order to provide these security features IEEE and IETF has provided many security standards and mechanism which serves as a countermeasure to these attacks. The present standard being followed in WLAN's to secure the network is IEEE 802.11i but still is not secure enough and vulnerable to various attacks including de-authentication, dis-association attacks, MIM, and DoS attacks. The aim of my thesis work is to secure the authentication process in order to prevent various DoS and Dos flooding attacks mounted by attackers.

WLAN is the wireless technology that uses electromagnetic waves, also known as radio waves to transmit data, voice, video, and graphics between the network entities thus getting rid of the use of cables to transfer data signals. The biggest advantage of wireless networks over wired is the concern for mobility and easy install of the network. The network entity can roam around within the coverage area of the base station, i.e., AP and can still remain connected to it and can easily communicate with other network entity. But since wireless networks use air as their mode for communication, securing data being communicated over networks requires utmost attention. For this many standards were defined and latest being IEEE 802.11i, but still it is not secure enough and many attacks can be mounted over it at various phases which includes de-authentication and dis-association attacks, DoS attacks, and many more. This motivated me to design a solution or a mechanism that can work to counter these attacks and is more secure. To protect image security over encryption is one of the better approaches for secure transmission over a wireless network.

Various methods [6–10] have been explored for image encryption. In spatial domain, pixels are directly manipulated by shuffling the position of the pixels. However, this can be resolved by many efforts. For that, transform domain is very popular to encrypt the images. Wavelet transform is one of the popular domains, where image is decomposed into low and high-frequency components [11–15]. These components are separately used for encryption. DES, RSA, and many other cryptography algorithms can be used for image encryption. But to high complexity, these algorithms are not so popular [16–18]. Other than, Arnold transform and pixel shuffling methods are very popular because of their simplicity and accuracy of decryption. Discrete orthonormal Stockwell Transform (DOST) also popular to give better directional information in comparison to wavelet transform. Many image processing applications [19–22] are also utilized the transform domains algorithms for image enhancement and restoration and these applications can be used for the purpose of image security as a preprocessing.

With this motivation, this paper has the following structure: Section 2 is about Conventional neural network (CNN), Sect. 3 gives brief overview of discrete orthonormal Stockwell transform (DOST), Sect. 4 gives information on the proposed algorithm employed for the encryption process, Sect. 5 represents the results and discussion and Sect. 6 concluded the paper.

2 Wireless Network Security

In today's world, wireless technology market is growing at a tremendous rate and wireless networks are being used by organizations, individuals, research institutes for transmitting data, voice, video, audio, graphics, and many more. Some of the wireless technologies being used include Wireless Fidelity (WiFi), WiMAX, Bluetooth, cellular technologies, and satellite technologies (VSAT). Though Wired Networks provide a reliable and a sophisticated platform to communicate data from one entity to other but sometimes it becomes very difficult to connect different nodes by means of cable networks when they are geographically very far from each other. In order to counter this, the concept of Wireless Distributed System (WDS) was defined in which several small wireless networks are connected by means of WDS links.

Nodes in the wireless network get connected and disconnected at random, i.e., dynamic nature, therefore, secure authentication of the nodes becomes an important issue which requires immediate attention. Therefore many standards were defined by IEEE in order to counter these problems and provide strong confidentiality and authentication features. IEEE 802.11i is the latest standard which is being used nowadays and provides all these characteristics but despite providing strong confidentiality and authentication feature it is still vulnerable to DoS and DoS flooding attacks including de-authentication and de-association attacks. Therefore, some new mechanism needs to be defined which can secure the authentication process and able to avoid these attacks.

IEEE 802.11 contains first two OSI layers, i.e., Physical and Data Link layer. Both layers are divided into 2 sub-layers based on their role and function they provide and a mechanism for reducing complexity of the architecture.

1. Physical Layer

Physical layer (PHY) has been divided into 2 sublayers: PMD and PLCP. PMD specifies various types of services for the physical medium to be used which includes Coding and manipulation techniques. PLCP layer is just an intermediary layer between the MAC and PMD and presents the packet to the upper MAC layer in a common and defined format so that the received packet does not depend on the PMD sublayer.

- Physical Layer Convergence Procedure

The PLCP layer [5] acts as an intermediary layer between the MAC layer and PMD sublayer. When MPDUs are to be communicated to the recipient, the main role of PLCP is to hide the PMD layer knowledge from MAC layer by mapping them into a frame format that can be easily transmitted by the PMD. Similarly, it also delivers the received incoming frames from medium to the upper MAC layer. PLCP layer modifies the frame format by appending a preamble field specific to the PHY used and some header fields to the original MPDU. These fields are required by physical layer transmitters and receivers in order to do their respective functions. This modified frame is referred as PLCP Service Data Unit (PSDU).

- Physical Medium Dependent

PMD sublayer [5] specifies the functions and services for a particular PMD. Hence for every PHY there is an associated PMD. Initially, three PHYs were defined by IEEE 802.11 standard committee, two of them were based on 2.4 GHz band and the remaining one was based IR radiation. Later on, three more PHYs were added: IEEE 802.11a based on 5.7 GHz spectrum and providing channel rate upto 54 Mbps; IEEE 802.11b based on 2.4 GHz spectrum and having max channel rate of 11 Mbps; IEEE 802.11g in 2.4 GHz band and having max channel rate of 54 Mbps.

2. Data Link Layer

Data Link layer [8, 9] is also divided into 2 sublayers: namely Medium MAC layer and Logical Link Control (LLC) layer. 802.11 make use of same 802.2 LLC layer and 48-bit addressing as used by other LANs allowing simple interconnection of wireless to IEEE 802.1x wired networks. MAC layer just provide Access Control related functions for different shared PHYs to support the LLC layer.

- 802.2 Logical Link Control

It represents the upper layer of Data Link layer and is just like an interface to the upper Network layer which can now be compatible with any kind of MAC [8]. Frame which is prepared or formed at LLC layer and passed to MAC layer is termed as LLC Protocol Data Unit (LPDU). In a network that supports multiple Network Layer protocols each of them will have Source Service Access point (SSAP) and Destination Service Access Point (DSSAP) where SAP simply represents a logical connection to the network layer. It defines two kinds of communication services namely: Connectionless (Type 1) and Connection-oriented (Type 2) and in case of Type 2 services LLC also take care of Sequence number of the received LPDUs. In case any of the LPDUs are lost during transmit, destination LLC makes request to the source in order to restart its transmission. LLC layer provides the following functions:

1. Multiplexing and de-multiplexing of information over the MAC layer.

2. Provides Flow and Error Control by splitting the data into different frames of some fixed length and sending them across the destination and specifies their order of re-assembling by appending related headers to the split messages.

- Medium Access Control layer

It is the lowest sublayer of the Data link layer [3]. It covers 3 areas:

- Access Control: Initially the IEEE 802.11 committee proposed 2 mechanisms for access control for a MAC algorithm: One was Distributed Access Protocol, in which the decision to transmit by means of carrier-sense technique was distributed over nodes like an Ethernet. Other was Centralized Access Protocol where there is a centralized entity known as decision maker which takes decision over transmission by the nodes. Finally, as a result of these two mechanisms, a new algorithm is known as Distributed foundation Wireless Access (DFWMAC) was developed which provides features of distributed access control protocol in addition to optional centralized control functions built over it. The lower sublayer of MAC layer is hence called as Distributed Coordination Function (DCF) and the upper sublayer is Point Coordination Function.
- Reliable Data Delivery: Now since the medium of communication of any wireless network is Air, transmission over it is always unreliable and prone to several interferences, noises, and errors which result in loss and corruption of several frames. Even using any of the error-correcting mechanism is still not able to guard against frame loss. However as wired LANs this situation is dealt at high layer by protocols like TCP. However, in order to deal with this reliability issue at MAC level, a new technique was developed in which to every packet being sent from source to destination a packet known as acknowledgement (ACK) is reverted by destination station towards the source. If the source station is not able to receive the ACK in response to send packet in some stipulated time, it deems it as loss or damage of the send packet and retransmits it. Now to further enhance the reliability a new 4-frame exchange mechanism was proposed. In this a station who wishes to send data packet first sends a RTS packet to destination. Destination if not involved in any other transmission or free response with clear to send (CTS) packet. After receiving this CTS frame, source station sends data frame to destination which acknowledges it by issuing ACK frame.
- Security: IEEE 802.11 provides mechanisms for both confidentiality and authentication. It provides mechanisms and algorithms like Wired Equivalent Privacy (WEP), Wi-Fi Protected Access (WPA), latest being 802.11i standards which makes wireless networks secure to external threats. It also provides 2 types of techniques for providing authentication features namely: Open System Authentication and Shared Key Authentication. In open system authentication process MAC control frames (authentication frame) are exchanged between the sender and recipient and thus provides no security while the later requires the 2 parties to share a secure shared key in order to communicate with each other.

Conventional Neural Network (CNN)
CNN uses the Siamese network, that uses two branches for every node, and includes three convolution layers and one max pooling layer which are similar that were used in [14]. With this, space consumption can be reduced, and by eliminating a completely connected layer, capacity is improved. It only requires 1.66 MB of memory.

The two groups refer to two types of uniform weight assignment tests, respectively "first patch 1 and second patch 0" and "first patch 0 and second patch 1." The likelihood of each class enables the potential of assigning each weight. When considering that the sum of two output probabilities is 1, in this situation the likelihood of each class only shows the weight assigned to its corresponding input patch. Using the method in the network is educated by high-quality image patches and their distorted versions. In the training cycle, according to the study in [14], the spatial size of the input patch is set to 16×16.

The development of workout examples is based on Gaussian multi-scale filtering and random sampling. The loss function of SoftMax is used as the target of optimization and we follow the stochastic gradient descent (SGD) algorithm to minimize it. The training process is carried out on the famous Caffe deep learning framework. It can be solved by transforming a completely connected layer into an analogous convolutionary layer containing two kernels of size $8 * 8 * 512$ due to overlapping patches, which could be developed for this dense prediction map.

3 Discrete Orthonormal Stockwell Transform (DOST)

The DOST [11] is a pared-down version of the fully redundant S Transform. In the sense of multiresolution, less temporal resolution is required for a lower frequency band based on the sampling theorem. As discussed in the previous chapter, the ST has redundantly stored an equal amount of data in each low-frequency band as in each high-frequency band, despite the fact that the Nyquist criterion indicates that these two bands have very different sampling requirements.

The 2-D ST is a separable transform, as is the 2-D DOST. For a better comparison between the coefficients, the plot of the coefficients is in log-scale. As we can see, the coefficients decay very quickly, which makes the DOST a powerful tool for image compression and other applications. Moreover, the DOST coefficients decay in a consistent way. As you can easily observe from the log-scale magnitude plot, there are still small "Lenas" on each corner of the plot. And even for the square and rectangular blocks inside the plot, where frequency bands with respect to different spatial axes of the image overlap, the stretched "Lenas" are still visible.

4 Proposed Architecture of Image Encryption

The overview of the proposed scheme is summarized in the following steps where block-wise Arnold map and pixel chaotic shuffle are used. CNN-based DOST is used to secure the images in a way so that after decryption edges should be better preserved.

The steps of proposed algorithm are processed as:

Step 1: Perform CNN over the input image for better edge preservation using three Layers.
Step 2: Approximation and detail parts are achieved using DOST.
Step 3: Apply Arnold transform for each DOST coefficients of Approximation parts where Arnold transformation is used, as given below:

$$\begin{bmatrix} x' \\ y' \end{bmatrix} = \begin{bmatrix} 1 & 1 \\ 1 & 2 \end{bmatrix} \times \begin{bmatrix} x \\ y \end{bmatrix} \bmod n$$

Step 4: Over the detail parts, DOST is performed and encryption is applied using pixel chaotic shuffle method, as given below:

(a) Initial values are selected to create chaotic variable sets.
(b) A sequence is generated of each dost coefficients.
(c) Transfer 2D matrix into 1D.
(d) Perform the shuffle function circularly at random times.

Step 5: Apply Inverse DOST to reconstruct the images.

5 Results of Experiment and Analysis

The results are obtained on images with size 512 × 512 using proposed method. In this paper, we use an image compression experiment to demonstrate the advantages of the DOST by analyzing the peak signal-to-noise ratio (PSNR), MSE, and Entropy difference.

The findings are shown in Figs. 1, 2, 3, and 4. Figures 1a, 2a, 3a, and 4a are the original images. Encrypted images are Figs. 1b, 2b, 3b and 4b and Decrypted images are Figs. 2c, 3c, and 4c.

PSNR, Mean Error, and Entropy Difference (ED) for original and decrypted images are calculated and given in Tables 1, 2, 3 and 4 to measure the efficiency of proposed methodology. A comparison with some typical existing schemes is done to assess the efficiency of the proposed technique. In PSNR the higher value produces better results. Although it reflects the best mean error and ED, this includes less value. From Tables 1, 2 and 3, it can be clearly analyzed that most of the cases, the proposed scheme gives better outcomes. Hence we can say that proposed methodology gives better outcomes in the terms of visual analysis and statistical analysis.

(a) (b) (c)

Fig. 1 **a** Original peppers: jellyfish, **b** encrypted image and **c** decrypted image

(a) (b) (c)

Fig. 2 **a** Original image: Lena, **b** encrypted image and **c** decrypted image

(a) (b) (c)

Fig. 3 **a** Original image: teeth, **b** encrypted image and **c** decrypted image

Fig. 4 **a** Original image: house, **b** encrypted image and **c** decrypted image

Table 1 Performance metrics on teeth image

Input images	PSNR	Mean error	Entropy difference (ED)
[1]	38.03	0.0614	0.4245
[2]	38.12	0.0526	0.8121
[3]	38.12	0.0932	0.3351
Proposed	39.91	0.0514	0.2351

Table 2 Performance metrics on lena image

Input images	PSNR	Mean error	Entropy difference (ED)
[1]	38.11	0.4256	0.4452
[2]	37.22	0.2352	0.3214
[3]	37.66	0.3468	0.5232
Proposed	39.23	0.1356	0.2335

Table 3 Performance metrics on pepper image

Input images	PSNR	Mean error	Entropy difference (ED)
[1]	41.13	0.0223	0.2356
[2]	40.12	0.0763	0.7889
[3]	40.77	0.0233	0.1123
Proposed	42.99	0.0131	0.0567

Table 4 Performance metrics on house image

Input images	PSNR	Mean error	Entropy difference (ED)
[1]	42.13	0.0423	0.6356
[2]	43.12	0.0663	0.6889
[3]	43.77	0.0733	0.5123
Proposed	45.99	0.0221	0.4567

6 Conclusions

The proposed method is used the concept of the CNN based DOST using Arnold and pixel shuffling method. A comparison is also done with some standard existing schemes. From experimental analysis, it can be clearly said that proposed method gives better results. The advantages of DOST and CNN also make strong the proposed method.

References

1. N. Haddaji, K. Nguyen, M. Cheriet, Towards end-to-end integrated optical packet network: empirical analysis. Opt. Switching Networking **27**, 18–39 (2018)
2. H. Diab, A.M. Elsemary, Secure image cryptosystem with unique key streams via hyper-chaotic system. Sig. Proc. **142**, 53–68 (2018)
3. X. Zhang, X. Wang, Multiple-image encryption algorithm based on mixed image element and permutation. Opt. Lasers Eng. **92**, 6–16 (2017)
4. G. Ye, X. Huang, An efficient symmetric image encryption algorithm based on an intertwining logistic map. Neurocomputing **251**, 45–53 (2017)
5. A. Belazi, A.A. Abd El-Latif, S. Belghith, A novel image encryption scheme based on substitution-permutation network and chaos. Signal Process. **128**, 155–170
6. X. Li, C. Li, I.-K. Lee, Chaotic image encryption using pseudo-random masks and pixel mapping. Signal Process. **125**, 48–63 (2016)
7. A. Daneshgar, B. Khadem, A self-synchronized chaotic image encryption scheme. Signal Process. Image Commun. **36**, 106–114 (2015)
8. A. Kanso, M. Ghebleh, An efficient and robust image encryption scheme for medical applications. Commun. Nonlinear Sci. Numer. Simul. **24**(1), 98–116 (2015)
9. E.A. Naeem, M.M. Abd Elnaby, N.F. Soliman, A.M. Abbas, O.S. Faragallah, N. Semary, M.M. Hadhoud, S.A. Alshebeili, F.E. Abd El-Samie, Efficient implementation of chaotic image encryption in transform domains. J. Syst. Softw. **97**, 118–127 (2014)
10. X. Zhang, Z. Zhao, J. Wang, Chaotic image encryption based on circular substitution box and key stream buffer. Signal Process. Image Commun. **29**(8), 902–913 (2014)
11. N.K. Pareek, V. Patidar, K.K. Sud, Diffusion–substitution based gray image encryption scheme. Digital Signal Process. **23**(3), 894–901 (2013)
12. W. Chen, X. Chen, Security-enhanced interference-based optical image encryption. Optics Commun. **286**, 123–129 (2013)
13. C. Zhu, A novel image encryption scheme based on improved hyperchaotic sequences. Optics Commun. **285**(1), 29–37 (2012)
14. Z. Liu, Xu. Lie, J. Dai, S. Liu, Image encryption by using local random phase encoding in fractional Fourier transform domains. Optik Int. J. Light Electron Opt. **123**(5), 428–432 (2012)
15. S. Deng, Y. Zhan, Di. Xiao, Y. Li, Analysis and improvement of a hash-based image encryption algorithm. Commun. Nonlinear Sci. Numer. Simul. **16**(8), 3269–3278 (2011)
16. Z. Liu, Xu. Lie, C. Lin, J. Dai, S. Liu, Image encryption scheme by using iterative random phase encoding in gyrator transform domains. Opt. Lasers Eng. **49**(4), 542–546 (2011)
17. J.W. Yoon, H. Kim, An image encryption scheme with a pseudorandom permutation based on chaotic maps. Commun. Nonlinear Sci. Numer. Simul. **15**(12), 3998–4006 (2010)
18. H. Yang, K.W. Wong, X. Liao, W. Zhang, P. Wei, A fast image encryption and authentication scheme based on chaotic maps. Commun. Nonlinear Sci. Numer. Simul. **15**(11), 3507–3517 (2010)

19. M. Diwakar, M. Kumar, Edge preservation based CT image denoising using Wiener filtering and thresholding in wavelet domain, in *2016 Fourth International Conference on Parallel, Distributed and Grid Computing (PDGC)* (IEEE, 2016), pp. 332–336
20. M. Diwakar, M. Kumar, CT Image noise reduction based on adaptive wiener filtering with wavelet packet thresholding, in *2014 International Conference on Parallel, Distributed and Grid Computing* (IEEE, 2014), pp. 94–98
21. M. Kumar, M. Diwakar, A new exponentially directional weighted function based CT image denoising using total variation. J. King Saud Univ. Comput. Inf. Sci. **31**(1), 113–124 (2019)
22. M. Diwakar, M. Kumar, CT image denoising based on complex wavelet transform using local adaptive thresholding and bilateral filtering, in *Proceedings of the Third International Symposium on Women in Computing and Informatics* (2015), pp. 297–302

Robust General Twin Support Vector Machine with Pinball Loss Function

M. A. Ganaie and M. Tanveer

Abstract Twin support vector machines (TWSVM) with hinge loss suffer from noise sensitivity and instability. To overcome these issues, pinball loss based general twin support vector machines (Pin-GTSVM) was recently proposed. However, TWSVM and Pin-GTSVM implement the empirical risk minimization principle. Also, the matrices in their dual formulations are positive semi-definite. To overcome these issues, we propose pinball loss based robust general twin support vector machines (Pin-RGTSVM). Pin-RGTSVM implements the structural risk minimization principle which embodies the marrow of statistical learning and pinball loss function makes it more robust for noisy datasets. Also, the matrices appear in the dual formulation of the proposed Pin-RGTSVM are positive definite. The incorporation of the structural risk minimization principle via introduction of the regularisation term leads to the improved generalization performance of the proposed Pin-RGTSVM. Numerical experiments and statistical evaluation on the real world benchmark datasets show the efficacy of the proposed Pin-RGTSVM.

Keywords Support vector machines · Pinball loss · Twin support vector machines · Hinge loss

1 Introduction

Support vector machines (SVM) [1–3] are powerful methods used in classification as well as regression problems. SVM is based on the maximum margin concept and shows better generalization performance as it implements the structural risk minimization (SRM) principle. SRM principle is at the core of the statistical learning and

M. A. Ganaie · M. Tanveer (✉)
Indian Institute of Technology Indore, Indore, MP 453552, India
e-mail: mtanveer@iiti.ac.in

M. A. Ganaie
e-mail: phd1901141006@iiti.ac.in
URL: http://www.iiti.ac.in

P. Kumar et al. (eds.), *Machine Learning for Intelligent Multimedia Analytics*,
Studies in Big Data 82,
https://doi.org/10.1007/978-981-15-9492-2_6

hence SVM and its variants have been successfully applied across different fields including detection of faces [4], categorization of text [5], classification of electroencephalogram signals [6], extraction of features [7, 8] and identification of chimera [9]. Despite its better generalization performance, SVM owns high computational complexity of the order of $O(l^3)$ where l is the size of the training dataset. To overcome this drawback of higher complexity, generalized eigenvalue proximal support vector machine (GEPSVM) [10] was proposed. GEPSVM builds two hyperplanes such that each hyperplane is proximal to its own class. Motivated by GEPSVM, twin support vector machines (TWSVM) [11] formulation was proposed for classification problems. Like GEPSVM, TWSVM builds two hyperplanes such that each hyperplane is proximal to its own class and farthest from the samples of other class. Unlike SVM wherein a single large quadratic programming problem (QPP) is solved, TWSVM solves two QPPs of smaller size as compared to SVM. Solving two smaller QPPs make TWSVM approximately 4 times faster than SVM. As TWSVM implements empirical risk minimization principle, twin bounded SVM (TBSVM) [12] implemented the structural risk minimization principle. TBSVM incorporated regularization term in the TWSVM formulation to make it well-conditioned for the real life scenarios and showed that the implementation of the SRM principle leads to better performance [12]. TWSVM has been adopted in other domains including clustering [13–15] and ensemble with other classification models [16, 17]. To reduce the complexity of TWSVM, several modifications have been proposed like least squares TWSVM [18], energy based least squares twin SVM [19], robust energy based least squares TWSVM (RELS-TSVM) [20] and angle based universum least squares twin support vector machine [21]. To make twin SVM more robust, TWSVM formulation have been improved in Newton method for implicit Lagrangian TWSVM [22], robust and sparse linear programming TWSVM [23], universum least squares twin parametric-margin support vector machine [24] and application of smoothing techniques for linear programming twin support vector machines [25]. To overcome the issues of class imbalance problems, reduced universum twin support vector machine for class imbalance learning (RUTSVM-CIL) [26] and regularized robust fuzzy least squares twin support vector machine for class imbalance learning [27] were proposed. Recent study [28] revealed that among the TWSVM classification models, RELS-TSVM is the best performing classification model. Twin support vector machines have also been extended to multiclass classification problems [29].

Among the several challenges in the classification problems, noise is the natural phenomenon to be present in the real world scenarios. Assuming that the noise level in the datasets is uniform may not be true in real world scenarios. TWSVM and its several variants are based on this assumption that the noise is uniform. To address the problem of noise, parametric-margin-ν-SVM [30] and twin parametric margin support vector machines [31] was proposed. In machine learning models, loss function is of vital importance for formulating the robust classification model. Different loss functions like 0–1 loss function, squared loss function etc have been used to meet different requirements. In standard SVM model, hinge loss function is used to penalty the data samples based on their distribution. Being the shortest distance function, hinge loss function suffers from the problems of instability and noise insensitivity.

To eliminate the aforementioned issues in SVM, pinball loss based SVM (Pin-SVM) [32] was proposed. Being a quantile distance function [33, 34], pinball loss function made the classification model noise insensitive. Truncated pinball loss SVM [35] proposed a flexible framework to balance the sparsity and feature noise insensitivity. As compared to standard SVM, Pin-SVM is theoretically and experimentally proved to be robust against the feature noise and more stable for resampling. Inspired by the noise insensitivity of the Pin-SVM and lower complexity of the TWSVM, twin parametric margin SVM with pinball loss [36] was proposed. However, using pinball loss function to attain noise insensitivity and stability leads to loss of sparsity. To retain sparsity as well as the advantages of noise insensitivity and stability, sparse pinball loss twin SVM (SPTWSVM) [37] used ϵ-insensitive zone pinball loss function. Unlike SPTWSVM, which implemented the empirical risk minimization principle, improved sparse pinball loss SVM [38] implemented structural risk minimization principle.

Inspired by [32, 36], pinball loss based general twin SVM (Pin-GTSVM) was proposed [39]. Unlike TWSVM which is noise sensitive and suffers from instability due to hinge loss function, Pin-GTSVM uses pinball loss function resulting in noise insensitivity and more stable for resampling. However, the drawback of TWSVM and Pin-GTSVM is that both implement the empirical risk minimization principle and the matrices in their dual formulations are positive semi-definite. Motivated by [32, 39], we propose pinball loss based robust general twin SVM (Pin-RGTSVM) model for classification problems. The several advantages of the proposed Pin-RGTSVM are given as follows:

- Pin-RGTSVM is more stable for re-sampling and less sensitive to noise.
- TWSVM and Pin-GTSVM minimizes the empirical risk while as the proposed Pin-RGTSVM implements the structural risk minimization principle which improves the generalization performance.
- The matrices in the dual formulation of TWSVM and Pin-GTSVM are positive semi-definite while as the matrices in the dual formulation of the proposed Pin-RGTSVM are positive definite.
- Experimental results and the statistical analysis show the efficiency of the proposed Pin-RGTSVM.

2 Related Work

In this section, we discuss the different kinds of loss functions and the formulations of the several related classification models which include support vector machines with pinball loss (Pin-SVM) [32], twin support vector machines [11] and general twin support vector machine with pin-ball loss (Pin-GTSVM) [39].

2.1 Loss Functions

Loss functions play an important role while designing the models for both classification and regression problems. Loss functions allow the model to measure the amount of goodness or badness of the model. Suppose T is a training set with l number of samples of the form (x_i, y_i) for $i = 1, \ldots, l$ where l is the size of the training set, $x_i \in \mathbb{R}^n$ represent the n-dimensional feature vector of the ith sample and $y_i \in \mathbb{R}$ denotes the corresponding label of the ith sample. Let the mapping function is given as $f : \mathbb{R}^n \rightarrow \mathbb{R}$. The functions used to measure the loss are given as follows:

- $0-1$ **loss function**:
  The $0-1$ loss function is given as:

$$L(x, y, f(x)) = \begin{cases} 0 & \text{if} \;\; y = f(x), \\ 1 & \text{if} \;\; y \neq f(x), \end{cases} \tag{1}$$

  where y is the original label and $f(x)$ is the predicted label.
- **Hinge loss function:**
  The hinge loss function is defined as:

$$L_{\text{hinge}}(x, y, f(x)) = \max(0, 1 - yf(x)), \tag{2}$$

  where y is the original label and $f(x)$ is the predicted label.
- **Pinball loss function:**
  The pinball loss function is defined as:

$$L_\tau(x, y, f(x)) = \begin{cases} -\tau(1 - yf(x)), & 1 - yf(x) < 0, \\ 1 - yf(x), & 1 - yf(x) \geq 0, \end{cases} \tag{3}$$

  where $\tau \in [0, 1]$, y is the original label and $f(x)$ is the predicted label.

2.2 Support Vector Machine with Pin-Ball Loss Function (Pin-SVM)

The objective function of the support vector machine with pin-ball loss function (Pin-SVM) [32] for non-linear case is given as follows:

$$\begin{aligned}
&\min_{w,b,\xi} \frac{1}{2}\|w\|^2 + c\sum_{i=1}^{l}\xi_i \\
&\text{s.t. } y_i(w^t\phi(x_i)+b) \geq 1-\xi_i, \\
&\quad\;\; y_i(w^t\phi(x_i)+b) \leq 1+\frac{\xi_i}{\tau},\quad i=1,2,\ldots,l,
\end{aligned} \tag{4}$$

where c is penalty parameter and $\xi = (\xi_1,\ldots,\xi_l)^t$ is a slack variable.

Pin-SVM provides noise insensitivity without incurring any additional cost as the complexity of Pin-SVM is a similar to the standard SVM. Note that, if $\tau \to 0$ then $\xi \geq 0$ i.e. Pin-SVM reduces to hinge loss SVM.

2.3 *Twin Support Vector Machine (TWSVM)*

TWSVM [11] is an efficient version of SVM as TWSVM is approximately 4-times faster than standard SVM. TWSVM generates two proximal planes such that each hyperplane is proximal to the samples of one class and farthest from the samples of other class. The objective functions of the TWSVM for non-linear case are given as follows:

$$\begin{aligned}
&\min_{w_1,b_1,\xi_1} \frac{1}{2}\|K(A,C^t)w_1+e_1b_1\|^2 + c_1e_2^t\xi_1 \\
&\text{s.t. } -(K(B,C^T)w_1+e_2b_1)+\xi_1 \geq e_2,\ \xi_1 \geq 0
\end{aligned} \tag{5}$$

and

$$\begin{aligned}
&\min_{w_2,b_2,\xi_2} \frac{1}{2}\|K(B,C^t)w_2+e_2b_2\|^2 + c_2e_1^t\xi_2 \\
&\text{s.t. } (K(A,C^T)w_2+e_1b_2)+\xi_2 \geq e_1,\ \xi_2 \geq 0,
\end{aligned} \tag{6}$$

where c_i are the penalty parameters, ξ_i are the slack variables, e_i are the vector of ones with appropriate dimensions and $i = 1, 2$, A represents the samples of positive class and B represents the samples of negative class, also $C = [A; B]$.

Taking the Lagrangian of (5), we have

$$\begin{aligned}
L = &\frac{1}{2}\|K(A,C^t)w_1+e_1b_1\|^2 + c_1e_2^t\xi_1 \\
&-\alpha^t(-(K(B,C^t)w_1+e_2b_1)+\xi_1-e_2)-\beta^t\xi_1
\end{aligned} \tag{7}$$

where α and β are the Lagrangian multipliers.

Applying KKT conditions, we have

$$K(A, C^t)^t(K(A, C^t)w_1 + e_1b_1) + K(B, C^t)^t\alpha = 0, \tag{8}$$
$$e_1^t(K(A, C^t)w_1 + e_1b_1) + e_2^t\alpha = 0, \tag{9}$$
$$c_1e_2 - \alpha - \beta = 0, \tag{10}$$
$$-(K(B, C^t)w_1 + e_2b_1) + \xi_1 \geq e_2, \xi_1 \geq 0, \tag{11}$$
$$\alpha^t(-(K(B, C^t)w_1 + e_2b_1) + \xi_1 - e_2) = 0, \beta^t\xi_1 = 0, \tag{12}$$
$$\alpha \geq 0, \beta \geq 0. \tag{13}$$

Rewriting (8) and (9), we have

$$[K(A, C^t)^t \;\; e_1^t][K(A, C^t) \;\; e_1][w_1, b_1]^t + [K(B, C^t)^t \;\; e_2^t]\alpha = 0 \tag{14}$$

Rewriting (14), we have

$$\begin{bmatrix} w_1 \\ b_1 \end{bmatrix} = -(S^tS)^{-1}R^t\alpha, \tag{15}$$

where $S = [K(A, C^t) \;\; e_1]$ and $R = [K(B, C^t) \;\; e_2]$.

Using (7) and the above given KKT conditions, the dual of (5), is given as

$$\max_{\alpha} \; e_2^t\alpha - \frac{1}{2}\alpha^tR(S^tS)^{-1}R^t\alpha,$$
$$\text{s.t.} \;\; 0 \leq \alpha \leq c_1. \tag{16}$$

Similarly, the dual of (6) is given as:

$$\max_{\beta} \; e_1^t\beta - \frac{1}{2}\beta^tL(N^tN)^{-1}L^t\beta,$$
$$\text{s.t.} \;\; 0 \leq \beta \leq c_2, \tag{17}$$

where $L = [K(A, C^t) \;\; e_1], N = [K(B, C^t) \;\; e_2]$ and the optimal hyperplane is given as:

$$\begin{bmatrix} w_2 \\ b_2 \end{bmatrix} = (N^tN)^{-1}L^t\beta. \tag{18}$$

2.4 General Twin Support Vector Machine with Pinball Loss (Pin-GTSVM)

Inspired by Pin-SVM [32] and TWSVM [11], Pin-GTSVM [39] was formulated. The objective functions of the Pin-GTSVM [39] for non-linear case are given as follows:

$$\begin{aligned} \min_{w_1,b_1,\xi_1} \quad & \frac{1}{2}\|K(A,C^t)w_1+e_1b_1\|^2+c_1e_2^t\xi_1 \\ \text{s.t.} \quad & -(K(B,C^t)w_1+e_2b_1)+\xi_1 \geq e_2, \\ & -(K(B,C^t)w_1+e_2b_1)-\frac{\xi_1}{\tau_2} \leq e_2 \end{aligned} \tag{19}$$

and

$$\begin{aligned} \min_{w_2,b_2,\xi_2} \quad & \frac{1}{2}\|K(B,C^t)w_2+e_2b_2\|^2+c_2e_1^t\xi_2 \\ \text{s.t.} \quad & (K(A,C^t)w_2+e_1b_2)+\xi_2 \geq e_1, \\ & (K(A,C^t)w_2+e_1b_2)-\frac{\xi_2}{\tau_1} \leq e_1, \end{aligned} \tag{20}$$

where c_i are penalty parameters, ξ_i are slack variables, e_i is the vector of ones of appropriate dimensions and $i=1,2$.

The Lagrange function of (19) is given as:

$$\begin{aligned} L = & \frac{1}{2}\|K(A,C^t)w_1+e_1b_1\|^2+c_1e_2^t\xi_1-\alpha^t(-(K(B,C^t)w_1+e_2b_1)+\xi_1-e_2) \\ & +\beta^t(-(K(B,C^t)w_1+e_2b_1)-\frac{\xi_1}{\tau_2}-e_2) \end{aligned} \tag{21}$$

Using necessary and sufficient KKT conditions, we have

$$K(A,C^t)^t(K(A,C^t)w_1+e_1b_1)+K(B,C^t)^t\alpha-K(B,C^t)^t\beta=0, \tag{22}$$

$$e_1^t(K(A,C^t)w_1+e_1b_1)+e_2^t\alpha-e_2^t\beta=0, \tag{23}$$

$$c_1e_2-\alpha-\frac{1}{\tau_2}\beta=0. \tag{24}$$

Since $\alpha \geq 0$, we can rewrite (24) as

$$-\tau_2c_1e_2 \leq (\alpha-\beta). \tag{25}$$

Rewriting Eqs. (22) and (23) in matrix form, we have

$$\begin{bmatrix} K(A, C^t)^t \\ e_1^t \end{bmatrix} [K(A, C^t) \;\; e_1] \begin{bmatrix} w_1 \\ b_1 \end{bmatrix} + \begin{bmatrix} K(B, C^t)^t \\ e_2^t \end{bmatrix} (\alpha - \beta) = 0. \tag{26}$$

Let $P = [K(A, C^t) \;\; e_1]$, $Q = [K(B, C^t) \;\; e_2]$ and $u_1 = \begin{bmatrix} w_1 \\ b_1 \end{bmatrix}$. Then, Eq. (26) can be written as

$$u_1 = -(P^t P)^{-1} Q^t (\alpha - \beta). \tag{27}$$

Using (21) and the above given KKT conditions, we have the dual of objective function (19) as

$$\begin{aligned} &\max_{(\alpha-\beta)} e_2^t(\alpha - \beta) - \frac{1}{2}(\alpha - \beta)^t Q(P^t P)^{-1} Q^t (\alpha - \beta) \\ &\quad \text{s.t.} - \tau_2 c_1 e_2 \leq (\alpha - \beta), \end{aligned} \tag{28}$$

where α and β are the Lagrange multipliers.

Likewise, the dual of the objective function (20) is given as follows:

$$\begin{aligned} &\max_{(\gamma-\theta)} \;\; e_1^t(\gamma - \theta) - \frac{1}{2}(\gamma - \theta)^t P(Q^t Q)^{-1} H^t (\gamma - \theta) \\ &\text{s.t.} - \tau_1 c_2 e_1 \leq (\gamma - \theta), \end{aligned} \tag{29}$$

where γ and θ are the Lagrange multipliers.

Once the objective functions (28 and 29) are optimized, the optimal separating hyperplanes are given as:

$$\begin{bmatrix} w_1 \\ b_1 \end{bmatrix} = -(P^t P)^{-1} Q^t (\alpha - \beta) \tag{30}$$

and

$$\begin{bmatrix} w_2 \\ b_2 \end{bmatrix} = (Q^t Q)^{-1} P^t (\gamma - \theta). \tag{31}$$

3 Proposed Method

TWSVM [11] and Pin-GTSVM [39] implement the empirical risk minimization principle. Also, matrices in their dual formulations are always positive semi-definite. To overcome these disadvantages, we propose a robust general twin support vector machine with pinball loss function (Pin-RGTSVM). The proposed Pin-RGTSVM implements the structural risk minimization principle and the matrices in the dual formulation of the proposed Pin-RGTSVM are always positive definite.

The formulation of proposed Pin-RGTSVM for linear and non-linear cases is given as follows:

3.1 Linear Case

The objective function of the proposed Pin-RGTSVM for linear case is given as follows:

$$\begin{aligned} \min_{w_1,b_1,\xi_1} \quad & \frac{c_3}{2}(\|w_1\|^2 + b_1^2) + \frac{1}{2}\|Aw_1 + e_1b_1\|^2 + c_1e_2^t\xi_1 \\ \text{s.t.} \; & -(Bw_1 + e_2b_1) + \xi_1 \geq e_2, \\ & -(Bw_1 + e_2b_1) - \frac{\xi_1}{\tau_2} \leq e_2 \end{aligned} \tag{32}$$

and

$$\begin{aligned} \min_{w_2,b_2,\xi_2} \quad & \frac{c_4}{2}(\|w_2\|^2 + b_2^2) + \frac{1}{2}\|Bw_2 + e_2b_2\|^2 + c_2e_1^t\xi_2 \\ \text{s.t.} \; & Aw_2 + e_1b_2 + \xi_2 \geq e_1, \\ & (Aw_2 + e_1b_2) - \frac{\xi_2}{\tau_1} \leq e_1, \end{aligned} \tag{33}$$

where c_i are penalty parameters, ξ_i are slack variables, e_i is the vector of ones of appropriate dimensions and $i = 1, 2$.

Taking Lagrangian of (32), we have

$$\begin{aligned} L = \frac{c_3}{2}(\|w_1\|^2 + b_1^2) + \frac{1}{2}\|Aw_1 + e_1b_1\|^2 + c_1e_2^t\xi_1 - \alpha^t(-(Bw_1 + e_2b_1) + \xi_1 - e_2) \\ + \beta^t(-(Bw_1 + e_2b_1) - \frac{\xi_1}{\tau_2} - e_2), \end{aligned} \tag{34}$$

where α and β are the Lagrange multipliers.

Using necessary and sufficient KKT conditions, we have

$$c_3w_1 + A^t(Aw_1 + e_1b_1) + B^t\alpha - B^t\beta = 0, \tag{35}$$

$$c_3b_1 + e_1^t(Aw_1 + e_1b_1) + e_2^t\alpha - e_2^t\beta = 0, \tag{36}$$

$$c_1e_2 - \alpha - \frac{1}{\tau_2}\beta = 0. \tag{37}$$

Since $\alpha \geq 0$, we can rewrite (37) as

$$-\tau_2 c_1 e_2 \leq (\alpha - \beta). \tag{38}$$

Rewriting Eqs. (35) and (36) in matrix form, we have

$$c_3 \begin{bmatrix} w_1 \\ b_1 \end{bmatrix} + \begin{bmatrix} A^t \\ e_1^t \end{bmatrix} [A \ \ e_1] \begin{bmatrix} w_1 \\ b_1 \end{bmatrix} + \begin{bmatrix} B^t \\ e_2^t \end{bmatrix} (\alpha - \beta) = 0. \tag{39}$$

Let $H = [A \ \ e_1]$, $G = [B \ \ e_2]$, $u_1 = \begin{bmatrix} w_1 \\ b_1 \end{bmatrix}$ and $\eta = (\alpha - \beta)$. Then, equation (39) can be written as

$$u_1 = -(H^t H + c_3 I)^{-1} G^t \eta. \tag{40}$$

Using (34) and the above given KKT conditions, we have the dual of the objective function (32) as

$$\begin{aligned} &\max_{\eta} \ e_2^t \eta - \frac{1}{2} \eta^t G (H^t H + c_3 I)^{-1} G^t \eta \\ &\text{s.t.} \ - \tau_2 c_1 e_2 \leq \eta. \end{aligned} \tag{41}$$

Likewise, the dual of the objective function (33) is given as follows:

$$\begin{aligned} &\max_{\zeta} \ e_1^t \zeta - \frac{1}{2} \zeta^t H (G^t G + c_4 I)^{-1} H^t \zeta \\ &\text{s.t.} \ - \tau_1 c_2 e_1 \leq \zeta, \end{aligned} \tag{42}$$

where γ and θ are the Lagrange multipliers and $\zeta = (\gamma - \theta)$.

Once the objective functions (41 and 42) are optimized, the optimal separating hyperplanes are given as:

$$\begin{bmatrix} w_1 \\ b_1 \end{bmatrix} = -(H^t H + c_3 I)^{-1} G^t \eta \tag{43}$$

and

$$\begin{bmatrix} w_2 \\ b_2 \end{bmatrix} = (G^t G + c_4 I)^{-1} H^t \zeta. \tag{44}$$

3.2 Non-linear Case

The objective functions of the proposed Pin-RGTSVM for non-linear case are given as follows:

$$\begin{aligned}
\min_{w_1,b_1,\xi_1} \quad & \frac{c_3}{2}(\|w_1\|^2+b_1^2)+\frac{1}{2}\|K(A,C^t)w_1+e_1b_1\|^2+c_1e_2^t\xi_1 \\
\text{s.t.} \quad & -(K(B,C^t)w_1+e_2b_1)+\xi_1 \geq e_2, \\
& -(K(B,C^t)w_1+e_2b_1)-\frac{\xi_1}{\tau_2} \leq e_2
\end{aligned} \tag{45}$$

and

$$\begin{aligned}
\min_{w_2,b_2,\xi_2} \quad & \frac{c_4}{2}(\|w_2\|^2+b_2^2)+\frac{1}{2}\|K(B,C^t)w_2+e_2b_2\|^2+c_2e_1^t\xi_2 \\
\text{s.t.} \quad & (K(A,C^t)w_2+e_1b_2)+\xi_2 \geq e_1, \\
& (K(A,C^t)w_2+e_1b_2)-\frac{\xi_2}{\tau_1} \leq e_1,
\end{aligned} \tag{46}$$

where c_i are penalty parameters, ξ_i are slack variables, e_i is the vector of ones of appropriate dimensions and $i = 1, 2$.

Taking Lagrangian of (45), we have

$$\begin{aligned}
L = & \frac{c_3}{2}(\|w_1\|^2+b_1^2)+\frac{1}{2}\|K(A,C^t)w_1+e_1b_1\|^2+c_1e_2^t\xi_1-\alpha^t(-(K(B,C^t)w_1 \\
& +e_2b_1)+\xi_1-e_2)+\beta^t(-(K(B,C^t)w_1+e_2b_1)-\frac{\xi_1}{\tau_2}-e_2),
\end{aligned} \tag{47}$$

where α and β are the Lagrange multipliers.

Using necessary and sufficient KKT conditions, we have

$$c_3w_1+K(A,C^t)^t(K(A,C^t)w_1+e_1b_1)+K(B,C^t)^t\alpha-K(B,C^t)^t\beta=0, \tag{48}$$

$$c_3b_1+e_1^t(K(A,C^t)w_1+e_1b_1)+e_2^t\alpha-e_2^t\beta=0, \tag{49}$$

$$c_1e_2-\alpha-\frac{1}{\tau_2}\beta=0. \tag{50}$$

Since $\alpha \geq 0$, we can rewrite (50) as

$$-\tau_2c_1e_2 \leq (\alpha-\beta). \tag{51}$$

Rewriting equation (48) and (49) in matrix form, we have

$$c_3 \begin{bmatrix} w_1 \\ b_1 \end{bmatrix} + \begin{bmatrix} K(A, C^t)^t \\ e_1^t \end{bmatrix} [K(A, C^t) \;\; e_1] \begin{bmatrix} w_1 \\ b_1 \end{bmatrix} + \begin{bmatrix} K(B, C^t)^t \\ e_2^t \end{bmatrix} (\alpha - \beta) = 0. \quad (52)$$

Let $P = [K(A, C^t) \;\; e_1]$, $Q = [K(B, C^t) \;\; e_2]$, $u_1 = \begin{bmatrix} w_1 \\ b_1 \end{bmatrix}$ and $\eta = (\alpha - \beta)$. Then, Eq. (52) can be written as

$$u_1 = -(P^t P + c_3 I)^{-1} Q^t \eta. \quad (53)$$

Using (47) and the above given KKT conditions, we have the dual of objective function (45) as

$$\max_{\eta} \; e_2^t \eta - \frac{1}{2} \eta^t Q (P^t P + c_3 I)^{-1} Q^t \eta$$
$$\text{s.t.} \; -\tau_2 c_1 e_2 \le \eta. \quad (54)$$

Likewise, the dual of the objective function (46) is given as follows:

$$\max_{\zeta} \; e_1^t \zeta - \frac{1}{2} \zeta^t P (Q^t Q + c_4 I)^{-1} H^t \zeta$$
$$\text{s.t.} \; -\tau_1 c_2 e_1 \le \zeta, \quad (55)$$

where γ and θ are the Lagrange multipliers and $\zeta = (\gamma - \theta)$.

Once the objective functions (54 and 55) are optimized, the optimal separating hyperplanes are given as:

$$\begin{bmatrix} w_1 \\ b_1 \end{bmatrix} = -(P^t P + c_3 I)^{-1} Q^t \eta \quad (56)$$

and

$$\begin{bmatrix} w_2 \\ b_2 \end{bmatrix} = (Q^t Q + c_4 I)^{-1} P^t \zeta. \quad (57)$$

4 Experimental Results

In this section, we discuss about the datasets used in the experimental analysis, selection of parameters, statistical analysis of the classification models. All the experiments were performed using Matlab $R2017a$ on a workstation with Intel Xeon processor (2.3 GHZ) and 128 GB of RAM on windows-10 platform. We used 5-fold cross validation to evaluate the performance of each model. All the experiments were performed using Gaussian kernel, $K(x_1, x_2) = \exp^{-\frac{\|x_1 - x_2\|^2}{\mu^2}}$, where μ is a kernel parameter. The other details are given as follows:

4.1 Dataset Details

To analyze the performance of the classification models, all the models were evaluated using 23 datasets taken from the UCI [40] and KEEL repository [41]. The details of the datasets are given in Table 1. Table 1 contains the information of number of samples in training and testing datasets, and number of features in each sample. For example, in brwisconsin 300 samples are used for training and 383 for testing the models, here each sample is of the dimension 10.

Table 1 Details of UCI [40] and KEEL datasets [41]

Dataset	Train size	Test size
Brwisconsin	(300 × 10)	(383 × 10)
Ecoli-0-1-4-7-vs-5-6	(150 × 7)	(182 × 7)
Ecoli-0-2-6-7-vs-3-5	(110 × 8)	(114 × 8)
Ecoli-0-3-4-6-vs-5	(100 × 8)	(105 × 8)
Ecoli-0-4-6-vs-5	(100 × 7)	(103 × 7)
Ecoli-0-6-7-vs-5	(110 × 7)	(110 × 7)
Ecoli4	(150 × 8)	(186 × 8)
Glass2	(100 × 10)	(114 × 10)
Glass4	(150 × 10)	(64 × 10)
Heart-stat	(130 × 14)	(140 × 14)
Monk2	(300 × 8)	(301 × 8)
Shuttle-6-vs-2-3	(100 × 10)	(130 × 10)
Shuttle-c0-vs-c4	(900 × 10)	(929 × 10)
Vehicle 1	(400 × 19)	(446 × 19)
Vehicle2	(400 × 19)	(446 × 19)
Votes	(200 × 17)	(235 × 17)
Vowel	(500 × 11)	(488 × 11)
Wpbc	(100 × 34)	(94 × 34)
Yeast-0-3-5-9-vs-7-8	(250 × 9)	(256 × 9)
Yeast-0-5-6-7-9-vs-4	(250 × 9)	(278 × 9)
Yeast2vs8	(250 × 9)	(233 × 9)
Yeast3	(500 × 9)	(984 × 9)
Yeast5	(500 × 9)	(984 × 9)

4.2 Parameter Selection

We followed grid search approach for finding the optimal parameters corresponding to given classification models. For all the classification models, the following range of parameters are used in the grid-search. $c_i = [2^{-5}, 2^{-4}, \ldots, 2^{5}]$, $\mu = [2^{-10}, 2^{-9}, \ldots, 2^{10}]$, $\tau_i = [0.5, 0.8, 1]$ and $i = 1, 2$. To reduce the computation, we used $c_1 = c_2, c_3 = c_4$ and $\tau_1 = \tau_2$. Also, to analyze the performance of the classification models on noisy datasets, we corrupted the datasets via feature noise [32] with zero mean Gaussian noise and standard deviation r, where r is chosen to be $[0, 0.05, 0.1]$ with $r = 0$ means noise free.

4.3 Results and Discussion

Table 2 represents the experimental results corresponding to the proposed Pin-RGTSVM and the other baseline models with noise free datasets. One can see that the proposed Pin-RGTSVM achieves lowest average rank among all the baseline models. Also, except Pin-TSVM all the baseline models achieved more than 93% average accuracy with the proposed Pin-RGTSVM achieving the maximum average accuracy. At different values of τ, Pin-GTSVM and proposed Pin-RGTSVM achieved lowest rank among all the models.

Table 3 represents the experimental results corresponding to the proposed Pin-RGTSVM and the other baseline models with noise level $r = 0.05$ added to the datasets. One can see that the proposed Pin-RGTSVM (with $\tau = 0.5$) achieved lowest average rank among all the baseline models. Except Pin-TSVM and the Pin-RGTSVM (with $\tau = 0.8$), all the baseline models achieved more than 93% average accuracy with the TWSVM achieving the maximum average accuracy followed by the proposed Pin-RGTSVM model (with $\tau = 0.5$) achieving 93.41% accuracy. At different values of τ, Pin-GTSVM and proposed Pin-RGTSVM (at $\tau = 0.5$) show better performance among all the baseline models.

Table 4 represents the experimental results corresponding to the proposed Pin-RGTSVM and the other baseline models with noise level $r = 0.1$ added to the datasets. One can see that the proposed Pin-RGTSVM (with $\tau = 0.5$) achieved lowest average rank among all the baseline models. Except Pin-TSVM and the Pin-GTSVM (with $\tau = 0.5, 0.8$), all the baseline models achieved more than 93% average accuracy with the proposed Pin-RGTSVM ($\tau = 0.5$) achieving the maximum average accuracy followed by the proposed Pin-RGTSVM model (with $\tau = 1$) achieving 93.46% accuracy. At different values of τ, Pin-GTSVM and proposed Pin-RGTSVM (at $\tau = 0.5$) show better performance among all the baseline models.

Table 2 Performance of the classification models with noise free datasets

Dataset	TWSVM(Acc, Time(s))	Pin-TSVM (Acc, Time(s))			Pin-GTSVM(Acc, Time(s))			Pin-RGTSVM(Acc, Time(s))		
		$\tau = 0.5$	$\tau = 0.8$	$\tau = 1$	$\tau = 0.5$	$\tau = 0.8$	$\tau = 1$	$\tau = 0.5$	$\tau = 0.8$	$\tau = 1$
Brwisconsin	(97.39, 0.025)	(97.91, 0.046)	(97.91, 0.052)	(97.91, 0.028)	(97.39, 0.047)	(97.91, 0.044)	(97.39, 0.043)	(96.87, 0.036)	(96.87, 0.038)	(96.87, 0.04)
Ecoli-0-1-4-7-vs-5-6	(97.8, 0.032)	(95.6, 0.031)	(95.6, 0.042)	(95.05, 0.026)	(98.35, 0.044)	(98.35, 0.017)	(98.35, 0.041)	(97.8, 0.022)	(98.35, 0.039)	(97.8, 0.04)
Ecoli-0-2-6-7-vs-3-5	(98.25, 0.014)	(92.11, 0.032)	(92.98, 0.022)	(92.11, 0.021)	(97.37, 0.021)	(98.25, 0.026)	(98.25, 0.022)	(98.25, 0.042)	(98.25, 0.039)	(98.25, 0.04)
Ecoli-0-3-4-6-vs-5	(98.1, 0.024)	(95.24, 0.02)	(95.24, 0.023)	(91.43, 0.031)	(99.05, 0.021)	(99.05, 0.022)	(99.05, 0.021)	(99.05, 0.038)	(99.05, 0.029)	(98.1, 0.04)
Ecoli-0-4-6-vs-5	(94.17, 0.019)	(88.35, 0.037)	(88.35, 0.009)	(88.35, 0.021)	(94.17, 0.042)	(94.17, 0.013)	(94.17, 0.032)	(94.17, 0.039)	(94.17, 0.039)	(94.17, 0.04)
Ecoli-0-6-7-vs-5	(96.36, 0.014)	(89.09, 0.058)	(90, 0.04)	(89.09, 0.04)	(96.36, 0.027)	(97.27, 0.024)	(96.36, 0.044)	(96.36, 0.04)	(96.36, 0.018)	(96.36, 0.04)
Ecoli4	(98.92, 0.016)	(93.55, 0.066)	(93.55, 0.022)	(93.55, 0.025)	(98.39, 0.026)	(98.39, 0.036)	(97.85, 0.042)	(97.85, 0.039)	(97.85, 0.025)	(97.85, 0.04)
Glass2	(91.23, 0.013)	(91.23, 0.036)	(91.23, 0.029)	(91.23, 0.021)	(91.23, 0.044)	(91.23, 0.024)	(91.23, 0.022)	(91.23, 0.038)	(91.23, 0.016)	(91.23, 0.03)
Glass4	(93.75, 0.013)	(92.19, 0.024)	(92.19, 0.017)	(92.19, 0.022)	(96.88, 0.024)	(93.75, 0.021)	(93.75, 0.045)	(96.88, 0.023)	(96.88, 0.041)	(96.88, 0.03)
Heart-stat	(83.57, 0.045)	(85, 0.029)	(84.29, 0.024)	(84.29, 0.025)	(82.86, 0.052)	(82.86, 0.043)	(82.86, 0.045)	(82.86, 0.039)	(82.86, 0.019)	(82.86, 0.04)
Monk2	(83.06, 0.032)	(63.12, 0.034)	(63.79, 0.036)	(63.79, 0.046)	(83.06, 0.038)	(83.06, 0.042)	(83.06, 0.047)	(83.72, 0.044)	(83.39, 0.042)	(83.06, 0.04)
Shuttle-6-vs-2-3	(99.23, 0.013)	(96.92, 0.023)	(96.92, 0.034)	(96.15, 0.033)	(100, 0.012)	(100, 0.016)	(100, 0.02)	(100, 0.016)	(100, 0.038)	(99.23, 0.04)
Shuttle-c0-vs-c4	(99.68, 0.437)	(99.78, 0.316)	(99.78, 0.221)	(99.78, 0.275)	(99.68, 0.452)	(99.68, 0.471)	(99.68, 0.432)	(99.68, 0.351)	(99.78, 0.375)	(99.68, 0.4)
Vehicle 1	(78.25, 0.067)	(69.73, 0.041)	(69.51, 0.043)	(69.28, 0.04)	(80.72, 0.068)	(80.94, 0.073)	(80.72, 0.088)	(80.49, 0.073)	(80.72, 0.065)	(80.72, 0.06)
Vehicle2	(97.98, 0.065)	(84.3, 0.038)	(84.3, 0.045)	(84.3, 0.038)	(98.43, 0.065)	(98.88, 0.073)	(98.43, 0.071)	(99.33, 0.069)	(99.1, 0.064)	(99.33, 0.06)
Votes	(95.32, 0.016)	(87.23, 0.037)	(88.09, 0.022)	(86.38, 0.027)	(95.74, 0.027)	(95.74, 0.042)	(95.74, 0.043)	(96.6, 0.021)	(95.32, 0.039)	(96.6, 0.04)
Vowel	(97.34, 0.095)	(92.01, 0.08)	(92.42, 0.085)	(92.42, 0.082)	(96.72, 0.164)	(96.93, 0.116)	(96.72, 0.137)	(97.54, 0.101)	(96.31, 0.132)	(97.13, 0.13)
Wpbc	(76.6, 0.012)	(78.72, 0.022)	(78.72, 0.021)	(78.72, 0.012)	(71.28, 0.023)	(71.28, 0.024)	(71.28, 0.02)	(72.34, 0.032)	(69.15, 0.039)	(72.34, 0.04)
Yeast-0-3-5-9-vs-7-8	(89.84, 0.036)	(88.28, 0.06)	(88.28, 0.048)	(88.28, 0.032)	(88.28, 0.042)	(88.28, 0.032)	(88.28, 0.036)	(89.84, 0.04)	(89.84, 0.047)	(89.84, 0.05)
Yeast-0-5-6-7-9-vs-4	(91.01, 0.031)	(91.37, 0.031)	(91.37, 0.039)	(91.37, 0.032)	(91.37, 0.044)	(91.37, 0.043)	(91.37, 0.036)	(91.37, 0.04)	(91.37, 0.037)	(91.37, 0.04)
Yeast2vs8	(97.85, 0.036)	(95.28, 0.053)	(95.28, 0.03)	(95.28, 0.041)	(96.57, 0.053)	(96.57, 0.046)	(96.57, 0.043)	(98.28, 0.037)	(98.28, 0.049)	(98.28, 0.04)
Yeast3	(94.61, 0.116)	(90.65, 0.075)	(90.65, 0.077)	(90.65, 0.073)	(92.78, 0.148)	(92.68, 0.132)	(92.78, 0.152)	(94.21, 0.105)	(92.68, 0.128)	(92.58, 0.12)
Yeast5	(97.46, 0.132)	(96.95, 0.069)	(96.95, 0.143)	(96.95, 0.09)	(98.27, 0.174)	(98.27, 0.169)	(98.17, 0.175)	(98.17, 0.12)	(98.07, 0.121)	(98.07, 0.12)
Average	93.38	89.33	89.45	89.07	93.26	93.26	93.13	93.6	93.3	93.42
Rank	5.09	7.41	7.22	7.67	4.8	4.35	5.02	4.04	4.59	4.8

*Acc denotes Accuracy

Table 3 Performance of the classification models with Gaussian noise $r = 0.05$ added to the datasets

Dataset	TWSVM(Acc, Time(s))	Pin-TSVM(Acc, Time(s))			Pin-GTSVM(Acc, Time(s))			Pin-RGTSVM(Acc, Time(s))		
		$\tau = 0.5$	$\tau = 0.8$	$\tau = 1$	$\tau = 0.5$	$\tau = 0.8$	$\tau = 1$	$\tau = 0.5$	$\tau = 0.8$	$\tau = 1$
Brwisconsin	(98.17, 0.033)	(97.91, 0.035)	(97.91, 0.024)	(97.91, 0.033)	(97.65, 0.043)	(97.65, 0.049)	(97.91, 0.044)	(96.61, 0.041)	(97.39, 0.043)	(97.65, 0.04)
Ecoli-0-1-4-7-vs-5-6	(97.8, 0.017)	(96.15, 0.021)	(95.6, 0.039)	(93.41, 0.022)	(98.35, 0.036)	(97.8, 0.044)	(98.35, 0.039)	(98.35, 0.045)	(97.8, 0.038)	(97.8, 0.04)
Ecoli-0-2-6-7-vs-3-5	(98.25, 0.017)	(92.11, 0.032)	(92.11, 0.02)	(93.86, 0.026)	(98.25, 0.037)	(98.25, 0.034)	(98.25, 0.024)	(98.25, 0.04)	(97.37, 0.03)	(96.49, 0.04)
Ecoli-0-3-4-6-vs-5	(98.1, 0.03)	(93.33, 0.025)	(91.43, 0.038)	(91.43, 0.02)	(96.19, 0.045)	(99.05, 0.013)	(98.1, 0.011)	(99.05, 0.033)	(96.19, 0.039)	(98.1, 0.03)
Ecoli-0-4-6-vs-5	(95.15, 0.027)	(88.35, 0.023)	(88.35, 0.023)	(91.26, 0.044)	(94.17, 0.044)	(94.17, 0.017)	(95.15, 0.024)	(96.12, 0.04)	(94.17, 0.04)	(95.15, 0.04)
Ecoli-0-6-7-vs-5	(96.36, 0.016)	(90.91, 0.022)	(89.09, 0.061)	(90, 0.022)	(95.45, 0.035)	(95.45, 0.024)	(96.36, 0.035)	(94.55, 0.041)	(96.36, 0.035)	(96.36, 0.04)
Ecoli4	(98.39, 0.028)	(91.94, 0.021)	(93.55, 0.022)	(91.94, 0.023)	(98.39, 0.037)	(98.92, 0.044)	(98.92, 0.044)	(98.39, 0.041)	(97.85, 0.04)	(97.31, 0.04)
Glass2	(91.23, 0.021)	(91.23, 0.015)	(91.23, 0.018)	(91.23, 0.032)	(91.23, 0.02)	(91.23, 0.039)	(91.23, 0.032)	(91.23, 0.028)	(91.23, 0.039)	(91.23, 0.04)
Glass4	(93.75, 0.015)	(92.19, 0.025)	(92.19, 0.028)	(92.19, 0.029)	(96.88, 0.044)	(96.88, 0.042)	(96.88, 0.03)	(95.31, 0.043)	(93.75, 0.04)	(93.75, 0.05)
Heart-stat	(83.57, 0.016)	(83.57, 0.025)	(83.57, 0.02)	(77.14, 0.014)	(80.71, 0.021)	(80, 0.038)	(80, 0.021)	(80, 0.039)	(82.86, 0.039)	(82.86, 0.04)
Monk2	(83.06, 0.035)	(64.45, 0.035)	(63.12, 0.055)	(63.79, 0.057)	(83.06, 0.044)	(82.06, 0.038)	(84.05, 0.043)	(83.06, 0.04)	(82.06, 0.04)	(82.72, 0.04)
Shuttle-6-vs-2-3	(99.23, 0.011)	(96.15, 0.022)	(96.15, 0.021)	(96.15, 0.08)	(100, 0.022)	(100, 0.022)	(100, 0.025)	(100, 0.029)	(100, 0.04)	(100, 0.04)
Shuttle-c0-vs-c4	(99.68, 0.413)	(99.78, 0.257)	(99.68, 0.246)	(99.78, 0.304)	(99.68, 0.48)	(99.68, 0.49)	(100, 0.41)	(99.68, 0.363)	(99.68, 0.382)	(99.68, 0.37)
Vehicle 1	(79.15, 0.063)	(69.51, 0.064)	(70.63, 0.039)	(69.73, 0.056)	(80.04, 0.068)	(79.6, 0.088)	(80.49, 0.068)	(81.39, 0.071)	(80.27, 0.075)	(81.61, 0.06)
Vehicle2	(97.98, 0.062)	(83.63, 0.041)	(83.41, 0.063)	(83.41, 0.107)	(98.21, 0.067)	(98.43, 0.082)	(98.88, 0.069)	(98.88, 0.055)	(98.21, 0.057)	(98.43, 0.06)
Votes	(94.89, 0.014)	(84.26, 0.035)	(82.55, 0.034)	(84.68, 0.048)	(96.17, 0.043)	(96.17, 0.028)	(95.74, 0.038)	(96.17, 0.043)	(96.17, 0.037)	(96.17, 0.04)
Vowel	(97.34, 0.115)	(93.24, 0.12)	(92.83, 0.128)	(92.01, 0.084)	(96.52, 0.157)	(97.75, 0.13)	(96.31, 0.149)	(97.13, 0.125)	(96.11, 0.128)	(97.13, 0.13)
Wpbc	(75.53, 0.01)	(78.72, 0.039)	(78.72, 0.022)	(78.72, 0.026)	(73.4, 0.02)	(70.21, 0.02)	(71.28, 0.022)	(73.4, 0.037)	(72.34, 0.041)	(64.89, 0.03)
Yeast-0-3-5-9-vs-7-8	(89.84, 0.039)	(88.28, 0.018)	(88.28, 0.028)	(88.28, 0.038)	(88.28, 0.033)	(88.28, 0.043)	(88.28, 0.037)	(89.84, 0.05)	(89.84, 0.051)	(89.84, 0.05)
Yeast-0-5-6-7-9-vs-4	(92.81, 0.036)	(91.37, 0.056)	(91.37, 0.017)	(91.37, 0.032)	(89.93, 0.042)	(91.73, 0.053)	(90.29, 0.038)	(91.73, 0.047)	(91.37, 0.038)	(90.29, 0.05)
Yeast2vs8	(97.85, 0.031)	(95.71, 0.049)	(95.28, 0.044)	(95.28, 0.045)	(96.57, 0.047)	(97.42, 0.046)	(97.42, 0.046)	(96.57, 0.043)	(96.57, 0.051)	(96.57, 0.04)
Yeast3	(94.92, 0.103)	(89.33, 0.077)	(89.84, 0.073)	(89.63, 0.116)	(93.8, 0.156)	(93.7, 0.142)	(93.29, 0.123)	(94.21, 0.114)	(93.6, 0.116)	(93.5, 0.12)
Yeast5	(97.46, 0.113)	(96.95, 0.116)	(96.95, 0.097)	(96.95, 0.13)	(97.97, 0.151)	(98.07, 0.164)	(98.27, 0.131)	(98.58, 0.12)	(98.17, 0.14)	(97.56, 0.15)
Average	93.5	89.09	88.86	88.7	93.08	93.15	93.28	93.41	93.02	92.83
Rank	4.04	7.35	7.8	7.89	5.04	4.59	4.02	3.85	5.33	5.09

*Acc denotes Accuracy

Table 4 Performance of the classification models with Gaussian noise $r = 0.1$ added to the datasets

Dataset	TWSVM(Acc, time(s))	Pin-TSVM(Acc, time(s))			Pin-GTSVM(Acc, time(s))			Pin-RGTSVM(Acc, time(s))		
		$\tau = 0.5$	$\tau = 0.8$	$\tau = 1$	$\tau = 0.5$	$\tau = 0.8$	$\tau = 1$	$\tau = 0.5$	$\tau = 0.8$	$\tau = 1$
Brwisconsin	(98.17, 0.032)	(97.91, 0.074)	(97.65, 0.089)	(97.91, 0.088)	(98.17, 0.043)	(97.65, 0.038)	(98.69, 0.042)	(97.91, 0.038)	(97.91, 0.039)	(98.17, 0.04)
Ecoli-0-1-4-7-vs-5-6	(97.8, 0.015)	(95.05, 0.029)	(95.6, 0.067)	(95.05, 0.062)	(98.35, 0.039)	(97.8, 0.028)	(98.35, 0.03)	(98.35, 0.041)	(98.35, 0.035)	(97.8, 0.02)
Ecoli-0-2-6-7-vs-3-5	(98.25, 0.013)	(92.11, 0.018)	(92.11, 0.036)	(91.23, 0.033)	(98.25, 0.042)	(98.25, 0.037)	(94.74, 0.024)	(98.25, 0.04)	(97.37, 0.033)	(98.25, 0.03)
Ecoli-0-3-4-6-vs-5	(98.1, 0.013)	(91.43, 0.032)	(91.43, 0.076)	(91.43, 0.031)	(98.1, 0.043)	(97.14, 0.018)	(98.1, 0.044)	(98.1, 0.033)	(97.14, 0.034)	(99.05, 0.03)
Ecoli-0-4-6-vs-5	(94.17, 0.027)	(91.26, 0.072)	(92.23, 0.028)	(91.26, 0.046)	(94.17, 0.039)	(97.09, 0.019)	(94.17, 0.023)	(96.12, 0.03)	(96.12, 0.036)	(96.12, 0.03)
Ecoli-0-6-7-vs-5	(96.36, 0.018)	(90.91, 0.024)	(90.91, 0.028)	(90, 0.023)	(97.27, 0.037)	(97.27, 0.025)	(97.27, 0.041)	(95.45, 0.037)	(96.36, 0.031)	(96.36, 0.02)
Ecoli4	(98.92, 0.025)	(88.17, 0.057)	(94.62, 0.042)	(91.4, 0.036)	(98.92, 0.035)	(97.31, 0.04)	(98.92, 0.027)	(97.85, 0.033)	(96.77, 0.042)	(97.85, 0.04)
Glass2	(91.23, 0.024)	(91.23, 0.033)	(91.23, 0.03)	(91.23, 0.045)	(85.96, 0.014)	(91.23, 0.025)	(91.23, 0.043)	(91.23, 0.04)	(91.23, 0.039)	(91.23, 0.02)
Glass4	(93.75, 0.175)	(92.19, 0.036)	(92.19, 0.037)	(92.19, 0.026)	(93.75, 0.044)	(93.75, 0.044)	(93.75, 0.041)	(95.31, 0.042)	(96.88, 0.032)	(95.31, 0.02)
Heart-stat	(85.71, 0.019)	(85.71, 0.039)	(62.14, 0.039)	(83.57, 0.045)	(81.43, 0.037)	(78.57, 0.042)	(82.14, 0.042)	(86.43, 0.038)	(79.29, 0.039)	(79.29, 0.03)
Monk2	(81.4, 0.033)	(64.12, 0.035)	(64.12, 0.088)	(65.45, 0.061)	(83.06, 0.039)	(82.06, 0.045)	(82.06, 0.044)	(82.39, 0.034)	(82.39, 0.027)	(82.72, 0.03)
Shuttle-6-vs-2-3	(99.23, 0.011)	(96.15, 0.051)	(97.69, 0.04)	(96.15, 0.026)	(100, 0.022)	(100, 0.025)	(99.23, 0.04)	(100, 0.041)	(100, 0.035)	(100, 0.02)
Shuttle-c0-vs-c4	(99.68, 0.328)	(99.78, 0.24)	(99.68, 0.381)	(99.68, 0.341)	(100, 0.482)	(99.68, 0.5)	(99.68, 0.422)	(99.68, 0.351)	(99.68, 0.397)	(99.68, 0.3)
Vehicle 1	(79.82, 0.064)	(71.97, 0.087)	(72.2, 0.048)	(68.83, 0.112)	(79.6, 0.065)	(78.92, 0.09)	(81.61, 0.071)	(78.25, 0.074)	(78.48, 0.075)	(78.25, 0.06)
Vehicle2	(97.53, 0.06)	(83.63, 0.077)	(83.41, 0.082)	(83.86, 0.052)	(98.65, 0.064)	(98.43, 0.085)	(98.43, 0.068)	(97.76, 0.052)	(98.21, 0.065)	(98.43, 0.07)
Votes	(96.17, 0.018)	(85.53, 0.067)	(81.28, 0.068)	(83.83, 0.042)	(96.6, 0.042)	(96.6, 0.036)	(96.6, 0.044)	(95.74, 0.034)	(96.6, 0.036)	(95.74, 0.03)
Vowel	(97.13, 0.108)	(92.42, 0.105)	(92.83, 0.174)	(92.21, 0.143)	(96.11, 0.129)	(96.31, 0.132)	(96.52, 0.147)	(97.13, 0.135)	(96.93, 0.147)	(96.31, 0.11)
Wpbc	(72.34, 0.01)	(78.72, 0.018)	(78.72, 0.045)	(78.72, 0.019)	(69.15, 0.024)	(70.21, 0.041)	(73.4, 0.043)	(75.53, 0.04)	(76.6, 0.021)	(78.72, 0.03)
Yeast-0-3-5-9-vs-7-8	(89.84, 0.066)	(88.28, 0.054)	(88.28, 0.064)	(88.28, 0.039)	(88.28, 0.035)	(88.28, 0.036)	(88.28, 0.045)	(89.84, 0.039)	(89.84, 0.05)	(89.84, 0.04)
Yeast-0-5-6-7-9-vs-4	(92.09, 0.048)	(91.37, 0.049)	(91.37, 0.024)	(91.37, 0.053)	(91.37, 0.044)	(90.29, 0.033)	(91.73, 0.046)	(91.73, 0.04)	(89.57, 0.04)	(92.09, 0.03)
Yeast2vs8	(97.85, 0.04)	(96.57, 0.019)	(95.28, 0.058)	(94.85, 0.062)	(97, 0.044)	(97, 0.037)	(96.57, 0.05)	(98.28, 0.04)	(97.42, 0.039)	(97.42, 0.04)
Yeast3	(93.29, 0.12)	(90.35, 0.155)	(90.04, 0.124)	(90.24, 0.101)	(93.09, 0.161)	(93.19, 0.152)	(93.09, 0.157)	(93.19, 0.122)	(93.9, 0.106)	(92.99, 0.1)
Yeast5	(97.46, 0.122)	(96.95, 0.153)	(96.95, 0.11)	(96.95, 0.169)	(97.76, 0.163)	(97.76, 0.149)	(96.95, 0.109)	(98.27, 0.117)	(98.07, 0.138)	(98.07, 0.15)
Average	93.32	89.21	88.35	88.94	92.83	92.82	93.11	93.6	93.27	93.46
Rank	4.3	7.59	8.15	8.11	4.41	5.33	4.7	3.89	4.48	4.04

*Acc denotes accuracy

4.4 Statistical Analysis

To evaluate the performance of the proposed Pin-RGTSVM and the other baseline models, we use Friedman test [42]. Let K number of classification models are evaluated on M number of datasets and R_j^i be the rank of the jth classifier on the ith dataset with the highest performing model being assigned the lower rank. In Friedman test, average ranks of the classifiers are compared for statistical evaluation. Under Null hypothesis, all the models are assumed to be of similar rank and hence similar performance. The Friedman statistic

$$\chi_F^2 = \frac{12M}{K(K+1)}\left[\sum_j R_j^2 - \frac{K(K+1)^2}{4}\right] \tag{58}$$

is distributed according to χ_F^2 with $(K-1)$ degrees of freedom under null hypothesis. However, χ_F^2 is undesirably conservative. Thus, better statistic is given by

$$F_F = \frac{(M-1)\chi_F^2}{M(K-1) - \chi_F^2} \tag{59}$$

which is distributed with $(K-1)$ and $(K-1)(M-1)$ degrees of freedom. As the Null hypothesis is rejected, Nemenyi post-hoc test [43] is used to perform pair wise significance tests of the K-classification models. Two classification models are said to be significantly different if the average ranks of the two classifiers differ by at least critical difference (CD)

$$\text{CD} = q_\alpha\sqrt{\frac{K(K+1)}{6M}} \tag{60}$$

where α is the level of significance and q_α is the studentized range statistic divided by $\sqrt{2}$.

In this paper, $\alpha = 0.05, M = 23$ and $K = 10$ which includes the models at different values of τ. Here, we evaluate the significance of the models at different levels of noise, noise level $r = 0$ (means noise free), and noisy datasets with $r = 0.05, 0.1$. The experimental results of the models with noise level $r = 0, 0.05, 0.1$ are given in Tables 2, 3 and 4, respectively. The analysis of each one is given as follows:

With simple calculations on the given average ranks of the models in Table 2, $\chi_F^2 = 42.3186$ and $F_F = 5.6534$. The critical value of $F(9, 198) = 1.935$ at $\alpha = 0.05$ level of significance. Hence, we reject the Null hypothsis. Thus, significant difference exists among the classification models. To evaluate the effectiveness, we use Nemenyi post hoc test. Based on the Nemenyi test, $CD = q_\alpha\sqrt{\frac{K(K+1)}{6M}} = 3.164\sqrt{\frac{10\times 11}{6\times 23}} = 2.83$. Thus, two models are said to be significanlty different if the average ranks of the two models differ by at least CD.

Based on the simple calculation, average rank differences of the statistically different models is given in Table 5 for noise free datasets. The blank entries denote that no

Table 5 Statistically significant differences among the pair of models in the row and column at $\alpha = 0.05$ with noise free datasets

	Pin-TSVM		
	($\tau = 0.5$)	($\tau = 0.8$)	($\tau = 1$)
TWSVM			
Pin-TSVM ($\tau = 0.5$)			
Pin-TSVM ($\tau = 0.8$)			
Pin-TSVM ($\tau = 1$)			
Pin-GTSVM ($\tau = 0.5$)			✓
Pin-GTSVM ($\tau = 0.8$)	✓	✓	✓
Pin-GTSVM ($\tau = 1$)			
Pin-RGTSVM ($\tau = 0.5$)	✓	✓	✓
Pin-RGTSVM ($\tau = 0.8$)			✓
Pin-RGTSVM ($\tau = 1$)			✓

The ✓ mark shows that the significant difference exists between the row method and the column method while as the blank entries signify that no significant difference exists. Also, the ✓ mark shows that the row method is better than the column method

significant difference exist among the models in the corresponding row and column while as the filled entries denote that significant difference exists and numbers denote the absolute differences among the average ranks of the models in the corresponding column and row. The non-blank entries denote that the row-method is significantly better than the column method. one can see that the Pin-GTSVM($\tau = 0.8$) and proposed Pin-RGTSVM ($\tau = 0.5$) are significantly better than the Pin-TSVM model at $\tau = 0.5, 0.8, 1$.

With simple calculations on the given average ranks of the models in Table 3, $\chi_F^2 = 56.97$ and $F_F = 8.35$. Since critical value of $F(9, 198) < F_F$, hence null hypothesis is rejected. We use Nemenyi test to evaluate the significant difference among the pairs of the classification models. After simple calculations, average rank differences of the statistically different models is given in Table 6 for noisy dataset with $r = 0.5$. one can see that TWSVM, Pin-GTSVM($\tau = 1$) and proposed Pin-RGTSVM ($\tau = 0.5$) are significantly better than the Pin-TSVM model at $\tau = 0.5, 0.8, 1$.

With simple calculations on the given average ranks of the models in Table 4, $\chi_F^2 = 68.40$ and $F_F = 10.85$. Since critical value of $F(9, 198) < F_F$, hence null hypothesis is rejected. We use Nemenyi test to evaluate the significant difference among the pairs of the classification models. After simple calculations, average rank differences of the statistically different models is given in Table 7 for noisy dataset with $r = 1$. one can see that proposed Pin-RGTSVM is significantly better than the Pin-TSVM model at $\tau = 0.5, 0.8, 1$.

From the above results, one can see that proposed Pin-RGTSVM achieves better or competitive average accuracy as compared to other baseline models. However, Nemenyi test fails to detect the significant difference among the proposed Pin-RGTSVM and other baseline models.

Table 6 Statistically significant differences among the pair of models in the row and column at $\alpha = 0.05$ with noise level $r = 0.5$

	Pin-TSVM		
	$(\tau = 0.5)$	$(\tau = 0.8)$	$(\tau = 1)$
TWSVM	✓	✓	✓
Pin-TSVM $(\tau = 0.5)$			
Pin-TSVM $(\tau = 0.8)$			
Pin-TSVM $(\tau = 1)$			
Pin-GTSVM $(\tau = 0.5)$			✓
Pin-GTSVM $(\tau = 0.8)$		✓	✓
Pin-GTSVM $(\tau = 1)$	✓	✓	✓
Pin-RGTSVM $(\tau = 0.5)$	✓	✓	✓
Pin-RGTSVM $(\tau = 0.8)$			
Pin-RGTSVM $(\tau = 1)$			

The ✓ mark shows that the significant difference exists between the row method and the column method while as the blank entries signify that no significant difference exists. Also, the ✓ mark shows that the row method is better than the column method

Table 7 Statistically significant differences among the pair of models in the row and column at $\alpha = 0.05$ with noise level $r = 1$

	Pin-TSVM		
	$(\tau = 0.5)$	$(\tau = 0.8)$	$(\tau = 1)$
TWSVM	✓	✓	✓
Pin-TSVM $(\tau = 0.5)$			
Pin-TSVM $(\tau = 0.8)$			
Pin-TSVM $(\tau = 1)$			
Pin-GTSVM $(\tau = 0.5)$	✓	✓	✓
Pin-GTSVM $(\tau = 0.8)$			
Pin-GTSVM $(\tau = 1)$	✓	✓	✓
Pin-RGTSVM $(\tau = 0.5)$	✓	✓	✓
Pin-RGTSVM $(\tau = 0.8)$	✓	✓	✓
Pin-RGTSVM $(\tau = 1)$	✓	✓	✓

The ✓ mark shows that the significant difference exists between the row method and the column method while as the blank entries signify that no significant difference exists. Also, the ✓ mark shows that the row method is better than the column method

4.5 Parameter Sensitivity Analysis

In proposed Pin-RGTSVM model, there are multiple parameters like c_1, c_2, c_3, c_4 and τ. Each parameter need to be tuned via cross validation to obtain the optimal performance. From Fig. 1, one can see that the different combination of the parameters lead to different performance. Therefore, it is important to determine the proper value of parameters for each dataset.

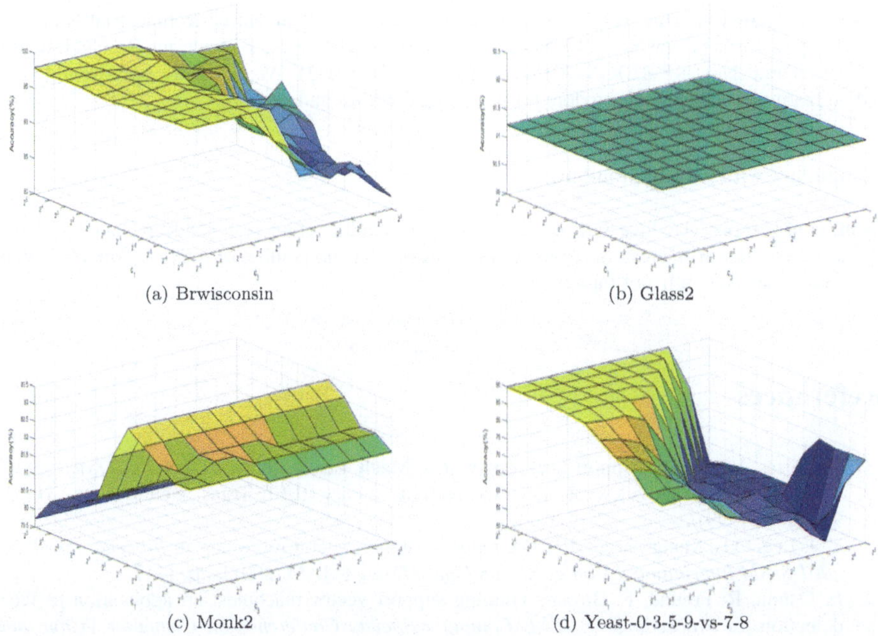

(a) Brwisconsin (b) Glass2

(c) Monk2 (d) Yeast-0-3-5-9-vs-7-8

Fig. 1 Effect of $c_1 = c_2$ and $c_3 = c_4$ on the performance of the proposed Pin-RGTSVM

5 Conclusion and Future Work

In this paper, we proposed an improved version of general twin support vector machines with pinball loss (Pin-GTSVM) known as robust general twin support vector machines (Pin-RGTSVM) with pinball loss function. Different from Pin-GTSVM and TWSVM, we added an extra regularization term to implement the structural risk minimization principle. Unlike TWSVM and Pin-GTSVM, the matrices in the dual formulation of the proposed Pin-RGTSVM are positive definite. Pinball loss function reduces the effect of noise and make it more robust for resampling, and the regularization term makes it theoretically sound as it involves the matrix inversions while optimizing the objective function. We evaluated the proposed Pin-RGTSVM and the other baseline models at different levels of noise. Experimental results and statistical analysis show the efficacy of the proposed Pin-RGTSVM. In future, we would like to extend this work to multilcass and large scale binary classification models. Another area of interest would be to design a fast algorithm for optimizing the objective functions with less complexity.

Acknowledgements This work is supported by the Council of Scientific & Industrial Research (CSIR), New Delhi (Grant No. 22(0751)/17/EMR-II) and Science and Engineering Research Board (SERB) (Grant Nos. SB/S2/RJN-001/2016, ECR/2017/000053). We gratefully acknowledge the Indian Institute of Technology Indore for providing facilities and support.

Compliance with Ethical Standards

Conflict of Interest Aradhya Saini declares she has no conflict of interest. Sandeep Daniel declares he has no conflict of interest. Satyam Saini declares he has no conflict of interest. Ankush Mittal declares he has no conflict of interest.

References

1. C. Cortes, V. Vapnik, Support-vector networks. Mach. Learn. **20**(3), 273–297 (1995)
2. V.N. Vapnik, An overview of statistical learning theory. IEEE Trans. Neural Netw. **10**(5), 988–999 (1999)
3. Y.-J. Lee, O.L. Mangasarian, RSVM: reduced support vector machines, in *Proceedings of the 2001 SIAM International Conference on Data Mining* (SIAM, 2001), pp. 1–17
4. E. Osuna, R. Freund, F. Girosit, Training support vector machines: an application to face detection, in *Proceedings of IEEE Computer Society Conference on Computer Vision and Pattern Recognition*. (IEEE, 1997), pp. 130–136
5. T. Joachims, Text categorization with support vector machines: learning with many relevant features. in *European Conference on Machine Learning* (Springer, 1998), pp. 137–142
6. B. Richhariya, M. Tanveer, EEG signal classification using universum support vector machine. Expert Syst. Appl. **106**, 169–182 (2018)
7. Y. Li, C. Guan, Joint feature re-extraction and classification using an iterative semi-supervised support vector machine algorithm. Mach. Learn. **71**(1), 33–53 (2008)
8. B. Richhariya, M. Tanveer, A. Rashid, A.D.N. Initiative et al., Diagnosis of alzheimer's disease using universum support vector machine based recursive feature elimination (usvm-rfe). Biomed. Sign. Proc. Control **59**, 101903 (2020)
9. M. Ganaie, S. Ghosh, N. Mendola, M. Tanveer, S. Jalan, Identification of chimera using machine learning. Chaos: An Interdisci. J. Nonlinear Sci. **30**(6), 063128 (2020)
10. O.L. Mangasarian, E.W. Wild, Multisurface proximal support vector machine classification via generalized eigenvalues. IEEE Trans. Pattern Anal. Mach. Intell. **28**(1), 69–74 (2005)
11. Jayadeva, R. Khemchandani, S. Chandra, Twin support vector machines for pattern classification. IEEE Trans. Pattern Anal. Mach. Intell. **29**(5), 905–910 (2007)
12. Y.-H. Shao, C.-H. Zhang, X.-B. Wang, N.-Y. Deng, Improvements on twin support vector machines. IEEE Trans. Neural Netw. **22**(6), 962–968 (2011)
13. B. Richhariya, M. Tanveer, A.D.N. Initiative et al., Least squares projection twin support vector clustering (lsptsvc). Inf. Sci. (2020)
14. Z. Wang, Y.-H. Shao, L. Bai, N.-Y. Deng, Twin support vector machine for clustering. IEEE Trans. Neural Netw. Learn. Syst. **26**(10), 2583–2588 (2015)
15. M. Tanveer, T. Gupta, M. Shah, Pinball loss twin support vector clustering. ACM Trans. Multimedia Comput. Commun. Appl. (TOMM) (2020)
16. M.A. Ganaie, M. Tanveer, P.N. Suganthan, Oblique decision tree ensemble via twin bounded SVM. Expert Syst. Appl. **143**, 113072 (2020)
17. M. Ganaie, M. Tanveer, LSTSVM classifier with enhanced features from pre-trained functional link network. Appl. Soft Comput. p. 106305 (2020)
18. M.A. Kumar, M. Gopal, Least squares twin support vector machines for pattern classification. Expert Syst. Appl. **36**(4), 7535–7543 (2009)

19. J.A. Nasiri, N.M. Charkari, K. Mozafari, Energy-based model of least squares twin support vector machines for human action recognition. Sign. Proc. **104**, 248–257 (2014)
20. M. Tanveer, M.A. Khan, S.-S. Ho, Robust energy-based least squares twin support vector machines. Appl. Intell. **45**(1), 174–186 (2016)
21. B. Richhariya, M. Tanveer, An efficient angle based universum least squares twin support vector machine for pattern classification. ACM Trans. Internet Technol. (TOIT) (2020)
22. M. Tanveer, Newton method for implicit lagrangian twin support vector machines. Int. J. Mach. Learn. Cyber. **6**(6), 1029–1040 (2015)
23. M. Tanveer, Robust and sparse linear programming twin support vector machines. Cognitive Comput. **7**(1), 137–149 (2015)
24. B. Richhariya, M. Tanveer, Universum least squares twin parametric-margin support vector machine, in *2020 International Joint Conference on Neural Networks (IJCNN)* (IEEE, 2020)
25. M. Tanveer, Application of smoothing techniques for linear programming twin support vector machines. Knowl. Inf. Syst. **45**(1), 191–214 (2015)
26. B. Richhariya, M. Tanveer, A reduced universum twin support vector machine for class imbalance learning. Pattern Recog. 107150 (2020)
27. M. Ganaie, M. Tanveer, P. Suganthan, Regularized robust fuzzy least squares twin support vector machine for class imbalance learning, in *2020 International Joint Conference on Neural Networks (IJCNN)*. (IEEE, 2020)
28. M. Tanveer, C. Gautam, P. Suganthan, Comprehensive evaluation of twin SVM based classifiers on UCI datasets. Appl. Soft Comput. **83**, 105617 (2019)
29. M. Tanveer, A. Sharma, P.N. Suganthan, Least squares KNN-based weighted multiclass twin SVM. Neurocomputing (2020)
30. P.-Y. Hao, New support vector algorithms with parametric insensitive/margin model. Neural Networks **23**(1), 60–73 (2010)
31. X. Peng, TPMSVM: a novel twin parametric-margin support vector machine for pattern recognition. Pattern Recogn. **44**(10–11), 2678–2692 (2011)
32. X. Huang, L. Shi, J.A. Suykens, Support vector machine classifier with pinball loss. IEEE Trans. Pattern Anal. Mach. Intell. **36**(5), 984–997 (2013)
33. R. Koenker, K.F. Hallock, Quantile regression. J. Econom. Perspect. **15**(4), 143–156 (2001)
34. A. Christmann, I. Steinwart, How SVMs can estimate quantiles and the median. in *Advances in Neural Information Processing Systems* (2008), pp. 305–312
35. X. Shen, L. Niu, Z. Qi, Y. Tian, Support vector machine classifier with truncated pinball loss. Pattern Recogn. **68**, 199–210 (2017)
36. Y. Xu, Z. Yang, X. Pan, A novel twin support-vector machine with pinball loss. IEEE Trans. Neural Networks Learn. Syst. **28**(2), 359–370 (2016)
37. M. Tanveer, A. Tiwari, R. Choudhary, S. Jalan, Sparse pinball twin support vector machines. Appl. Soft Comput. **78**, 164–175 (2019)
38. M. Tanveer, T. Rajani, M.A. Ganaie, Improved sparse pinball twin SVM, in *2019 IEEE International Conference on Systems, Man and Cybernetics (SMC)* (IEEE, 2019), pp. 3287–3291
39. M. Tanveer, A. Sharma, P.N. Suganthan, General twin support vector machine with pinball loss function. Inf. Sci. **494**, 311–327 (2019)
40. D. Dua, C. Graff, UCI machine learning repository (2017). [Online]. Available: http://archive.ics.uci.edu/ml
41. J. Alcalá-Fdez, A. Fernández, J. Luengo, J. Derrac, S. García, L. Sánchez, F. Herrera, Keel datamining software tool: data set repository, integration of algorithms and experimental analysis framework. J. Mult-Valued Log. Soft Comput. **17** (2011)
42. J. Demšar, Statistical comparisons of classifiers over multiple data sets. J. Mach. Learn. Res. **7**(Jan), 1–30 (2006)
43. P. Nemenyi, Distribution-free multiple comparisons, in *Biometrics*, vol. 18, no. 2. International Biometric Soc 1441 I ST, NW, Suite 700, Washington, DC 20005-2210, 1962, p. 263

Noise Resilient Thresholding Based on Fuzzy Logic and Non-linear Filtering

Shreya Goyal, Gaurav Bhatnagar, and Chiranjoy Chattopadhyay

Abstract This chapter proposes a novel image thresholding technique resilient to noise based on the fuzzy logic and the non-linear filtering. The rationale of the proposed technique is to use adaptive Kuwahara filtering, which essentially suppresses the noise while retaining the image texture information. Finally, an optimized threshold selection process is formalized using the fuzzy logic properties. The performance of the proposed technique is then evaluated by extensive experiments on different types of synthetic, real-world images, and floor plan images. Finally, the superiority of the proposed technique is carried out by detailed comparison with the existing methods.

Keywords Segmentation · Thresholding · Non-linear filtering · Adaptive Kuwahara filter · Fuzzy logic

1 Introduction

In image processing, thresholding is an important research area, which is essentially employed to segment the images to extract the desired objects. In general, image segmentation can be defined as a technique in which an image is classified into categories or regions which may correspond to different objects in the image or different parts of the image. Every pixel in the image will be categorized into one of these categories. A good segmented image has each pixel of same object being classified into the same region and its neighboring pixel which do not belong to the same object should classify into other regions. It is a very critical step in most of

S. Goyal (✉) · G. Bhatnagar · C. Chattopadhyay
Indian Institute of Technology Jodhpur, Jodhpur, India
e-mail: goyal@iitj.ac.in

G. Bhatnagar
e-mail: goravb@iitj.ac.in

C. Chattopadhyay
e-mail: chiranjoy@iitj.ac.in

P. Kumar et al. (eds.), *Machine Learning for Intelligent Multimedia Analytics*, Studies in Big Data 82,
https://doi.org/10.1007/978-981-15-9492-2_7

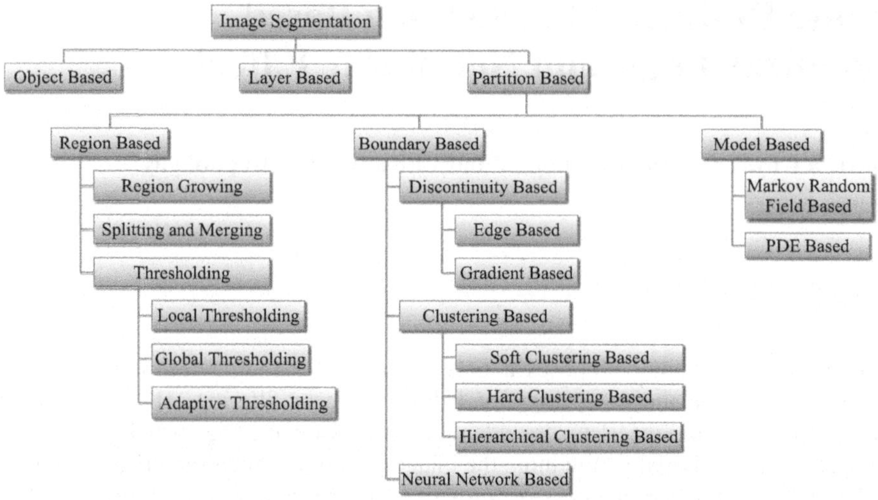

Fig. 1 A broad classification of image segmentation techniques

the applications because if segmentation is done properly then the further stages of the application are quiet simpler. Therefore, in a nutshell, it is a decisive stage for numerous image-processing and computer vision applications such as shape recognition [1], document image analysis [2], infrared image segmentation [3, 4], quality inspection of materials [5], Synthetic Aperture Radar (SAR) image segmentation [6], and medical image applications [7, 8].

In the recent past, numerous segmentation techniques have been proposed. The underlying philosophies and requirements of these distinct segmentation techniques are diverse in nature. Based on this diversity, a broad classification of segmentation techniques can be seen in Fig. 1. The most initial classification of these techniques will be based on object, layer, and partitions. In the former, an image is segmented into the objects with exact boundaries. Here, the object can be geometrical or text. The layer-based techniques segment the image with respect to mask layers, which are assumed to be the demarcation in the constitution of the given image. It has been observed that these techniques are sophisticated and quite complicated and thus have limited applicability. In contrast, partitions based techniques work on finding different attributes of different partitions of an image. These attributes include histogram, color, transform coefficients, gradients and edges. These are the simplest techniques and are bifurcated into three categories: (1) Region-Based, (2) Boundarybased, and (3) Model-based techniques. Region based techniques generally rely on the similarity properties of the image and the segmentation is done on the basis of alikeness among different regions. The most common techniques under this category include thresholding, region growing, splitting, and merging. Thresholding techniques can be classified into single- or bi-level thresholding. In the former, a single threshold is required to segment the image whereas more than one thresholds are required to segment the image in the later. A variety of thresholding techniques have been proposed over the years. A thorough survey and studies of these techniques can be seen in [9–11]. The threshold selection process in single- or bi-level thresholding can further be classified into global process and local process. In global thresholding process, the threshold value(s) is(are) obtained according to the characteristics of the whole image. In contrast, the threshold value(s), in local thresholding techniques, is(are) obtained using the local activity information of the images. Among all these, most notable approaches are based on the between-class variance (Otsu's method) [12], entropy [13–17], clustering [18], gradient [19], maximum correlation [20], fuzzy logic [2, 21–26] and the fuzzy version of the maximum entropy [27]. Otsu method utilizes the discriminant analysis to find the maximum separability of two classes. It is based on a very simple idea of finding a threshold to maximize the intra-class variance [12]. Majority of other techniques exploit the entropy of the distribution of the gray levels in an image. The underlying idea is to consider the image foreground and background as two different signals and the image is said to be optimally thresholded when the sum of the entropies of the two signals reach its maximum. Thereafter, different definitions of entropy are deployed to find the optimally thresholded image. For instance, authors have used Shannon's entropy in [13], cross entropy in [14], Renyi's entropy in [15] and Tsallis entropy in [16]. In [17] authors have defined the.

thresholding problem as the change in the correlation of observation on specification of the foreground and background classes. The optimum threshold is then obtained by maximizing the correlation among the observed classes. Next class of techniques is fuzzy logic based thresholding techniques. These techniques assume the fuzzy membership as indicator of the fact that how strongly a gray value is associated to the background or to the foreground. In [20], the authors have used the fuzzy compactness of the S-function for membership evaluation. In contrast, a measure of fuzziness based on the relationship between fuzzy set and its complement are exploited for thresholding in [2, 21]. In [22], authors have tried to relate each pixel of the image to different output centroids via a fuzzy membership function to select an optimal threshold. The methods [23] and [24] employed the neural networks and co-occurrence matrix to select an optimal threshold on fuzzy measure.

The boundary-based techniques rely on the discontinuity among the pixels. The idea is to collect all the regions using the unanticipated changes in the intensity values. This property can generally be obtained using edges, points, corners, and lines in the image. There are few advanced techniques in this category, which rely on neural networks (NN) and clustering. These approaches assumed segmentation as unsupervised learning task and form the clusters based on the principle of maximizing(minimizing) the intra(inter)-class similarity. Clustering-based techniques can be bifurcated in hard, soft and hierarchical clustering based techniques. In hard clustering, image pixels are clustered in an exclusive way while pixels are clustered based on the exhibit membership values to more than one cluster in soft clustering [32–34]. Further, two-dimensional informational is collected and translated into the progressive grouping in the Hierarchical clustering based techniques [35, 36]. The major drawback of these techniques is that different similarity measures lead to varied outcomes [36]. In contrast, NN-based techniques first map each pixel to a neuron, and then dynamic equations are used to obtain the edges using minimum energy [37–40]. A detailed review on NN based segmentation techniques can be seen in [41]. Another set of techniques are based on different models. The basic idea is to incorporate local information with the geometrical state of the object to obtain a comprehensive representation in contrast with distinct objects of the image. The only requirement is the availability of the exact shape of the object. The most prominent techniques in this category are Markov random field (MRF) and Partial Differential Equation (PDE) based techniques. MRF based techniques work in an iterative fashion wherein the coarse regions are segmented initially and the finer regions are then processed. In some of the cases, the edge detection process must be accompanied to locate the edges prominently [42–44]. In contrast, PDE based techniques try to construct a curve or active contours conditioned with the refrainment of an image to locate the objects. The curve essentially locates the boundary of the objects by observing the internal and external forces while tracing the image [45–48].

However, all these techniques fail to deal with the noise corrupted images. This is due to the fact that the gray-levels of objects are dispersed and merged due to the noisy distortions. Specifically, under the low SNR levels the achievements of the above-mentioned methods' drop considerably. To mitigate this drawback a novel thresholding technique is proposed based on the fuzzy logic and non-linear filtering. The core idea is to use adaptive Kuwahara filter prior to the thresholding selection to suppress the noise in image while retaining the texture information. The use of adaptive Kuwahara filter ensures a tradeoff so that no smaller window would have suppressed the noise and no larger window would have suppressed the texture information required for the better thresholding result. It essentially obtains an optimal window size to ensure the best results and to ensure that all the texture information will be retained even if a large sliding window will be deployed in the adaptive filtering. Finally, the threshold selection process based on fuzzy logic picks an optimal threshold. Two different fuzzy membership functions, viz., 1) trapezoidal and 2) triangular membership function are considered for the threshold selection process. Performance of the proposed method is evaluated on the number of different types of experimental images. The superiority of the proposed technique is then confirmed using the comparison with existing thresholding techniques. The strength of proposed work is further confirmed by its usage in floor plan analysis as an application. The main contributions of the proposed work can be summarized as follows:

(1) An image segmentation technique is proposed in this work specifically defined for noisy images.
(2) It surveys the literature of image segmentation thoroughly.
(3) It is the very first attempt, as per the best of authors' knowledge, to integrate adaptive Kuwahara filters with segmentation strategies.
(4) The superiority of the proposed work is carried out by extensive experiments on different synthetic and real images.
(5) Explores comparative analysis elaborately to substantiate the superior performance of proposed work.

The rest of this chapter is organized as follows. The proposed technique is presented in Sect. 2 followed by a detailed description of the experimental settings and the results obtained on the experimental images in Sect. 3. The application of the proposed framework in floor plan analysis is illustrated in Sect. 4 and the concluding remarks are finally given in Sect. 5.

2 Proposed Thresholding Technique

In this section, some of the motivating factors in the design of proposed thresholding technique are discussed. The process to suppress the noise in the image while retaining the texture information using the adaptive Kuwahara filtering is first presented, and then the optimal threshold selection process is devised. Without loss of generality, assume that f represents the original image of size $M \times N$. The detailed description of the proposed technique is as follows.

2.1 Adaptive Kuwahara Filtering

The adaptive Kuwahara filter is a non-linear filter which essentially adjusts the window size of the filter first and then makes color intensity smooth and removes small noise while maintaining the edges [28]. For this filter, the window size is dynamically changing according to the local properties of the image. The need for using dynamic window size arouses from the fact that too small window cannot handle all the noise while too large window can dissolve smaller details of the images along with the noise. Therefore, an optimal window size will ensure that the noise is removed while retaining the contours of the image. The complete adaptive Kuwahara filtering process can be summarized as follows [28].

(a) The starting point is to select an initial window size. Without loss of generality, let us assume 3×3 as the initial window size.
(b) Divide the filter window of size $2a + 1 : a \in \mathbb{R}$ around the pixel (x, y) into four basic areas denoted as ξ_k, where $k \in \{0, 1, 2, 3\}$. Schematically, these basic areas can be defined as

$$\xi_k(x, y) = \begin{cases} [x, x + a] \times [y, y + a] \text{ if} k = 0 \\ [x - a, x] \times [y, y + a] \text{ if} k = 1 \\ [x - a, x] \times [y - a, y] \text{ if} k = 2 \\ [x, x + a] \times [y - a, y] \text{ if} k = 3 \end{cases} \tag{1}$$

where $\times$ is the Cartesian product and initially $a = 1$. It must be noted that the pixels lie on the borders of two regions belonging to both regions so there is a slight overlap between the regions.

(c) For each of these areas, the values of mean (μ_k) and variance (σ_k^2) are calculated ass

$$\mu_k = \frac{1}{(n+1)\times(n+1)} \sum_{(x,y)\in\xi_k} \varphi(f(x,y)) \tag{2}$$

$$\sigma_k^2 = \frac{1}{(n+1)\times(n+1)} \sum_{(x,y)\in\xi_k} [\varphi(f(x,y)) - \mu_k]^2 \tag{3}$$

(d) Now, the size of the window is increased (i.e., $a \rightarrow a+1$) and repeat the previous step to get μ_{k+1} and σ_{k+1}^2 for each basic area.

(e) If the new variance is smaller than the previous one, then the mean and variance of the kth area will take the new values, i.e.,

$$m_k = m_{k+1} \text{ and } \sigma_k^2 = \sigma_{k+1}^2 \text{ if } \sigma_k^2 < \sigma_{k+1}^2 \tag{4}$$

(f) Repeat the steps 4 and 5 until the size of window reaches to its maximum or the variance of the newly enlarged area is greater than the previously calculated variance ($\sigma_k^2 > \sigma_{k+1}^2$).

(g) Once the requirement of the previous step is achieved, the final variances of the basic areas are compared to generate the filtered image. The resulting value of the output pixel is the average value of the basic area for which the variance is smallest, i.e.,

$$\widetilde{f}(x,y) = m_p \text{ such that } \sigma_p^2 = \sigma_{\min}^2 = \min_{k\in\{0,1,2,3\}} \sigma_k^2 \tag{5}$$

It is clear from Eq. 5, that the output pixel is the average value of the basic area having the lowest variance. It is evident that the variance is a good indicator of the image activity. More precisely, small variance indicates the flat regions whereas larger variance indicates the texture regions of the image. Therefore, the output image has the smoothness along with the image activity information.

2.2 *Proposed Thresholding Process*

The proposed thresholding process can be summarized as follows:

(a) Apply adaptive Kuwahara filter on the source image (f) to suppress the noise and retain the edges in the image. The final image formed is ($\widetilde{f}$).

(b) Extract L centroids, for the clusters ($Ck : k = 0, 1, 2, ..., L$) for the image f. The value of L is decided by the user and this clustering is done by k-mean clustering algorithm.

(c) A membership function $\phi_k(x)$ is defined to obtain the membership values for each pixel for every cluster. This membership function is fuzzy membership function. The membership function used here is triangular and trapezoidal membership functions. The triangular membership function is described by three parameters *a*, *b* and *c* and is defined as

$$\phi^{\Delta}(x; a, b, c) = \max\left\{\min\left(\frac{x-a}{b-a}, \frac{c-x}{c-b}\right), 0\right\} \tag{6}$$

where the parameters *a* and *c* locate the 'feet' of the triangle and the parameter *b* locates the 'peak' of the triangle. In contrast, the trapezoidal membership function is described by four parameters *a, b, c* and *d* and is defined as:

$$\phi^{\Delta}(x; a, b, c, d) = \max\left\{\min\left(\frac{x-a}{b-a}, 1, \frac{d-x}{d-c}\right), 0\right\} \tag{7}$$

where the parameters *a* and *d* locate the 'feet' of the trapezoid and the parameter *b* and *c* locate the 'shoulder' of the trapezoid. A number of other membership functions such as S-shaped, Z-shaped, Gaussian, Bell-shaped, sigmoidal membership functions can be used. However, the aforesaid membership functions are the basic and computationally inexpensive [29]. In general, the triangular and trapezoidal membership functions perform well and a theoretical justification can be seen in [30].

(d) The membership value for each pixel is assigned using the membership functions, denoted by $\phi_k^{\theta}(\tilde{f}(x, y))$, where $\theta \in \{\Delta, \cup\}$. The fuzzy membership function not only classifies a pixel in to a cluster, but also calculates membership value for all the pixels. Thus, $\phi_k^{\theta}(\tilde{f}(x, y))$ is a collection of membership values represented as:

$$\phi_k^{\theta}(\tilde{f}(x, y)) = \left[\phi_1^{\theta}(\tilde{f}(x, y))\ \phi_2^{\theta}(\tilde{f}(x, y))\ \cdots\ \phi_k^{\theta}(\tilde{f}(x, y))\right] \tag{8}$$

such that the condition of consistency is satisfied, i.e.,

$$\sum_{k=1}^{L} \phi_k^{\theta}(\tilde{f}(x, y)) = 1 \tag{9}$$

(e) Generate an initial thresholded image as follows.

$$\Upsilon(x, y) = \arg\max_{L}\{\phi_k^{\theta}(\tilde{f}(x, y))\} \tag{10}$$

Though Υ is the thresholded image, but it doesn't have local activity information of the image and therefore Υ is highly sensitive to the centroid selection process. In other words, an optimal search will produce an optimal thresholded image, but there is no universal way to decide an optimal search process which will efficiently work for all image modalities. Therefore, the initial thresholded image has to be improved considering the local activity information of the image.

(f) At this point, the initial thresholded image Υ is improved by considering local information. The basic idea is to consider the membership values of the pixels in the neighborhood N_δ centered at (x, y). The membership values are considered because the image is no longer in the spatial domain and the local activity can be processed by some aggregation method. In this work, local aggregation is performed by the adaptive median operator and the center value is replaced by it.

$$\widetilde{\Upsilon}(x, y) = \underset{(a,b)\in N_\delta(x,y)}{\text{aMedian}} \left[\phi_k^\theta\left(\widetilde{f}(a, b)\right)\right] \tag{11}$$

The use of adaptive median operator will essentially obtain an optimal value of δ to ensure best results. In other words, it will search for the optimal neighborhood size before aggregating local information. This is essential because the pixels from flat regions are having same membership values and the local activity may not be obtained with a single value of δ.

(g) Find the optimal threshold by defuzzifying the modified membership values. For defuzzification, centroid defuzzifier is used, which essentially computes the center of gravity of the membership values. Mathematically, it can be defined as:

$$T^{opt} = \frac{\sum_{(x,y)} \left[\widetilde{\Upsilon}(x, y)\, \widetilde{f}(x, y)\right]}{\sum_{(x,y)} \widetilde{\Upsilon}(x, y)} \tag{12}$$

(h) In the final step, segment the image into object and background using the optimal threshold T^{opt}.

3 Experiments and Results

In this section, the experimental results on a variety of images are depicted to illustrate the performance of the proposed thresholding techniques. The performance and effectiveness of the proposed technique are assessed by a number of experiments on different images using MATLAB platform. The used dataset consists of synthetic and real-world images, each having a manually generated ground-truth image. These images contained small or large objects, text, objects with clear or fuzzy boundaries, and can be either noisy or smooth. The main motivation for these experiments is to compare the results of proposed technique with state-of-the-art techniques. Therefore, a more elaborated performance comparison is accomplished with six different methods namely the Otsu's [12], the Clustering-based [18], the

Entropy-based [13], the Fuzzy c-means (FCM) [23], the Fuzzy thresholding (FTH) [26], and the Gradient-based [19] methods.

In order to demonstrate the significant performance of the proposed technique quantitatively, misclassification error measure is considered. This measure considers the image segmentation as a pixel classification problem and measures the degree of pixels misclassification. It essentially reflects the percentage of background pixels wrongly classified to foreground, and vice versa (foreground pixels wrongly classified to background). Without loss of generality, let us assume that the ground truth and thresholded images are represented by G and R respectively. Mathematically, this measure is expressed as [31]

$$\eta(G, R) = 1 - \frac{|B_G \cap B_R| + |F_G \cap F_R|}{|B_G| + |F_G|} \tag{13}$$

where B_G and F_G denote the background and foreground of the ground-truth images respectively, B_R and F_R denote the counterparts in the resulting segmented images, and $| \cap |$ represents the cardinality of a set. The value of η varies from zero for perfectly segmented image to one for a poorly segmented image.

3.1 Performance on Synthetic Images

In this section, the performance of the proposed thresholding technique is tested using the synthetic images (having two intensities, one is 0 and another is 255) for which the optimal threshold can be generated directly. The original synthetic images are of size 256×256 and are shown in the Figs. 2a and 3a, respectively. The Gaussian noise having zero mean and variance 0.1, 0.2, 0.3, 0.4 and 0.5, respectively, is then added to the synthetic images followed by the optimal threshold generation. The visual results of the proposed and the existing thresholding techniques are shown in Figs. 2, 3 and the quantitative performance measures for each technique are listed in Table 1. As is apparent from Figs. 2, 3 and Table 1, the performance of the existing methods degrades as the level of noise increases. The performance of all the methods is similar when the Gaussian noise is having the variance of 0.1. In this case, the misclassification error is nearly zero for all the methods. The significant difference is observed in the performance as the variance of the Gaussian noise is increased. For thresholding methods presented in [12, 13, 18, 19, 23], a lot of pixels are misclassified in the presence of the noise. In contrast, all the pixels are classified correctly by the proposed technique even in the presence of the Gaussian noise with variance 0.5. The worst performance is for the entropy-based method [13] and this was expected because the first-order entropy is unable to seize the variations present in the image. In contrast, the performances of clustering based [18], Fuzzy c-means based [23], Gradient-based [19] and Otsu techniques [12] are comparatively better due to their classification-based optimal threshold selection criterion. Among all the

Fig. 2 Results on Synthetic image 1 (Syn1): (a) Original Synthetic image; Synthetic image with zero mean and variance (a1) 0.1 (a2) 0.2 (a3) 0.3 (a4) 0.4 (a5) 0.5; Results of thresholding based on (b1,b2,b3,b4,b5) Otsu's method (c1,c2,c3,c4,c5) Clustering; (d1,d2,d3,d4,d5) Entropy; (e1,e2,e3,e4,e5) Fuzzy c-mean; (f1,f2,f3,f4,f5) Fuzzy thresholding; (g1,g2,g3,g4,g5) Gradient; Proposed technique with (h1,h2,h3,h4,h5) Trapezoidal; (i1,i2,i3,i4,i5) Triangular membership function

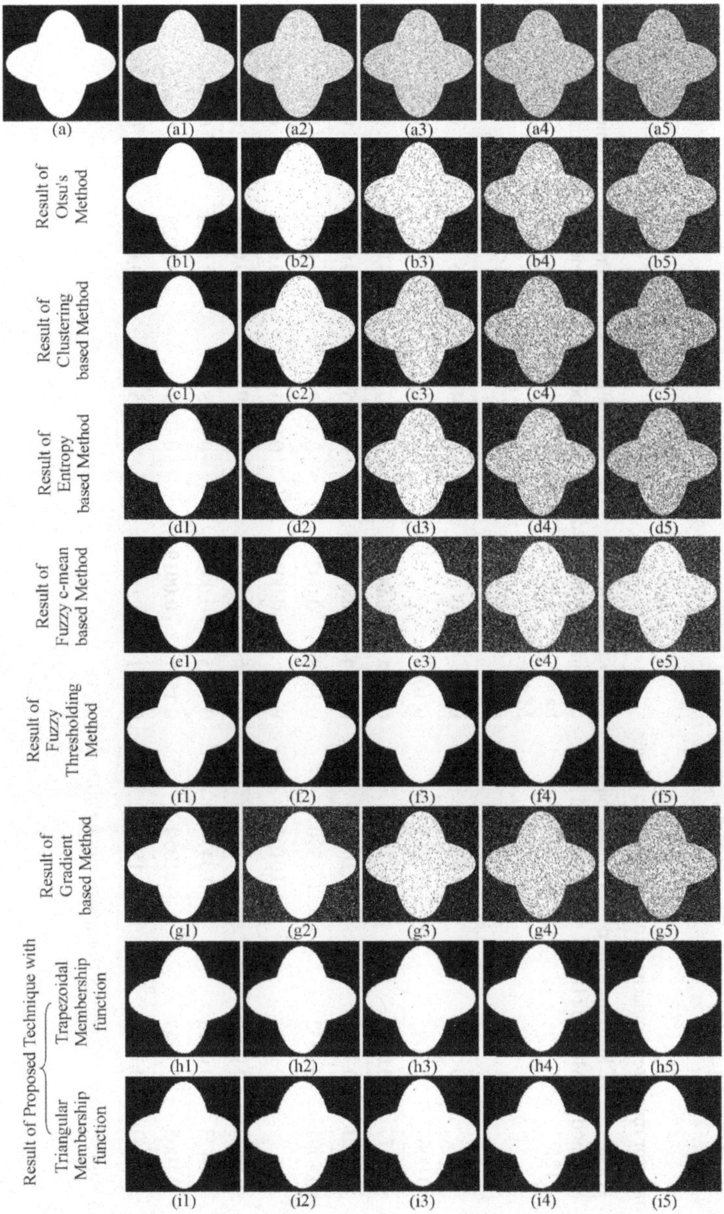

Fig. 3 Results on Synthetic image 2 (Syn2): (a) Original Synthetic image; Synthetic image with zero mean and variance (a1) 0.1 (a2) 0.2 (a3) 0.3 (a4) 0.4 (a5) 0.5; Results of thresholding based on (b1,b2,b3,b4,b5) Otsu's method (c1,c2,c3,c4,c5) Clustering; (d1,d2,d3,d4,d5) Entropy; (e1,e2,e3,e4,e5) Fuzzy c-mean; (f1,f2,f3,f4,f5) Fuzzy thresholding; (g1,g2,g3,g4,g5) Gradient; Proposed technique with (h1,h2,h3,h4,h5) Trapezoidal; (i1,i2,i3,i4,i5) Triangular membership function

Table 1 Misclassification error measure comparison for the synthetic images

Image	Noise with zero mean and variance	Otsu's Method [12]	Clustering-based Method [18]	Entropy-based Method [13]	Fuzzy c-mean based Method [23]	Fuzzy Thresholding Method [26]	Gradient-based Method [19]	Proposed Technique with the membership function	
								Trapezoidal	Triangular
Syn1	0.1	0	0	0.0138	0	0	0.0164	0	0
	0.2	0.0067	0.0097	0.0151	0.1004	0	0.0922	0.0019	0.0019
	0.3	0.0477	0.0525	0.0486	0.1008	**0.0011**	0.0507	0.0027	0.0027
	0.4	0.1061	0.1098	0.1069	0.1464	**0.0021**	0.1085	0.0034	0.0037
	0.5	0.1591	0.1632	0.1590	0.1898	0.0023	0.1603	0.0030	**0.0022**
Syn2	0.1	0	0	0.0163	0	0.0031	0.0163	0.0026	**0.0029**
	0.2	0.0057	0.0090	0.0186	0.0226	**0.0029**	0.0939	0.0037	0.0041
	0.3	0.0473	0.0498	0.0475	0.1055	0.0039	0.0508	0.0038	**0.0036**
	0.4	0.1065	0.1069	0.1050	0.1501	**0.0019**	0.1091	0.0022	0.0025
	0.5	0.1567	0.1599	0.1572	0.1891	0.0034	0.1595	**0.0023**	0.0027

The values highlighted in bold signifies the best performance among all techniques state-of-the-art methods, the fuzzy thresholding technique [26] has an exceptional performance which leads to a cleaner thresholded image which is as accurate as the results of proposed technique. The proposed technique with triangular membership function performs overall better in the presence of the noise. Therefore, the proposed technique can be used efficiently to process noisy images having different levels of noises.

3.2 Performance on Real-World Images

Generally, it is easy to find an optimal threshold for the synthetic images due to the less number of gray-levels. In contrast, the distribution of gray-levels in a real image is complex in nature, which makes the optimal threshold calculation sloppy. Therefore, the performance of the proposed thresholding technique is tested using the real-world images in this section. Five different real gray-scale images are selected to demonstrate the robustness of the proposed technique. These images are "Blocks", "Gear", "Hippo", "Rocks" and "Top", respectively. All the experimental images and associated manually generated ground truth images are depicted in Fig. 4. The visual results of all the techniques are presented in Fig. 4 and the quantitative performance measure η for each technique is listed in Table 2. It is clear from Table 2, the proposed thresholding technique has the lowest average performance of 5.29% with the lowest standard deviation of 3.92%. In contrast, the higher average performance and the higher standard deviation of existing techniques essentially show the superior performance of the proposed technique. Clearly, for all experimental images, the proposed method has performed better when compared to the existing methods. However, for the "Blocks" and "Top" images, the accuracy of the proposed method is slighter lower (but comparable to) than Gradient-based method [19]. For these two

Fig. 4 **a**, **b**, **c**, **d**, **e** Real-world images; **f**, **g**, **h**, **i**, **j** corresponding ground-truth images

Table 2 Misclassification error measure comparison for real-world images

Image name	Otsu's method [12]	Clustering-based method [18]	Entropy-based method [13]	Fuzzy c-mean based method [23]	Fuzzy thresholding method [26]	Gradient-based method [19]	Proposed technique with the membership function	
							Trapezoidal	Triangular
Blocks	0.1899	0.4276	0.1601	0.1444	0.1861	**0.0565**	0.1066	0.1119
Gear	0.0338	0.0850	0.0242	0.0194	0.0365	0.4584	**0.0317**	0.0340
Hippo	0.0673	0.0651	0.4828	0.0975	0.0638	0.1515	**0.0265**	0.0359
Rocks	0.0087	0.0080	0.3238	0.0874	0.0081	0.1229	**0.0072**	0.0077
Top	0.2232	0.2313	0.1201	0.0682	0.2200	**0.0376**	0.0924	0.1107
Mean	0.1046	0.1634	0.2222	0.0834	0.1029	0.1654	**0.0529**	0.0601
Standard deviation	0.0860	0.1512	0.1623	0.0407	0.0843	0.1523	**0.0392**	0.0430

The values highlighted in bold signifies the best performance among all techniques images, the results of proposed thresholding techniques are having some false detections, as apparent from Fig. 5(g1,h1,g4,h4). In contrast, Gradient-based method [19] is having the worst result among all the techniques for the "Gear" image, as depicted in Fig. 5(f2). For this image, Gradient-based method only classifies few pixels correctly and leads to a visually unintelligible output. Overall, the worst performance is for the entropy-based method, which is well justified visually from Fig. 5(c1,c2,c3,c4,c5) where a lot of false detections are clearly ostensible. In contrast, same trend can be observed from Table 2 where entropy-based method is having higher values. The Fuzzy thresholding method [26], which is also resilient to the noise (see Sect. 3.1 for a detailed discussion) performs moderately and introduces false detection in the results and therefore is not applicable for the real-world images.

From the experiments on the synthetic and real-world images, it can be concluded that the proposed techniques are capable of selecting better optimal threshold for both the clean and noisy images with different levels. Overall, the performance of the proposed technique with trapezoidal membership function is better due to the use of non-linear filtering and fuzzy logic properties, which essentially captures the background variations while retaining the edge information and incorporates them in the optimal threshold selection criterion.

4 Application to Floor Plan Image Segmentation

Image segmentation is having numerous applications ranging from remote sensing to autonomous vehicles to medical technologies [41, 49–52]. To judge the practical efficiency of the proposed work, we have considered a relatively new application of floor plan segmentation. Floor plans are the line drawing representation of interiors of any house, office, or any other building. These schematic representations depict the placement of rooms, walls, decors, and other entities in an indoor setting. There are different components of a floor plan, for example, walls, doors, windows, rooms, etc. which help users to identify the structure of the building. However, these complex drawings have notations and symbols which are used by engineers and architects and are difficult to be read by a common user. Hence, in order to automate the understanding of these engineering drawings, several processes have to be devised which involve techniques from the domain of image processing and computer vision. Image segmentation is one such technique, which is used for extracting specific components from the image into consideration. Hence, in the context of floor plans, image segmentation is used for wall, décor, door identification, and classification. However, image segmentation technique fails to generate good results if the image is poorly illuminated or consists of noise of any form. Hence, there is a requirement of technique which is used as a pre-processing step for floor plan components identification. Also, in case of the very old floor plans, for an old building or any historical monument, the images are the scanned document images of a very old paper based document. These documents get a yellow tint with time or get deteriorated with humidity and whether conditions. Hence, in order to recover the accurate floor plan image and

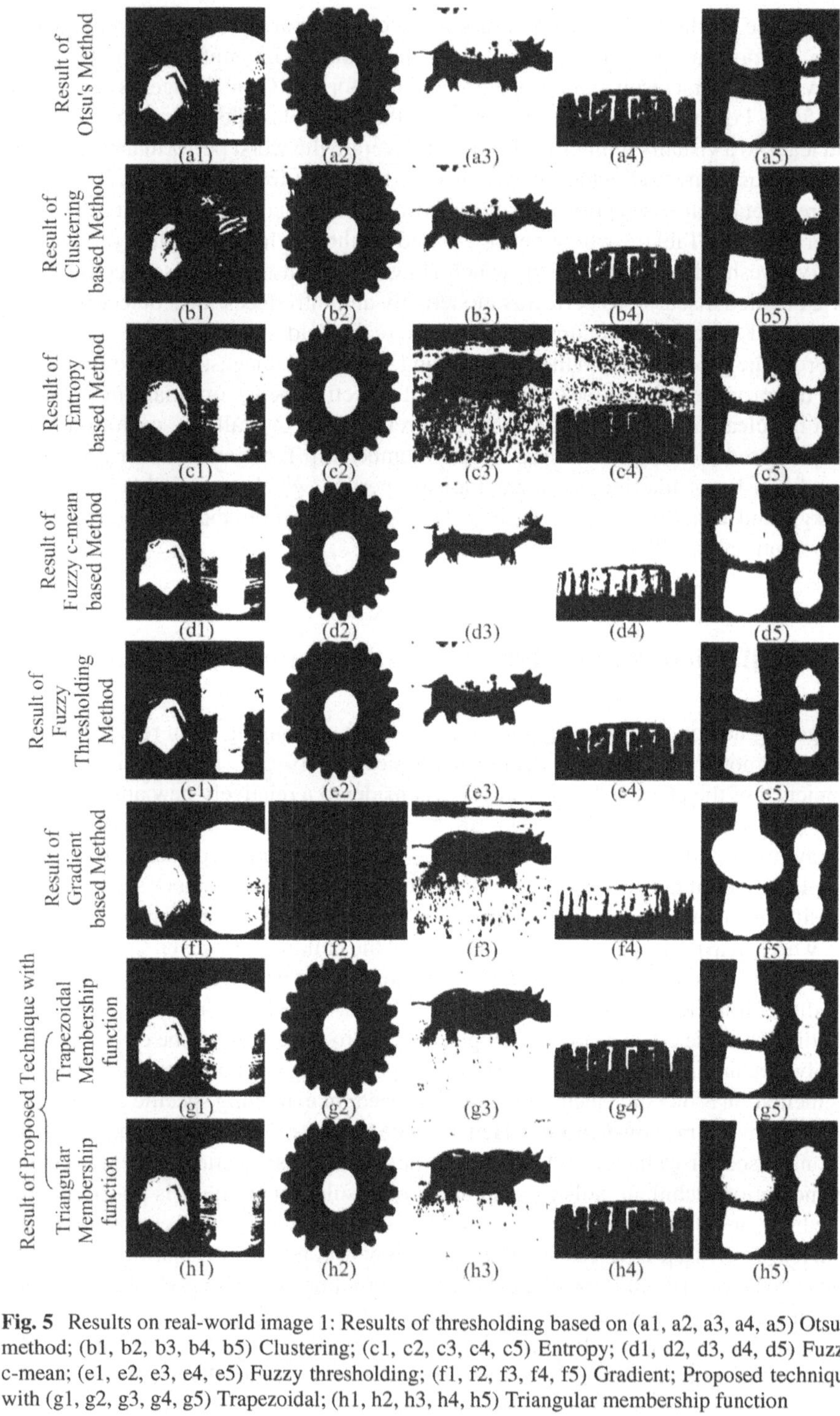

Fig. 5 Results on real-world image 1: Results of thresholding based on (a1, a2, a3, a4, a5) Otsu's method; (b1, b2, b3, b4, b5) Clustering; (c1, c2, c3, c4, c5) Entropy; (d1, d2, d3, d4, d5) Fuzzy c-mean; (e1, e2, e3, e4, e5) Fuzzy thresholding; (f1, f2, f3, f4, f5) Gradient; Proposed technique with (g1, g2, g3, g4, g5) Trapezoidal; (h1, h2, h3, h4, h5) Triangular membership function

line drawings, it is required to employ some pre-processing technique which could accurately recover the line drawing beneath the poor quality of the image.

The technique proposed in this chapter has application as the pre-processing step for the floor plan image analysis, especially for the old floor plan images. The proposed noise resilient image thresholding techniques, also retains the edges in the image along with choosing the correct threshold for the image thresholding and removing the noise. The retention of edges is an important aspect in floor plan image analysis, since edges depict different types of walls and décor. The noisy and old floor plans have a wide distribution of grey levels, hence finding an optimum threshold manually, is a challenge. The proposed technique helps in finding an optimum level of threshold in floor plan images along with removing noise and retaining edges. The original and segmented floor plan images are depicted in Fig. 6. It is clear that the results shown below on the old and noisy floor plans show the successful segmentation step and they can be taken to the next step for further processing. Here the segmented results were obtained considering the triangular membership function,

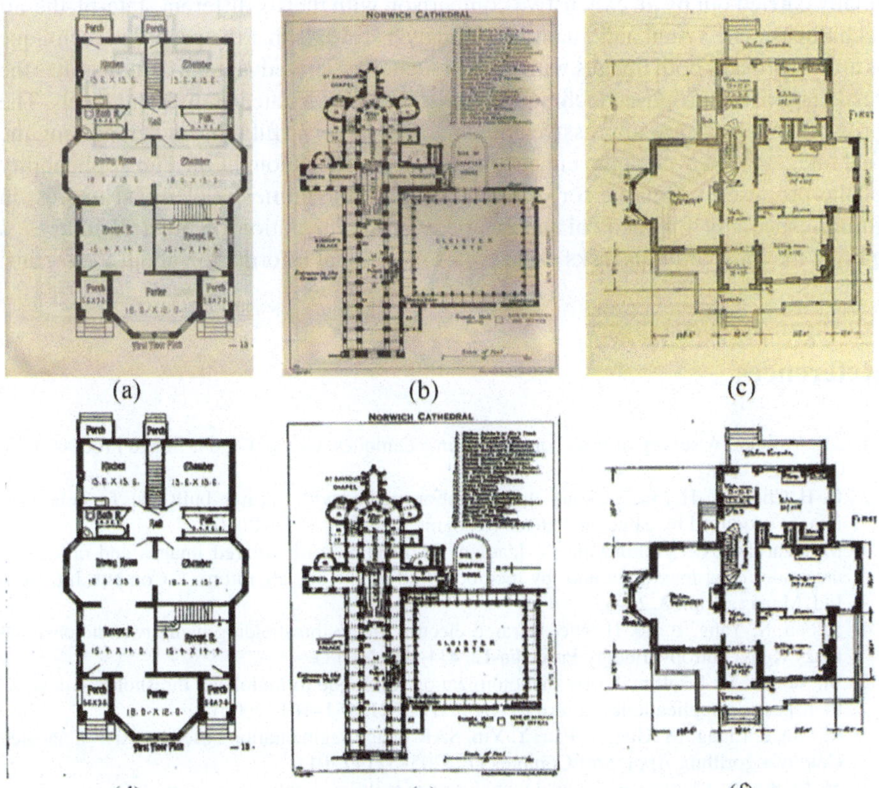

Fig. 6 Results on floor plan segmentation: **a, b, c** Noisy floor plan images; Segmented results using proposed technique with Triangular membership function

which was giving the best result among the considered membership functions in the analysis done in Sect. 3. Therefore, it can be concluded that the proposed segmented framework is well utilized in the floor plan analysis.

5 Conclusions

In this paper, an effective noise resilient image thresholding technique is developed using fuzzy logic and the non-linear filtering. The core idea is to use non-linear filter to inhibit the noise while preserving the image texture information. In this work, adaptive Kuwahara filter is used as the non-linear filter. Finally, a process based on the fuzzy logic properties is devised to obtain an optimal threshold value to perform image thresholding. The experiments on various synthetic and real-world images coming from wide-range applications essentially prove the practicability and efficiency of the proposed technique. The superiority of the proposed technique is finally carried out by an exhaustive comparison with the six different state-of-the-art techniques. The visual and quantitative analysis reveals that the proposed technique is able to process both images with or without noise. This advantage demonstrates the resiliency of the proposed technique against the noise located at different levels. The provided analyses are impressive and emerge as a powerful tool for segmenting the real images. However, some enhancements can still be thought of. The applicability of the proposed technique for different imaging modalities is of thrust interest in future especially for medical images because segmentation of medical images is among the most arduous tasks as it reveals the critical information about the organs.

References

1. J.S. Weszka, A survey of threshold selection techniques. Comp. Graph. Image Process. **7**(2), 259–265 (1978)
2. C.-H. Chou, W.-H. Lin, F. Chang, A binarization method with learning-built rules for document images produced by cameras. Pattern Recognit. **43**, 1518–1530 (2010)
3. M. Azarbad, A. Ebrahimzade, V. Izadian, Segmentation of infrared images and objectives detection using maximum entropy method based on the bee algorithm. J. Comput. Inf. Syst. Ind. Manag. Appl. **3**, 26–33 (2011)
4. S. Fan, S. Yang, P. He, H. Nie, Infrared electric image thresholding using two-dimensional fuzzy Renyi entropy. Energy Procedia **12**, 411–419 (2011)
5. M. Sezgin, R. Tasaltin, A new dichotomization technique to multilevel thresholding devoted to inspection applications. Pattern Recognit. Lett. **21**, 151–161 (2000)
6. M. Ma, J. Liang, M. Guo, Y. Fan, Y. Yin, SAR image segmentation based on Arti- ficial Bee Colony algorithm. Appl. Soft Comput. **11**, 5205–5214 (2011)
7. M. Maitra, A. Chatterjee, A novel technique for multilevel optimal magnetic resonance brain image thresholding using bacterial foraging. Measurement **41**, 1124–1134 (2008)
8. N.P. Sathya, R. Kayalvizhi, Optimal segmentation of brain MRI based on adaptive bacterial foraging algorithm. Neurocomputing **74**, 2299–2313 (2011)

9. P.K. Sahoo, S. Soltani, A. Wong, A survey of thresholding techniques. Comput. Vision Graph. Image Process. **41**(2), 233–260 (1988)
10. M. Sezgin, Survey over image thresholding techniques and quantitative performance evaluation. J. Electron. Imag. **13**, 146–168 (2004)
11. W. Lievers and A. Pilkey, An evaluation of global thresholding techniques for the automatic image segmentation of automotive aluminum sheet alloys. Mater. Sci. Eng.: A, **381**, 134–142 (2004)
12. N. Otsu, A threshold selection method from gray-level histograms. IEEE. Trans. Syst. Man. Cybern. **9**, 62–66 (1979)
13. J. Kapur, P.K. Sahoo, A. Wong, A new method for gray-level picture thresholding using the entropy of the histogram. Comput. Vis. Graph. Image Process. **29**, 273–285 (1985)
14. C.H. Li, C. Lee, Minimum cross entropy thresholding. Pattern Recognit. **26**, 617–625 (1993)
15. P. Sahoo, C. Wilkins, J. Yeager, Threshold selection using Renyi's entropy. Pattern Recognit. **30**, 71–84 (1997)
16. M.P. Albuquerque, I. Esquef, A.G. Mello, Image thresholding using Tsallis entropy. Pattern Recognit. Lett. **25**, 1059–1065 (2004)
17. H. Cheng, J.-R. Chen, J. Li, Threshold selection based on fuzzy c-partition entropy approach. Pattern Recognit. **31**, 857–870 (1998)
18. S.H. Kwon, Threshold selection based on cluster analysis. Pattern Recognit. Lett. **25**, 1045–1050 (2004)
19. Z. Wang, A new approach for segmentation and quantification of cells or nanoparticles. IEEE. Trans. Ind. Infor. **12**, 962–971 (2016)
20. J.-C. Yen, F.-J. Chang, S. Chang, A new criterion for automatic multilevel thresholding. IEEE Trans. Image Process. **4**, 370–378 (1995)
21. J. Derrac, S. García, D. Molina, F. Herrera, A practical tutorial on the use of nonparametric statistical tests as a methodology for comparing evolutionary and swarm intelligence algorithms. Swarm Evol. Comput. **1**, 3–18 (2011)
22. S. Dey, I. Saha, S. Bhattacharyya, U. Maulik, Multi-level thresholding using quantum inspired meta-heuristics. Knowl. Based Syst. **67**, 373–400 (2014)
23. W.A. Hussein, S. Sahran, S.N. Huda, S. Abdullah, A fast scheme for multilevel thresholding based on a modified bees algorithm. Knowl. Based Syst. **101**, 114–134 (2016)
24. K. Hammouche, M. Diaf, P. Siarry, A comparative study of various meta-heuristic techniques applied to the multilevel thresholding problem. Eng. Appl. Artif. Intell. **23**, 676–688 (2010)
25. X.-Y. Wang, J. Bu, A fast and robust image segmentation using FCM with spatial information. Digital Signal Processing **20**, 1173–1182 (2010)
26. S. Aja-Fernández, A.H. Curiale, G.V.-S. Ferrero, A local fuzzy thresholding methodology for multiregion image segmentation. Knowl. Based Syst. **83,** 1–12 (2015)
27. H.D. Cheng, X.H. Jiang, Y. Sun, J.J. Wang, Color image segmentation: advance and prospects. Pattern Recogn. **34**, 2259–2281 (2001)
28. K. Bartyzell, Adaptive Kuwahara filter. Sig. Ima. Video Proce., **10**, 663–670 (2016)
29. D. Wu, Twelve considerations in choosing between Gaussian and trapezoidal membership functions in interval type-2 fuzzy logic controllers, in *IEEE International Conference on Fuzzy Systems*, Brisbane, pp. 1–8 (2012)
30. A. Barua, L.S. Mudunuri, O. Kosheleva, Why trapezoidal and triangular membership functions work so well: Towards a theoretical explanation. J. Uncertain Syst. **8**, 164–168 (2014)
31. M. Sezgin, B. Sankur, Selection of thresholding methods for nondestructive testing applications, in *Proceedings International Conference on Image Processing*, vol. 3, pp. 764–767 (2001)
32. A. Kouhi, H. Seyedarabi, A. Aghagolzadeh, Robust FCM clustering algorithm with combined spatial constraint and membership matrix local information for brain MRI segmentation. Expert Syst. Appl. **146** (2020)
33. Q. Wang, X. Wang, C. Fang, W. Yang, Robust fuzzy c-means clustering algorithm with adaptive spatial & intensity constraint and membership linking for noise image segmentation. Appl. Soft Comput. **92** (2020)

34. J. Jia, B. Liu, L. Jiao, Soft spectral clustering ensemble applied to image segmentation. Front. Comput. Sci. China **5**, 66–78 (2011)
35. J. GalbiatiHéctor, A. Becerra, Dynamic Image Segmentation Method Using Hierarchical Clustering, Progress in Pattern Recognition, Image Analysis, Computer Vision, and Applications, CIARP 2009, Lecture Notes in Computer Science, vol. 5856, pp. 177–184 (2009)
36. I.A. Pestunov, S.A. Rylov, V.B. Berikov, Hierarchical clustering algorithms for segmentation of multispectral images. Optoelectron. Instrument. Proc. **51**, 329–338 (2015)
37. X. Yu, Fuzzy infrared image segmentation based on multilayer immune clustering neural network. Optik **140**, 959–963 (2017)
38. Y. Duan, F. Liu, L. Jiao, P. Zhao, L. Zhang, SAR Image segmentation based on convolutional-wavelet neural network and markov random field. Pattern Recogn. **64**, 255–267 (2017)
39. B. Skourt, A. El Hassani, A. Majda, Lung CT image segmentation using deep neural networks. Procedia Comput. Sci. **127**, 109–113 (2018)
40. R. Haq, A. Hotca, A. Apte, A. Rimner, J.O. Deasy, M. Thor, Cardio-pulmonary substructure segmentation of radiotherapy computed tomography images using convolutional neural networks for clinical outcomes analysis. Phys. Imaging Radiation Oncol. **14**, 61–66 (2020)
41. F. Sultana, A. Sufian, P. Dutta, Evolution of image segmentation using deep convolutional neural network: a survey. Knowl.-Based Syst., 201–202 (2020)
42. S.P. Awate, S. Garg, R. Jena, Estimating uncertainty in MRF-based image segmentation: A perfect-MCMC approach. Med. Image Anal. **55**, 181–196 (2019)
43. Z. Liu, H. Kodamana, A. Afacan, B. Huang, A GMM-MRF Based Image Segmentation Approach for Interface Level Estimation **52**, 28–33 (2019)
44. C. Hu, W. Fan, J. Du, Y. Zeng, Model-Based segmentation of image data using spatially constrained mixture models. Neurocomputing **283**, 214–227 (2018)
45. D.N. Liu, R. Hou, W.Z. Wu, J.W. Hua, X.Y. Wang, B. Pang, Research on infrared image enhancement and segmentation of power equipment based on partial differential equation. J. Vis. Commun. Image Representation **64** (2019)
46. Y. Wang, Q. Yuan, C. He, Indirect diffusion based level set evolution for image segmentation. Appl. Math. Model. **69**, 714–722 (2019)
47. D. Yeo, C.-O. Lee, Variational shape prior segmentation with an initial curve based on image registration technique. Image Vis. Comput. **94** (2020)
48. R. Kumar, S. Srivastava, R. Srivastava, A fourth order PDE based fuzzy c-means approach for segmentation of microscopic biopsy images in presence of Poisson noise for cancer detection. Comput. Methods Programs Biomed. **146**, 59–68 (2017)
49. X. Zhang, P. Xiao, X. Feng, Object-specific optimization of hierarchical multiscale segmentations for high-spatial resolution remote sensing images. ISPRS J. Photogram. Remote Sens. **159**, 308–321 (2020)
50. W. Dan, H. Guoqing, L. Chengzhi, Multi-path connected network for medical image segmentation. J. Vis. Commun. Image Represent. (2020). https://doi.org/10.1016/j.jvcir.2020.102852 (In Press)
51. Y. Yang, R. Wang, C. Feng, Level set formulation for automatic medical image segmentation based on fuzzy clustering. Signal Process. Image Commun. **87**, article id. 115907 (2020)
52. M. Lan, Y. Zhang, L. Zhang, B. Du, Global context based automatic road segmentation via dilated convolutional neural network. Inf. Sci. **535**, 156–171 (2020)

Deep Learning Methods for Audio Events Detection

Giuseppe Ciaburro

Abstract Deep learning is a branch of machine learning that is based on the use of algorithms whose purpose is the modeling of high-level abstractions on data. It is part of a family of targeted techniques learning methods to represent data. Convolutional neural networks (CNN) represent a type of neural network in which the connection pattern between neurons is inspired by the structure of the visual cortex in the animal world. The individual neurons present in this part of the brain respond to certain stimuli in a restricted region of observation, called the receptive field. CNNs are designed to recognize visual patterns directly in images represented by pixels and require a preprocessing quantity that is zero or very limited. They can recognize extremely variable patterns, such as freehand writing and images representing the real world. Typically, a CNN consists of several alternating layers of convolution and pooling followed by one or more final levels fully connected in the case of classification, or by a certain number of levels of upsampling in the regression case. In this chapter, we will see how to identify audio events in a complex sound scenario using convolutional neural networks. An automatic identification system for sound events can prove extremely useful in different contexts: safety of public events, home acoustic monitoring, bioacoustics monitoring, acoustic monitoring for healthcare, and more generally for the needs of smart cities. The variability of sound events determines continuous changes in frequency content that can hardly be traced with the instruments commonly used in traditional acoustics. Convolutional neural networks (CNN) are useful for learning the shift-invariant filters that are essential for modeling audio events.

Keywords Deep learning · Machine learning · Sounds classification · Data mining · Artificial intelligence · Convolutional neural networks · Pattern recognition · Audio events detection · Spectral signature · Mel-frequency cepstral coefficients

---

G. Ciaburro (✉)
Department of Architecture and Industrial Design, University of Campania Luigi Vanvitelli, Caserta, Italy
e-mail: giuseppe.ciaburro@unicampania.it

P. Kumar et al. (eds.), *Machine Learning for Intelligent Multimedia Analytics*, Studies in Big Data 82,
https://doi.org/10.1007/978-981-15-9492-2_8

1 Introduction

The use of the internet and of social networks has now invaded our lives in a completely transparent way, becoming a central part of many of our daily activities, be they working or recreational [1]. The success of this technology is mainly due to the availability of large quantities of multimedia content that users all over the world share. The sources of these data are extremely diversified, ranging from universities, research centers, and institutional organizations to the ordinary citizen eager to make his contribution [2]. Most of the material available on the Internet is placed on a positive perspective for perfectly legitimate purposes related to the exchange of information for private or commercial use. However, as is the case with any other communication technology, especially in the initial stages of its development, the Internet has also conveyed a certain volume of information of potentially harmful or illegal content, or has lent itself to abuses aimed at carrying out criminal activities [3].

To stem this phenomenon, it was necessary to resort to the widespread use of tools for analyzing the contents of multimedia data published and disseminated via the Internet [4]. In a first phase, this check was carried out by people who, under the pseudonym of moderator, checked the content of the multimedia resources before they were published, and then passed on to the police who tried with the same approach to identify criminal organizations dedicated to illegal activities. Given the amount of data that is published and shared today, this approach is inadequate. It was therefore necessary to study automatic multimedia content control systems with the creation of so-called intelligent multimedia systems [5].

The multimedia data foresee the simultaneous use of different communication tools including text, graphics, images, animations, sounds, which integrate each other. The simultaneous control of such data becomes at least complicated by any software that provides for the search for a particular object among all those present in the source: It also assumes that the object being searched is well known, which is not always the case [6].

In this context, a new technology based on artificial intelligence has relatively quickly made its way, which through a machine learning process has proven to be suitable for controlling the multimedia content of data on a large scale. As is now known, algorithms based on machine learning can extract knowledge from a database in a similar way to what is done by a human operator. This procedure takes place through the recognition of patterns that are difficult to identify even for an expert eye. In this way, multimedia content can be automatically processed in search of potentially harmful or even illegal content. But as said, machine learning can extract knowledge from the data allowing not only to identify specific contents but to identify new trends that allow us to identify new scenarios and foresee potentially dangerous situations [7].

Among the algorithms based on Machine Learning, convolutional neural networks have proved particularly suitable for the management of multimedia content such as text, images, and sounds: Recent applications have concerned the use of convolutional neural networks for the identification and classification of sounds. In this chapter, we will learn how to use convolutional neural networks to identify audio events in complex sound scenarios. The usefulness of an audio event recognition system is evident in contexts where safety becomes crucial for the management of daily life [8].

2 Machine Learning-Based Model

Machine Learning indicates a set of methods with which artificial intelligence is trained so as to be able to carry out unscheduled activities, but above all it represents a learning methodology from previous experience that operates in analogy to human intelligence, correcting itself and therefore improving through mistakes made and making autonomous decisions [9].

The term Machine Learning is inextricably linked with the principles underlying mathematical optimization, which provides methods, theories, and domains of application. In fact, many machine learning problems are formulated as problems of minimization of a certain loss function applied to sets of examples (training sets). Specifically, it expresses the difference between the values predicted by the model during training and the expected values. The goal is therefore to teach the network the ability to correctly predict the expected values on a set of instances do not present in the training set (test set) by minimizing the loss function. The algorithms based on machine learning can synthesize new knowledge starting from a knowledge base previously acquired [10].

2.1 Artificial Neural Networks (ANN)

A neural network is a computational model capable of implementing any function which, starting from a certain number of input data, produces an output through an internal mechanism inspired by the recognition model of the human brain. Neural networks are therefore made up of central blocks called neurons, organized into layers which, starting from the so-called input layer, can process the information produced by the intermediate hidden layers to produce the desired output [11]. They can implement any function, what limits their possibilities can be essentially attributed to three factors:

- have enough annotated data, on which to train them appropriately
- use algorithms that allow you to set all the parameters of the function to be implemented, effectively and efficiently
- have technology that can perform the required operations in a sustainable time.

An artificial neural network is formed by an interconnected group of small computational units called neurons; whose operating principle is based on biological neural networks [12]. A neuron i is described by a mathematical function represented by the following equation:

$$y_i = f(x_i) \tag{1}$$

The function f is generally non-linear and is called the activation function. In the previous equation the term xi can be expressed through the following equation:

$$x_i = \sum (w_{ji} + b_i) \tag{2}$$

The terms in the previous equation are defined as follows:

- w_{ji} = weights of the connections
- b_i = bias

These parameters are crucial for the success of neural network-based algorithms, as they contain the information extracted in the learning process. A parameter is changed, be it the weight or the deviation of the net, to approach more and more correct and precise prediction models. The set of weights and the deviation value constitute the bias and represent precisely the information that the neuron learns during the training phase and that it retains for subsequent reuse [13]. An example of how neurons are connected to each other is shown in Fig. 1.

In Fig. 1 each circle represents a neuron, defined by f, while each arrow is a connection of weight w_{ji}. Figure 1 is a typical feedforward neural network architecture and is characterized by layers, composed of multiple neurons, called: input layer, hidden layer, and output layer.

What characterizes artificial neural networks is their ability to approximate continuous functions with a single layer. In a generic multiclass classification problem, the classes are not linearly separable, and it is necessary to formulate a linear function that is able to perform the corresponding linear separation. Approaching the problem with an artificial neural network it is not necessary to define any function but only the so-called hyperparameters. In this way, the model learns to classify the elements

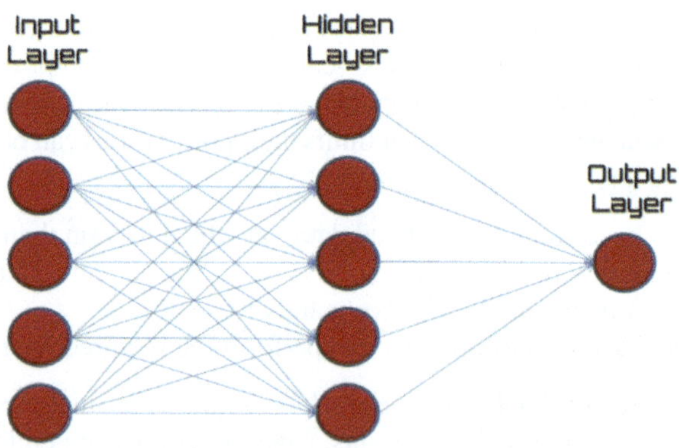

Fig. 1 Feedforward neural network architecture

once the weights have been determined using the error backpropagation algorithm. Typical hyperparameters are number of layers, number of neurons, type of activation functions [14].

2.2 *Deep Neural Networks (DNN)*

Deep Neural Networks (DNN) are neural networks characterized by many hidden layers, unlike standard neural networks where there can also be only one hidden layer. The importance of DNN lies in the fact that they reflect even more faithfully the visual recognition model of the human brain since, being made up of a large number of layers, they are able, level after level, to gradually recognize more and more complex information aggregating the simplest information recognized at previous levels. This allows DNNs to reach very high levels of precision which nowadays, in different areas, are even higher than human ones [15].

Initially, it was impossible to train DNN as the training algorithms were too slow making DNN unusable. For this reason, algorithms were introduced which instead of calculating the gradients of the cost function to be minimized for all inputs of the training set, performed the calculation only for some inputs, chosen in a stochastic way. What contributed most to the dissemination of DNNs was the introduction of the backpropagation algorithm for the training phase. This allowed in fact to replace the multitude of forward steps, which within a neural network had to be performed to calculate the necessary gradients with only two steps: a forward for the calculation of the network outputs at each layer and a backward which, starting from the network outputs and therefore layer after layer up to the input, it simultaneously calculates all the partial derivatives used in the calculation of the necessary gradients [16].

2.3 *Convolutional Neural Networks*

The convolutional neural networks (CNN) are the most widely used algorithm of Deep learning. It is used for the processing of data characterized by a grid topology: a CNN can highlight local relationships starting from adjacency structures present in the data. This process of knowledge extraction takes place through automatic and adaptive learning of patterns from low to high level [17].

CNNs represent the main model used in the field of computer vision and in general in applications that require object recognition and artificial vision.

CNNs are based on an architecture inspired by the biological structure of the visual cortex, in which there is a hierarchy of two basic types of cells: simple and complex cells. Simple cells react to primitive patterns found in subregions of the visual field, called receptive fields, while complex cells synthesize information from the former to identify more complex structures (). Likewise, the neurons present in a convolutional layer are connected to sub-regions of the previous level, and not affected by signals located outside that area. The receptive fields can also overlap: the neurons of a CNN therefore produce spatially correlated results.

It is, therefore, possible to identify the main difference with a fully connected neural network: while in this type of network the number of parameters to be learned increases with increasing input size, a convolutional neural network reduces the number of parameters thanks to the reduced number of connections, shared weights and sub-sampling [18].

A convolutional neural network can have dozens or hundreds of layers, each of which is made up of different filters used to detect different features and build the corresponding feature maps.

In the case of images, in fact, filters are applied to each input (called kernels) and, through a convolution operation, feature maps are generated. The latter will be used as input for the next layer. The filters of the first layers look for very simple features (e.g., edges), to take on more and more complex shapes, capable of uniquely defining the object [19].

The architecture of a convolutional neural network includes several blocks, such as, for example, convolutional layers, pooling layers, and fully connected layers ().

The architecture of the convolutional neural network can be divided into two parts: A first part of feature detection, which then deals with the extraction of features through operations such as convolution, pooling, and ReLU. This is accompanied by a classification part that generates the predicted output with fully connected layers and operations, for example, of softmax [20].

The network parameters are optimized in the same way as described for back-propagation neural networks: the function signal is propagated forward in the network to the final layer, where the gradient of the loss function is computed and back-propagated to allow the weights update by means of a gradient descent algorithm [21].

Let us now deal in detail with the different layers of a CNN, highlighting the characteristics and functions they perform within the architecture of the network.

2.3.1 Convolution Layers

Convolution is a particular type of linear operation in which a filter, or kernel, is applied to a multidimensional array of numbers, called the input tensor, i.e., a mask of multidimensional parameters limited in height and width but which extends for the whole depth of the input volume.

The kernel scans the input tensor in height and width: in each position, the scalar product between the kernel and the scanned input portion is calculated, to obtain the output value in the current position. The output volume obtained once the process is finished is called the activation map or feature map [22].

To understand the function performed by a convolutional layer, let us take an example: Let us make the case where the entrance to the network is an image. In the simplest case of gray-level image, described by function $I(x, y)$, a two-dimensional K kernel will be used to implement the convolution operation ():

$$C(x, y) = (K * I)(x, y) == \sum_m \sum_n I(m, n) K(x - m, y - n) \quad (3)$$

Here,

- $C(x, y)$ is the result of the convolution
- K is the kernel
- $*$ is the symbol of convolution
- $I(x, y)$ is the input passed to the convolutional layer
- x, y are respectively the number of rows and columns of the input tensor
- m, n are respectively the number of rows and columns of the output tensor.

The procedure can be repeated using different filters to obtain the desired number of activation maps, each of which will represent different characteristics of the input tensor extracted from the specific filter [23].

It is now possible to better understand how the key feature of the convolution operation lies in the sharing of the weights of the filters alongwith the positions of the input tensor: this allows not only to increase the efficiency of the neural model, reducing the number of weights to learn, but it makes the pattern of local characteristics extracted from the mask invariant at translation.

Taking up again the interesting analogy with the visual system, the result of the convolution operation can be interpreted as a three-dimensional structure of neurons in which each unit is connected to a receptive field of the previous level: along the depth of the volume, each neuron will cover the same field of view but will learn to detect different input characteristics. The parameters of the filters (the weights and the bias characterizing each kernel) represent the parameters that can be learned in the convolutional layer during the training phase of the network [24].

The following hyper-parameters must instead be set before the training phase:

- **Filter size**: height (h) and width (w) of the kernels, which therefore define the size of the receptive field of each neuron.
- **Number of filters**: corresponds with the depth of each hidden layer, thus defining the number of feature maps.
- **Stride**: controls the distance between two successive locations of the kernel, i.e., the step with which the filter moves by scanning the input in height and width. It therefore determines the overlap of the accommodation fields.
- **Padding**: by adding a border to the input tensioner in different ways (padding at zero, at 1, with flip, etc.) it allows you to adjust the height and width of the volume exiting the convolution. In the case of zero padding, a border of zeros is added such as to obtain as output a volume that has the same dimensions as the input tensor.

2.3.2 Rectified Linear Units

The convolution layer is typically followed by a non-linear activation function: only the activated features are passed to the next layer. The most used function is the rectified linear unit (ReLU), which performs a rectification operation on each element by mapping the negative values to zero and keeping the positive values unchanged:

$$f(x) = \begin{cases} x, & \text{if } x \geq 0 \\ 0, & \text{if } x < 0 \end{cases} \tag{4}$$

It has been shown that the use of the ReLU function allows faster and more effective training than traditional units such as the hyperbolic tangent [25].

2.3.3 Pooling Layer

The pooling layer performs a non-linear sub-sampling operation that reduces the transverse dimension of the activation maps, leaving their depth unchanged, with the aim not only of simplifying (reducing the parameters) the output of the previous convolutional layer, but also to introduce translational invariance with small displacements and distortions and, consequently, providing greater robustness with respect to the localization of the features [26].

Taking up the analogy with the visual cortex, the pooling unit was inspired by behavior of complex cells: by capturing a growing visual field, they can learn spatial hierarchies of feature patterns, being less sensitive to slight shifts in the position of the salient features.

The most popular form of the pooling operation is max pooling, which performs a non-linear subsampling by dividing the input into rectangular regions and returning the maximum value within each window.

It is a layer that does not have appreciable weights but for which a set of hyper-parameters must be set before training, such as filter size, stride, and padding. If the size of the filter r and stride are equal, the operation reduces the size of the input tensor, for each channel, by a factor r. The max-pooling technique leads to faster convergence and improves the generalization process of the model [27].

So far, we have dealt with layers that allow us to extract features from our input data. Now we will analyze the layers that allow us to operate the classification.

2.3.4 Fully Connected Layer

The activation maps output from the last feature extraction block is transformed into a vector and connected to one or more fully connected layers (fully-connected layers, FC layers) in which each value of the input vector is connected to all the values of the next layer. The last fully connected layer will generate a vector of size K equal to the number of classes in which the input must be classified [28].

2.3.5 SoftMax Layer

The activation function applied to the last FC layer generally differs from those used in the previous levels and will be selected based on the task required from the network. In the multi-class learning paradigm, the softmax function is used, which normalizes the K real values obtained from the last FC layer in probability of belonging to the K classes in question [29].

2.3.6 Classification Layer

The last layer of the convolutional neural network assigns the class to which it belongs. In the multi-class classification approach, the probabilities calculated in the softmax layer feed the cross-entropy loss function [30].

2.4 Architecture of a CNN

After analyzing each component of a CNN in detail, it is time to see the general structure of it. For example, starting from images as input levels, there will be a series of convolutional levels interspersed with a ReLU level and, if necessary, the standardization and pooling levels. Finally, there will be a series of FC layers before the output layer. In Fig. 2 it is possible to analyze an example of CNN architecture.

The basic idea is to start with a large image and continuously reduce the data step by step until you get a single result. The more the convolution steps, the more the neural network will be able to understand and elaborate complex functions.

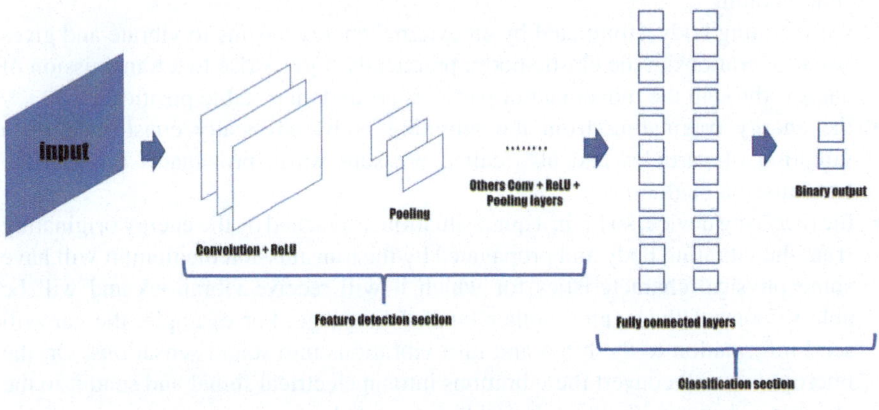

Fig. 2 Typical architecture of a convolutional neural network

3 Introduction to Sound Physics

Sound is a variation of air pressure or any other elastic medium, compared to a constant value: Sound is made up of vibrations. When this variation is repeated cyclically a certain number of times over a period, we perceive a sound. For a sound to propagate it needs a means of transporting it; the air is one of these since its particles transmit the vibration generated by the sound source and propagate it in space. This means that any medium, solid, liquid, or gaseous, can carry sound, influencing, among other things, its speed [31].

The set of devices through which we can perceive sound make up the so-called sound chain, as follows:

- sound source: device or object capable of generating variations in acoustic pressure or vibrations
- transmission medium: system that allows the propagation of vibrations
- receiving device: device or object that can perceive vibrations.

As examples of sound sources, we can cite the most disparate: the human voice, an acoustic musical instrument, an electro-acoustic transduction device (speaker) and, more generally, any vibrating body.

As examples of the transmission medium we can mention: the air, in which we commonly listen to sounds, the water, in which we will certainly have heard sounds and, more generally, any elastic medium.

As examples of receiving device: the ear, an electroacoustic transducer (microphone) and, more generally, any vibrating body capable of perceiving vibrations.

The dynamics of the production, propagation, and reception of sound is synthetically the following:

- the vibrating body and the receiving device are "immersed" within the transmission medium
- the vibrating body, stimulated by an external energy, begins to vibrate and gives rise to vibrations in the elastic body; practically it gives rise to a transmission of energy, through the movement of particles around their stable position
- the energy originating from the vibrating body, physically constituted by a vibration of particles and also called pressure wave, propagates through the transmission medium
- the receiving device, so far in a quiet situation, is reached by the energy originating from the vibrating body and propagated by the transmission medium; it will have some physical characteristics for which it will receive vibrations and will be able to convert them into another type of language. For example, the ear will send information to the brain and turn vibrations into sound sensations. Or, the microphone will convert the vibrations into an electrical signal and send it to the electronic device with which it will be connected.

Transmission media typically have their own viscosity, which turns into energy loss: the further the signal moves away from the source, the more it loses intensity, becoming gradually less and less powerful. The receiving devices will have physical characteristics for which not all types of acoustic vibrations will be received by them: The human ear can only receive some sounds, within a very precise frequency band. The microphones, similarly, depending on the physical electrical characteristics, will be able to receive some types of vibrations and not others [32].

The sounds are distinguished in pure sounds and complex sounds. Pure sounds have a sinusoidal waveform. A sinusoid is characterized by its frequency (the number of cycles per second), and by its amplitude (the height of each peak). Complex sounds are all those that have complicated shapes and that can be obtained as the sum of a certain number of pure sounds. Among other things, the oscillations are called periodic if they repeat in the same way in equal time intervals.

Sound waves can be generated by different types of objects, vibrating bodies that alternately moving forward and backward alternately cause compression and rarefaction of the surrounding air. The global effect is a pressure wave that spreads. In all cases, what is highlighted is an oscillatory motion. The study of physics has associated this phenomenon with that of the harmonic oscillator, having the following motion equation:

$$x(t) = A \sin(2\pi f t + \varphi) \tag{5}$$

Here,

- $x(t)$ is the position on the abscissa axis at the instant t
- A is the amplitude of the oscillation, understood as the maximum distance from the equilibrium point and is measured in meters
- f is the frequency of the vibration, i.e., it measures how many times in time the oscillation completes a complete cycle; it is measured in Hertz
- φ is the initial phase, the value of the angle of the sine function for $t = 0$ and is measured in radians
- $(2\pi ft + \varphi)$ is the instantaneous phase and is measured in radians

Another way of writing the harmonic motion equation is by using the pulsation, defined by the following equation:

$$\omega = 2\pi f \tag{6}$$

The pulsation ω which is measured in rad/s, transforms Eq. (5) into the following form:

$$x(t) = A\sin(\omega t + \varphi) \tag{7}$$

Another important element to be introduced is the period T. It is defined as the time interval within which the oscillation completes a complete cycle. Recalling that the frequency f is the number of cycles per second, it follows that it is linked to the period T by the following equation:

$$f = \frac{1}{T} \tag{8}$$

The physical device that generates the sound, therefore, depending on its physical characteristics, will be such as to enter vibration and highlight a certain amplitude, frequency, and phase in the excitation produced. It is like saying that the generated sound will have those same characteristics.

A harmonic oscillator represents a borderline case, since there are a whole series of factors that characterize real systems. It is practically impossible to eliminate friction and viscosities from the mechanical systems involved in the production of sound. The most likely oscillators are, therefore, the so-called damped oscillators, those which model friction and damping through a time constant τ. The amplitude of the oscillation will no longer be constant over time but will decrease according to a factor equal to $e^{-\frac{t}{\tau}}$. In these hypotheses, Eq. (5) is modified leading to the equation that defines the damped harmonic oscillator, as follows:

$$x(t) = Ae^{-\frac{t}{\tau}}\sin(2\pi f t + \varphi) \tag{9}$$

The frequency will also no longer be constant but will vary over time.

Finally, it is appropriate to consider the case of the forced oscillator. It mathematically models the physical situation in which the excitement that initiates the oscillator vibration is no longer simply impulsive; it is not a simple push, but it is a force of intensity which is also periodic. After a certain period, called transient, the oscillator will oscillate with the same frequency as the external force and with an amplitude proportional to the external force applied. For this reason, the oscillations are called forced. The phenomenon by which oscillation is enhanced in the oscillator's own frequency, or in the near frequencies, is called resonance. To enhance the oscillation, we mean to ensure that the amplitude of oscillation is multiplied by a positive factor and, therefore, is accentuated. This occurs precisely when the exciter element has a frequency equal to that of the oscillator, this frequency is also called natural [33].

The waveforms analyzed so far are simply sinusoids characterized by amplitude, frequency, pulsation, period, and phase. Sounds are complex, not easily identifiable in a simple sinusoidal harmonic form, and much more difficult to analyze. From the analysis of the functions, however, we know that if a function is continuous and periodic of period T it can be developed as the sum of harmonic functions. This operation is called Fourier series development (). This means that it is sufficient for a waveform to be periodic to be the sum of pure harmonic forms. The frequency f_n of the nth harmonic term will be equal to $f_n = f_0 n$ where f_0 is the frequency of the first term of the sum and is called the fundamental frequency of the original wave.

Qualitatively we can say that carrying out the harmonic analysis of a periodic function consists of determining n harmonics, each with its own characteristics of frequency, phase, and amplitude and making a comparison with the original form. In this regard, we can speak of the harmonic spectrum as the set of amplitudes of the harmonics that form the wave.

3.1 Sound Frequency Analysis

The most immediate and easy-to-interpret sound analysis is the one concerning the trend over time of the signal, in which the latter is represented in a coordinate system having the time on the x axis and the amplitude on the y-axis of the signal. This analysis allows us to evaluate how the audio signal evolves over time, however limiting the analysis to only the sound power that gives us information on the energy to which the receptor is exposed: Nothing tells us about the frequency content of the sound [34].

The analysis in the frequency domain allows us to deepen the perspective from which we observe the phenomenon, allowing us to extract more information from the audio signal. The graph that is obtained in this case is the frequency spectrum, which has the axis of the frequencies in a specific interval as abscissa and a value associated with the contribution that each frequency provides to the entire signal. The spectrum, therefore, allows to identify the frequencies present in the signal. To pass from the time domain to the frequency domain, a mathematical transformation of the original signal is required.

3.1.1 The Fourier Transforms

The Fourier transform (FT) is the most known transformation in frequency, due to the idea of the French mathematician Jean Baptiste Joseph Fourier [35]. A generic signal can be expressed as the sum of simple waves, each characterized by a specific frequency and a coefficient that modulates its amplitude. The Fourier transform is obtained through the following equation:

$$X(f) = \int_{-\infty}^{-\infty} x(t)e^{-2\pi ft}\mathrm{d}t \tag{10}$$

The exponential term present in the integral can be rewritten through the Euler formula in the following way:

$$e^{-2\pi ft} = \cos(2\pi ft) + j\sin(2\pi ft) \tag{11}$$

For each value of f this entails multiplying the entire signal by a complex expression consisting of sine and cosine at the frequency f and calculating the integral.

The idea is that if $x(t)$ contains a component at a certain frequency f^, this component coincides with the sinusoidal terms at the frequency f^, providing as a product a curve having a relatively consistent amplitude. With the integration, operation all this is summarized in a single value, which is marked in the spectrum with a certain amplitude at f^. If the Fourier transform gives a consistent value for $f = f$, the signal $x(t)$ is said to have a spectral component dominant at the frequency f^. Conversely, if the integration result is relatively small, it means that the wave at frequency f^

contained in the signal $x(t)$ has reduced amplitude. If the result is zero, the $x(t)$ signal does not contain that type of wave.

3.1.2 Octave Band Filters

One method of frequency analysis of stationary signals involves the use of bandpass filters, that is to say, a series of devices each of which allows the passage of only a certain range of frequencies, excluding the components of the sound at major and minor frequencies. With a measuring instrument at the output of each filter, it is possible to measure the level that belongs to the frequency range [36].

A filter is a two-port network capable of transferring signals with frequencies included within predetermined frequency ranges (passbands) and eliminate signals at other frequencies (dark bands).

It is possible to represent a band-pass filter with a zone in which the gain is almost constant and equal to 0 dB (effective band) and with two zones, external to the first, in which the gain is negligible. The effective band is between two extreme frequencies f_1 and f^2, called cut-off frequencies, placed at half energy with respect to the pass band; by definition $G(f_1) = G(f_2) = -3$ db. f_c is defined as the center band frequency and is such that $G(f_c) = 0$ dB.

Theoretically, an ideal filter should have a rectangular pulse as the gain curve, but since the device is made of passive components, the rising and falling edges can never be perfectly vertical.

Basically, there are two band spectra: the constant band spectrum, in which all the bands have the same amplitude, and the constant percentage band spectrum, in which each band is twice as wide as the previous one.

3.2 Sound Event Detection (SED)

Sound event detection (SED) consists of two activities: First, it aims to recognize the type of sound events present in an audio stream by processing the so-called audio tags. Subsequently, it aims to identify the start and end times of the event, this activity is called localization. Sound event detection is relatively recent topic for audio signal processing techniques, particularly in the field of acoustic surveillance. The aim is to recognize sound events that could highlight us of possible threats to the safety of people or things or could offend people's sensitivity [37].

The sound classes attributable to dangerous events include gunshots, explosions, or screams, those instead attributable to offensive behavior are bad language, racist insults, and offenses to religious sensitivity. Most of the sound recognition algorithms described in the literature are based on the extraction of acoustic characteristics and the recognition of statistical schemes.

To perform a correct recognition of audio events, it is necessary to have sample data on which to perform analyzes. The difficulty lies in separating the sources to be identified from the background noise. Mesaros et al. [38] collected a set of binaural recordings in a database to facilitate the ambient sound recognition. The recordings were made in 15 different acoustic environments, ranging from residential areas to domestic environments. The database also contains annotations for individual sound events to facilitate detection of sound events. For each record, the start, offset, and label of the sound events are marked. The authors in this work performed a classification of an acoustic scenario using a supervised algorithm using mel frequency cepstral coefficients and Gaussian mixture models.

Heittola et al. [39] they developed a sound event detection system for natural multi-source environments. The authors used a sound source separation front-end to detect sound events in various daily contexts. The procedure involves a pre-processing of the audio by means of non-negative matrix factorization and separation into four individual signals. Each class of sound event is represented by a hidden Markov model trained using cepstral frequency coefficients mel extracted from the audio. Each separate signal is used individually for the extraction of the characteristics and therefore the segmentation and classification of the sound events by means of the Viterbi algorithm. Separation allows detection of up to four overlapping events. The proposed system shows a significant increase in the accuracy of event detection compared to a system capable of producing a single sequence of events.

Adavanne et al. [40] studied automatic detection of sound events (SED), using spatial and harmonic functions in a model based on the recurrent neural network (RNN). They used multi-channel audio recordings to recognize overlapping sound events, which is technically difficult using mono channel audio. The audio events used to test the proposed system were retrieved from the TUT sound events detection 2016 database (Mesaros et al. 2016). They also presented a comparison between the multichannel and the single-channel model.

McFee et al. [41] used training labels that indicate the presence of that type of audio event, but not the temporal location within the recording. The reason that prompted the authors to choose this strategy is due to the onerous work of collecting the data labels necessary to train a supervised classification algorithm. This saves huge resources by indicating only the presence of the event. To allow the aggregation of the labels, adaptive pooling was used with automatic adaptation to the characteristics of the sound sources. To test the technology, the authors used a model based on convolutional neural ethics.

Vesperini et al. [42] used a CNN architecture based on the capsule neural network (CapsNet) for polyphonic sound detection. CapsNet has the purpose of solving the problems of scarce robustness of similar transformations typical of CNN. CapsNet extracts a deeper understanding of the objects contained in the image, encoding the units that form the object within vectors of instantiation parameters, which represent the position of the unit. This result is obtained thanks to the capsules, which are neurons capable of representing the units present in the image. The authors use connect the capsules via dynamic routing which encourages learning of relationships between whole parts and improves detection performance in a polyphonic context.

4 Sound Event Recognition Using Convolutional Neural Network

The extraction of characteristics from sounds is a process that requires computational effort that can sometimes prove to be extremely expensive. In addition, the most suitable features capable of discriminating between the different sources depend on the nature of the sounds and can also vary significantly. To overcome these problems, deep learning can be adopted, in this way, we can take advantage of the layered structure of these models to extract characteristics from the input of sound representation given with greater abstraction on each level. In this way, we are also able to improve the learning ability of the network which is particularly suitable for identifying the extremely complex relationship between the sound representation and the labels of the target sound events.

4.1 Methodology

In this work, we analyzed the sounds obtained through recordings in complex urban scenarios. Recordings were made relating to different contexts: outdoor and indoor environments. For each scenario, periods were extracted in which specific events occurred: screams, bangs, crashes. Furthermore, periods were extracted in which no specific events occurred, or at least attributable to the types mentioned above. Figure 3 shows waveforms and spectrogram related to these events.

Identifying these sounds in this context becomes very simple: Both the waveform and the spectrogram highlight the features that characterize each event. This becomes difficult when the acoustic scenario becomes complex, as these sounds spread in an environment characterized by a background noise that partially obscures them.

A system based on a threshold of the acoustic sound level does not allow us to solve the problem. This is because sounds in an urban environment have comparable sound levels and often pleasant sounds make sound levels greater than unwanted ones: For example, think of the music so loved by all of us that it can reach very high sound levels.

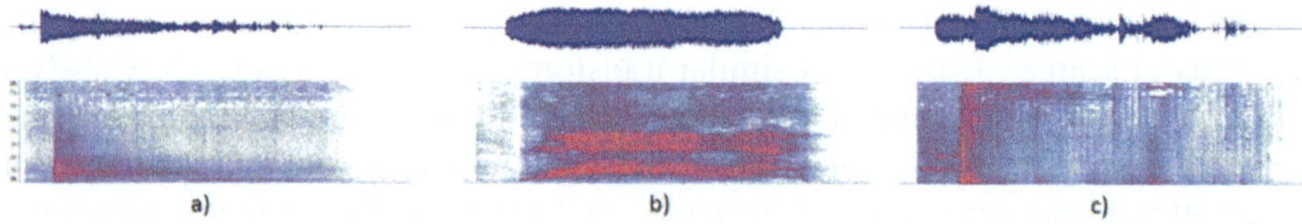

Fig. 3 Waveforms and a spectrogram for three sound events: **a** gunshot; **b** screams; **c** car crash. A spectrogram is the graphic representation of the intensity of a sound as a function of time and frequency

We have seen that the features of the sounds we have isolated are well highlighted in a spectrogram. We could then use this tool to identify the presence of an event in a complex acoustic scenario. As we said, in a complex acoustic scenario, in addition to the event we want to identify, there is the background noise that characterizes the spectrogram. Due to the background noise, the recognition of the event by a visual analysis of the spectrogram is very difficult, at least for the human eye. As well as potentially causing errors, this procedure proves to be long and expensive.

The spectrogram is an image, and in the Convolutional Neural Networks section, we said that these models can recognize specific patterns and return interesting results. We also said that this technologist frees us from the problem of extracting the features that allow us to characterize the phenomenon. This means that we can pass the spectrograms to a model based on convolutional neural networks and obtain the recognition of specific events within sounds representative of complex urban scenarios.

4.2 Convolutional Neural Network-Based Algorithm

The architecture of CNN is composed of an input level, two pairs of convolutional and pooling levels, two fully connected hidden levels, and an output level with four classes.

The CNN-based model must recognize audio events in a complex acoustic scenario. To do this we must first prepare data that we will use in the training phase of the network: We have isolated fragments of the recordings and labeled them according to the event contained in it. Four types of events have been labeled: scream, gunshot, crash, no event.

Properly labeled audio signals must be processed to produce the spectrograms. The mel-scaled spectrogram was developed using the librosa Python library (). The mel filter performs a nonlinear transformation in the frequency domain. At lower frequencies, the behavior is linear, while in higher frequencies the behavior is logarithmic. In this way, we try to simulate the human reception of the sound, which is dependent on the frequency of the received sound.

The preprocessed data was subsequently divided into two sets: training set, and test set. In this way, the data used in the training phase will not be reused for the evaluation of the model's performance. This procedure becomes crucial in the development of a machine learning-based model as it protects us from any overfitting problems.

Subsequently, the model was trained to regulate weights and bias to minimize the classification errors. For the adjustment of the weights, the Stochastic Gradient Descent was used which is an iterative algorithm to minimize a loss function. Finally, the model obtained was used to classify the sounds contained in the test set. Figure 4 contains the flowchart of the recognition procedure of audio events in complex urban scenarios.

The results obtained were evaluated using the accuracy metric. Accuracy returns the percentage of correct classifications on all observations presented to the model.

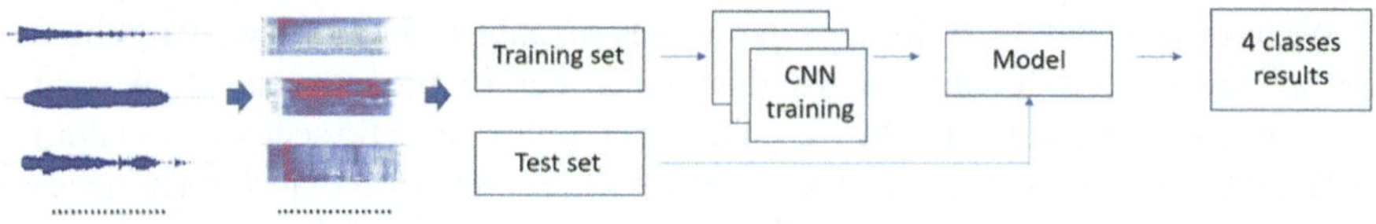

Fig. 4 CNN-based model flowchart

The algorithm based on the convolutional neural networks used in this study returned an accuracy of 0.89 confirming the quality of the elaborated procedure to identify sound events in a complex sound scenario. The accuracy of a prediction indicates how close the expected value of a quantity is to the real value of that quantity. In our case, the real value is available since we measured the accuracy of the model on test data that was correctly labeled. A classification model is a mathematical function that uniquely determines the class to which a statistical unit belongs, based on the values observed for the variables of interest. Its predictive accuracy depends on the ability to correctly classify new units, regardless of the class they come from. A result like that obtained (0.89) tells us that the model can correctly classify 89 cases out of 100 that have been presented to it. Furthermore, it should be noted that this performance was obtained on a sample equally represented in the different classes.

The accuracy returned by the CNN-based algorithm proved to be in line with the results obtained from previous studies that used the recognition of schemes in different areas, indeed it showed superior performance. Piczak [43] used CNN to classify short audio clips of ambient sounds. A deep model consisting of 2 convolutional levels with maximum pooling and 2 fully connected levels is trained on a low-level representation of the audio data by means of segmented spectrograms. The results of the model returned an accuracy between 0.65 and 0.80, depending on the data set used to confirm the importance of the data in the correct classification of the audio sources. Salamon et al. [44] used CNN to classify ambient sound. The precision returned by the model was 0.75–0.80, the performance was improved by performing a data increase operation. Bardou et al. [45] used CNN to classify lung sounds. The authors extracted the functionality of the local binary model (LBP) from the visual representation of the audio files using spectrograms: The accuracy returned by the model was 0.80.

5 Conclusion

In this chapter, we analyzed how to identify audio events in a complex sound scenario using convolutional neural networks. An automatic identification system for sound events can prove extremely useful in different contexts. The variability of sound events determines continuous changes in frequency content that can hardly be traced with the instruments commonly used in traditional acoustics. Convolutional neural networks (CNN) are useful for learning the displacement invariant filters that are essential for modeling audio events. A CNN-based model returns a system capable of recognizing audio events in multimedia objects. In this way, it will be possible to develop automatic systems to recognize audio events that could inform us of possible threats to the safety of people or things or could offend the sensitivity of people.

References

1. C. Haythornthwaite, Social networks and Internet connectivity effects. Inf. Community Soc. **8**(2), 125–147 (2005)
2. L. Atzori, A. Iera, G. Morabito, M. Nitti, The social internet of things (siot)–when social networks meet the internet of things: Concept, architecture and network characterization. Comput. Netw. **56**(16), 3594–3608 (2012)
3. S. Livingstone, L. Haddon, Risky experiences for children online: charting European research on children and the internet. Child. Soc. **22**(4), 314–323 (2008)
4. P. Horster (ed.), *Communications and Multimedia Security II*. Springer (2016)
5. F. Thissen, *Screen Design Manual: Communicating Effectively Through Multimedia*. Springer Science & Business Media (2003)
6. R. Steinmetz, K. Nahrstedt, *Multimedia Systems*. Springer Science & Business Media (2013)
7. C. Sammut, G.I. Webb (eds.), *Encyclopedia of Machine Learning*. Springer Science & Business Media (2011)
8. B. Prasad, S.M. Prasanna (eds.), *Speech, Audio, Image and Biomedical Signal Processing Using Neural Networks*, vol. 83 (Springer, Heidelberg, 2007)
9. B. Clarke, E. Fokoue, H.H. Zhang, *Principles and Theory for Data Mining and Machine Learning* (Springer Science & Business Media, 2009)
10. F. Hutter, L. Kotthoff, J. Vanschoren, *Automated Machine Learning* (Springer, New York, NY, USA, 2019)
11. R. Rojas, *Neural Networks: A Systematic Introduction* (Springer Science & Business Media, 2013)
12. G. Dreyfus, *Neural Networks: Methodology and Applications* (Springer Science & Business Media, 2005)
13. I.N. Da Silva, D.H. Spatti, R.A. Flauzino, L.H.B. Liboni, S.F. dos Reis Alves, *Artificial Neural Networks* (Springer International Publishing, Cham, 2017), p. 39
14. S.C. Wang, Artificial neural network, in *Interdisciplinary Computing in Java Programming* (Springer, Boston, MA, 2003), pp. 81–100
15. J. Kacprzyk, W. Pedrycz (eds.), *Springer Handbook of Computational Intelligence* (Springer, 2015)
16. R. Miikkulainen, J. Liang, E. Meyerson, A. Rawal, D. Fink, O. Francon, B. Hodjat, et al., Evolving deep neural networks, in *Artificial Intelligence in the Age of Neural Networks and Brain Computing* (Academic Press , 2019), pp. 293–312
17. H.H. Aghdam, E.J. Heravi, *Guide to Convolutional Neural Networks* (Springer, New York, NY, 2017)
18. C.C. Aggarwal, *Neural Networks and Deep Learning*, vol. 10 (Springer, Heidelberg, 2018), pp. 978 983
19. L. Lu, Y. Zheng, G. Carneiro, L. Yang, Deep learning and convolutional neural networks for medical image computing, in *Advances in Computer Vision and Pattern Recognition* (2017)
20. T. Elsken, J.H. Metzen, F. Hutter, *Simple and Efficient Architecture Search for Convolutional Neural Networks* (2017). arXiv preprint arXiv:1711.04528
21. H. Sun, X. Chen, Q. Shi, M. Hong, X. Fu, N.D. Sidiropoulos, Learning to optimize: training deep neural networks for wireless resource management, in *2017 IEEE 18th International Workshop on Signal Processing Advances in Wireless Communications (SPAWC)* (IEEE, 2017, July), pp. 1–6

22. J. Gehring, M. Auli, D. Grangier, D. Yarats, Y.N. Dauphin, Convolutional sequence to sequence learning, in *Proceedings of the 34th International Conference on Machine Learning*, vol. 70 (2017, August), pp. 1243–1252, JMLR.org
23. G. Huang, Z. Liu, L. Van Der Maaten, K.Q. Weinberger, Densely connected convolutional networks, in *Proceedings of the IEEE Conference on Computer Vision and Pattern Recognition* (2017), pp. 4700–4708
24. Y. Pang, M. Sun, X. Jiang, X. Li, Convolution in convolution for network in network. IEEE Trans. Neural Netw. Learn. Syst. **29**(5), 1587–1597 (2017)
25. A.F. Agarap, Deep Learning Using Rectified Linear Units (relu) (2018). arXiv preprint arXiv: 1803.08375
26. J. Gu, Z. Wang, J. Kuen, L. Ma, A. Shahroudy, B. Shuai, T. Chen, Recent advances in convolutional neural networks. Pattern Recogn. **77**, 354–377 (2018)
27. N. Akhtar, U. Ragavendran, Interpretation of intelligence in CNN-pooling processes: a methodological survey, in *Neural computing and applications* (2019), pp. 1–20
28. G. Deshpande, P. Wang, D. Rangaprakash, B. Wilamowski, Fully connected cascade artificial neural network architecture for attention deficit hyperactivity disorder classification from functional magnetic resonance imaging data. IEEE Trans. Cybern. **45**(12), 2668–2679 (2015)
29. G. Du, C. Tian, Z. Li, D. Zhang, Y. Yin, Y. Ouyang, Efficient softmax hardware architecture for deep neural networks, in *Proceedings of the 2019 on Great Lakes Symposium on VLSI* (2019, May), pp. 75–80
30. N. Sharma, V. Jain, A. Mishra, An analysis of convolutional neural networks for image classification. Proc. Comput. Sci. **132**, 377–384 (2018)
31. H. Kuttruff, *Acoustics: An Introduction*. CRC Press (2007)
32. T. Rossing (ed.), *Springer Handbook of Acoustics* (Springer Science & Business Media, 2007)
33. M.A. Breazeale, M. McPherson, Physical acoustics, in *Springer Handbook of Acoustics* (Springer, New York, NY, 2014), pp. 215–246
34. J. Foreman, *Sound Analysis and Noise Control* (Springer Science & Business Media, 2012)
35. R.N. Bracewell, R.N. Bracewell, *The Fourier Transform and Its Applications*, vol. 31999 (McGraw-Hill, New York, 1986)
36. M.J. Crocker (ed.), *Handbook of Noise and Vibration Control* (John Wiley & Sons, 2007)
37. T. Heittola, A. Mesaros, A. Eronen, T. Virtanen, Context-dependent sound event detection. EURASIP J. Audio Speech Music Process. **2013**(1), 1 (2013)
38. A. Mesaros, T. Heittola, T. Virtanen, TUT database for acoustic scene classification and sound event detection, in *2016 24th European Signal Processing Conference (EUSIPCO)* (IEEE, 2016, August) pp. 1128–1132
39. T. Heittola, A. Mesaros, T. Virtanen, A. Eronen, Sound event detection in multisource environments using source separation, in *Machine Listening in Multisource Environments* (2011)
40. S. Adavanne, G. Parascandolo, P. Pertilä, T. Heittola, T. Virtanen, Sound Event Detection in Multichannel Audio Using Spatial and Harmonic Features (2017). arXiv preprint arXiv:1706.02293
41. B. McFee, J. Salamon, J.P. Bello, Adaptive pooling operators for weakly labeled sound event detection. IEEE/ACM Trans. Audio Speech Language Process. **26**(11), 2180–2193 (2018)
42. F. Vesperini, L. Gabrielli, E. Principi, S. Squartini, Polyphonic sound event detection by using capsule neural networks. IEEE J. Selected Topics Signal Process. **13**(2), 310–322 (2019)
43. K.J. Piczak, Environmental sound classification with convolutional neural networks, in *2015 IEEE 25th International Workshop on Machine Learning for Signal Processing (MLSP)* (IEEE, 2015, September),pp. 1–6
44. J. Salamon, J.P. Bello, Deep convolutional neural networks and data augmentation for environmental sound classification. IEEE Signal Process. Lett. **24**(3), 279–283 (2017)
45. D. Bardou, K. Zhang, S.M. Ahmad, Lung sounds classification using convolutional neural networks. Artif. Intell. Med. **88**, 58–69 (2018)

A Framework for Multi-lingual Scene Text Detection Using K-means++ and Memetic Algorithms

Neelotpal Chakraborty, Averi Ray, Ayatullah Faruk Mollah, Subhadip Basu, and Ram Sarkar

Abstract Recent years have witnessed an exponential surge in interest to explore the domain of scene text detection as well as analysis in natural scene images. However, owing to the complexities arising due to various factors, it can be said that existing techniques may fail at times while attempting to detect text components. This paper presents a system wherein an image is taken as input and its color components are extracted at first. Next the intensity values from each color channel are grouped together using K-means++ clustering algorithm. Memetic algorithm is then applied to get an optimal set of candidate components from the color maps while eliminating the background. The spurious components are removed on the basis of their dimension and entropy measure. This system is experimentally evaluated on two standard datasets namely MLe2e and KAIST, and on an in-house dataset of 400 images, all having multi-lingual texts. The results obtained are comparable with some state-of-the-art methods.

Keywords Scene text · Multi-lingual · Text detection · Text localization · K-means++ clustering · Memetic algorithm

1 Introduction

Text usually represents a set of descriptions or information about some entities, events or both. Ever since the invention of writing, humans have relied on data obtained from various textual sources, to develop knowledge about any subject. Thus, texts have held a certain significance in human life, since ancient times. Derived from human abstraction and manipulation [1], textual content in natural scenes directly

N. Chakraborty (✉) · A. Ray · S. Basu · R. Sarkar
Department of Computer Science and Engineering, Jadavpur University, Kolkata 700032, India
e-mail: chakraborty.neelotpal@gmail.com

A. F. Mollah
Department of Computer Science and Engineering, Aliah University, Kolkata 700160, India

P. Kumar et al. (eds.), *Machine Learning for Intelligent Multimedia Analytics*, Studies in Big Data 82,
https://doi.org/10.1007/978-981-15-9492-2_9

carry high-level semantic information. Hence, it entitles text occurring in natural scene images to be claimed as an important source of information.

The past few decades have witnessed an exponential surge in several areas of information retrieval. Of these, the problem of detecting text in natural scene images is very challenging due to the presence of various complexities. Image analysis for scene text detection [2] finds its applications in computer vision and visual-aided technologies that include general public assistance like street signboard recognition, image to text conversion, and tourist guidance [3]. Thus, researchers across the globe are intrigued and continuously persevering to ameliorate the hurdles posed by an image, in order to detect and localize the text regions. The complexities usually involved in scene text detection arise due to factors affecting image quality like motion blur, noise out of poor camera quality, uneven intensities [4] and so on. Sample images across the datasets considered, are shown in Fig. 1 having different complexity factors.

Scene text in general, depicts a text containing region in an image which is captured by any camera based handheld device either in an outdoor or indoor environment.

Fig. 1 Sample images containing multi-lingual texts taken from standard public datasets: **a** KAIST, **b** MLe2e, and **c** in-house dataset. A system's robustness can be marred by some complexity factors like noise, blur, and heterogeneous intensity variation as evident in the sample images

Scene text detection from camera captured images is a computer vision [18] task which gained a lot of significance after the emergence of smart phones with good quality camera. Texts in scene images may have variations with respect to shape, font, color, and position. Text detection implies searching and retrieval of any textual region present in an image and later surrounding or localizing the desired region with a rectangular bounding box [19]. Text detection can be accomplished using either rule-based methods or by implementing some intelligent techniques or both in combination.

The presence of multi-lingual [4] text in images is considered as an added complexity in this domain. For instance, India is a culturally diverse country and has several places where signboards, hoardings, and posters with texts in multiple languages are prevalent. Not all benchmark mechanisms are suitable for detecting texts in varied languages. This is mainly due to structural and geometric dissimilarities among texts written in different languages [20]. This newly identified complexity becomes an addition to the already existing challenges faced in this domain. This eventually necessitates the development of a robust system versatile enough for detecting and localizing multi-lingual scene texts efficiently.

The proposed method involves a modular approach where the input image is resized to suitable dimension and K-means++ [5] algorithm is applied to the histogram of intensities in Red (R), Green (G) and Blue (B) color channels individually. The choice of the cluster centers is optimized using Memetic Algorithm (MA) [6]. This module clusters the intensities across the RGB color space such that the text components remain intact. The candidate components from every cluster then pass through a hybrid filtering mechanism where the spurious candidates are significantly reduced on the basis of their dimension and entropy measure. Finally, morphological operation [3] is applied such that any candidate text components emerge in word form with spurious regions further discarded.

The performance evaluation of the proposed system is done using standard multi-lingual datasets namely, KAIST [7, 8] and MLe2e [9, 10], and the performance scores are compared with that of some state-of-the-art methods. The system is also assessed against an in-house dataset which comprises 400 images containing texts in English, Bangla, and Hindi. A few sample images from the different datasets considered are shown in Fig. 1. A relatively good performance is demonstrated and is observed to surpass some of the latest techniques.

2 Recent Developments

Although the domain of scene text detection is not new and the state-of-the-art performance has significantly improved, still the performance falls short of the desired outcome. This is due to complexities like scene texts with multiple orientations and multiple languages that have raised the bar for enhancing the versatility of standard methods used for detecting scene texts. Altogether, there is a need for further improvement in performance so as to achieve efficient detection and localization of text regions in various scene images as evident from the methods described below.

A major breakthrough came with the advent of two techniques: Maximally Stable Extremal Region (MSER) detection and Stroke Width Transform (SWT). Matas et al. introduced the concept of MSER in [11] for detecting blobs of co-variant regions having near-similar intensity values. These regions exhibit an extremal property of the intensity variation between the region and the boundary at its outer side. The MSERs represent sets of components in an image, closed under the image coordinates' affine transformation, and not affected by the affine transformation of intensity. A text component is usually considered stable due to low-intensity variation among its pixels. To increase its robustness against blur, a Canny edge based MSER is developed in [2] and SWT is used later to minimize the number of spurious MSERs. This method of SWT was first proposed by Epshtein et al. [3] where a local image ensembled with geometric analysis is applied to retrieve text components. Stroke width is defined as the perpendicular distance between two adjacent pixels of a component, with the same orientation. By utilizing logical and dynamic geometric analysis, regions with nearly same stroke width can be combined to form bigger component that may represent a word. SWT associates dense estimation, determined at each pixel, with non-local scope since stroke width relies upon the data often contained in distant pixels.

The limitations of this MSER descriptor, however, still remain since it is sensitive to optimization of some parameters like region area selection and intensity threshold. Attempts in [12, 13] are made to study and develop relations shared by these parameters with text object properties like height, width, intensity variation, image histogram and dimension. Overall, the aim was to make MSER more dynamic. In the work presented in [12], an input image is transformed into its grayscale form and its histogram is obtained. From this histogram the number of sharp peaks is determined by analyzing the histogram of a local region in the grayscale image using a window sliding across the grayscale image. The width of this sliding window is determined by the ratio between width of the grayscale image and the logarithmic value of the width of the grayscale image with base as 2. The height of the sliding window is determined in the same manner with respect to the height of the input image. The count of sharp peaks thus calculated, serves as the threshold of intensity variation of the MSER descriptor. Again, deciding a region area range plays an important role in marking the desired MSERs. This is done by calculating the minimum area as the ratio of the dimensions of the grayscale image to the dimensions of the sliding window and, maximum area as the product of the dimensions of the grayscale image with the logarithmic value of the color in bit depth of the input image with base as 2. The MSER components thus obtained are treated as candidates that undergo refinement by fulfilling certain criteria based on the geometric properties like eccentricity and solidity that a text component is assumed to hold, such that the spurious components are minimized. The residue components are then morphologically closed such that word-level localization can be performed, since in case of multi-lingual texts, every word in a text region may have language variation. An experimental study is performed in the work present in [13] where a certain number of images are randomly selected from every dataset considered and their dimensions are recorded. Then, the MSER descriptor is applied on these images by manually setting the parameters like threshold for variation in intensity and the threshold for the minimum area as well as the maximum area for determination of the range of the region area. The values of these parameters are set in a manner such that only those regions are covered in the input image that contains textual information. Since certain spurious regions are covered as well, the closest approximation is chosen for the parameters of the MSER descriptor. These parametric values are recorded against the corresponding images taken under consideration. The images in the pool under consideration, are categorized into two groups: images that have large text regions and images that have small text regions. In case of images that have texts covering large area of region, the text components are not covered as a single MSER component. Rather these text components break into a number of small blobs. On the other hand, in case of images having texts covering small region of area, although the text components get covered as a single blob, the problem that arises is that the area covered by the MSER blob is observed to be greater than what is necessary. To understand and resolve these issues, the areas covered by the text regions are determined and plotted against the dimensions of the input image such that a suitable relation curve is devised. In order to fit maximum possible coordinate points, initially a polynomial form of relation is realized. However, as the degree of the polynomial increases, the feasibility of such relation becomes questionable since the non-monotonous nature of any polynomial relation and its inclination towards force fitting all the plotted points leads to generating complex relations. Hence, the minimum area of region is formulated as a first degree polynomial relation between the image dimensions and the dimensions of the text region. The maximum area of region is observed to resemble a power curve relation between the image dimensions and the dimensions of the text region.

A language invariant text proposal generator is presented in [7] where information from multiple color channels is utilized by a Random Forest classifier to differentiate textual regions from the rest. Here, text regions are scaled down to relatively small as well as fixed size for depiction of certain patterns by their image gradients irrespective of their geometric structures and types of language. Then features constructed on the basis of means and standard deviations of these image gradients are then used to train a Random Forest in order to retrieve text regions via several image scales and color channels. A sliding window having fixed size is used to look for candidate regions for grouping the horizontally aligned ones into candidate text lines. These candidate lines are then ranked on the basis of the Random Forest responses where horizontally extended candidates get greater scores owing to their likeliness to be the desired text lines. Stroke based features prove to be valuable in scene text detection thereby leading to its usage in [8] to train a multi-layer non-linear neural network for detecting multi-lingual scene texts. The learning of stroke features, which are of the most representative and discriminative is assumed to be capable of capturing the essential substructures of the candidate strokes. Firstly, an unsupervised algorithm serving as a multilayered network convolutional kernels, learns the basic stroke features. Then, this trained model is deployed for learning advanced-level abstract features of the text regions.

Scene text detection and localization is taken a step further by treating this as a one-class problem in the work presented by Mukhopadhyay et al. in [4]. In this work, candidate components are extracted from an input image using the combined mechanism of MSER and SWT. These components are then localized and fed to a trained one-class classifier for classifying the text regions from the spurious ones. This one class classifier is based on a binary class Support Vector Machine (SVM) classification model where the classifier can be trained with instances that belong to only one class. Here, the candidate text regions are taken to be the instances of a single class and features are obtained from these regions using a standard frequency-based Gabor filter to train the one-class classifier. This setup of one class classification enables the entire scene text detection scheme to improve its precision. However, its recall drastically falls since many regions containing textual information are falsely classified as spurious or non-text regions.

The combination of MSER and SWT has become quite popular in multi-lingual scene text processing mechanisms where the first and challenging stage is to determine those components that can serve as text candidates. The work in [9] utilizes this combination to generate candidate text lines and extracted to be resized into dimensions that is suitable for preserving the stroke parts. These stroke parts are scanned by a sliding window in order to get representation of essential features of the strokes in the candidate text lines. These features of the stroke parts are then fed to a trained Naïve-Bayes Nearest Neighbor (NBNN) classification model for determining the required text regions. An improvement in this method is made in [10] where conjoined networks are ensembled to collectively learn the discriminative features of the stroke parts of the candidate text lines and their relative position of significance in any patch-based classification scheme. This weighted model of NBNN classifier is substituted by a patch-based classification scheme so as to integrate it with the process of training the classification module by utilizing a conjoined network ensemble. In this method, the input image is depicted as a pool of local descriptors extracted from patches using a sampling strategy. These local features thus obtained, are then passed into a global classifier scheme, that decides for the input image.

The method proposed in [14] first converts an input image into a grayscale image and then applies edge enhanced MSER to extract a number of homogeneous blobs that may or may not be text components. Then, certain geometric properties of these components, like area, perimeter, compactness, and occupancy are calculated so as to eliminate the spurious components on the basis of the thresholds set for each geometric property. The number of spurious components is further reduced by the filtering process based on SWT criterion which is calculated on the basis of the standard deviation of stroke width value of all the components. The residue components are morphologically dilated such that they form a word or a text line. Still, a number of unwanted regions remain along with the text regions which is taken care of by passing these regions through an Optical Character Recognition (OCR) system. The major drawback with this approach is the use of an OCR mechanism, which is till date language specific. As a result, the overall scene text detection system can work on only those texts written in a language specific to that for which OCR is used. Again, edge enhanced MSER is used in [15] to extract stable components from the grayscale version of an input image. At first, the number of spurious components gets reduced on the basis of geometric properties like aspect ratio, area and skeleton length, and then analysis of connected components is done for combining the candidate regions. Further refinement is performed on the basis of the property that text usually appears on a surrounding that has distinct intensity. To realize this observation, the corresponding image patches for the detected MSER components are determined. To separate out the spurious regions from the textual information, these image patches are binarized using Otsu's method and acquire a region drawn from the intersection between the binarized patch and the corresponding MSER blob. The candidate regions are then passed through a filtering module where some standard criteria like Histogram of Oriented Gradients (HOG) and Pyramidal HOG (PHOG) are calculated. The HOG and PHOG values are observed to be lower for text regions compared to that of the spurious ones. Elimination of spurious regions is also undertaken by analysis based on SWT property and entropy of the candidate region. Although, the precision of this method enhances, its recall is acutely compromised. This is due to usage of too many elimination parameters where many text regions do not fulfil the criteria put forth in the excessively tight filtering module.

An alternate technique of two-level binning is developed by Dutta et. al in [16] to counter the limitations of MSER. In this method, certain gray levels are binned together using a background-foreground analysis, such that the text components remain intact unlike text MSERs that have potentially greater chances of getting broken. Usually, in any given image, the foreground components comprise pixel values similar to one another and differ significantly with that of their corresponding background. This observation is exploited to derive a relation that calculates the range of gray levels within which a particular bin can be represented. Then, the components falling under a particular bin are analyzed for their mutual alignment such that the aligned components are combined to form a word candidate. However, with increasing variation among the intensity values, the number of generated bins becomes unmanageable which in turn increases the computational cost.

The method proposed in [18] applied the MSER cum SWT combination to generate text candidates that undergo a Generative Adversarial Network (GAN) based refinement. Then, a Convolution Neural Network (CNN) based model is used for language identification of the detected scene texts. The work in [19] proposed a new feature descriptor which is based on distance transform of a text region. Here a scene image is binarized and the components are distance transformed. Then four pixels that are far away from both horizontal as well as vertical direction with equidistant values are determined and area of the rectangular box bounding the text region is measured. The ratio between areas held by each equidistant pixel and the rectangular bounding box is determined and treated as feature descriptor which is then fed to a standard classifier for text/non-text classification. A light-weight scene text spotter is developed in [20] to address the problem of cluttered environment in scene images. An end-to-end trainable deep neural network is designed that utilizes information of oriented region proposals at local, global, and context level for identifying text instances. This helps in localizing texts in scene images having background clutters responsible for broken and distorted text components. The issue of misclassification triggered due inter-class interference is resolved to some extent by analysis of inter-class separability and intra-class compactness.

The methods described above are overall robust to an extent, but the high presence of complexities mentioned previously, in an input image, limit their robustness. Hence, to overcome the limitations of the existing methods, in the present work, a method has been proposed where K-means++ algorithm is used for clustering of gray levels of an input image and an MA is used for optimizing the centroids or seed points obtained, for determining appropriate clusters so as to generate unbroken text components.

3 K-means++ Algorithm

Clustering usually refers to grouping of similar data into classes that are unknown. The dissimilarities are analyzed on the basis of attributes describing the objects. Often, distance-based measurements are utilized in determining such clusters. Clustering finds its applications in several areas related to data mining, statistics, biology, and machine learning [21]. There are some popular clustering algorithms like K-means that efficiently group among data points. K-means [22] is a very simple and popular clustering technique that aims to minimize the average squared distance between points belonging to the same cluster. However, its accuracy is unassured and it is sensitive to the initialization of the centroids or the mean points. This may lead to a centroid initialized as "far-off" point, to end up with no points associated with it. Simultaneously, more than one cluster might end up linked with a single centroid. Alternatively, two or more centroids might be initialized into the same cluster leading to poor clustering. To overcome these drawbacks, K-means was updated as K-means++ in the work [5], which ensures better initialization of the centroids and enhances the clustering quality. Cluster analysis finds its usage in a number of real-time applications like including market research, pattern recognition, data analysis, and image processing [21]. It is also employed in assisting the classification of documents for retrieving information on the Web.

K-means++ algorithm initially chooses the values or "seeds" for the K-means clustering procedure to avoid its occasional poor clustering. The foundation of this approach lies in spreading out the K initial cluster centers. The first center is randomly chosen from the set of data points being clustered, after which each resulting cluster center is selected out of the remaining data points having probability proportional to the square of its distance from the closest standing cluster center of the point. The steps for K-means++ are as follows:

1. An intensity value is chosen as one uniform center at random from among from among all intensities in the color histogram.

2. For each data point x, the distance between x and the nearest cluster center that has already been chosen which is called the distance between two histogram peaks or two called chromosome, is computed as $D(x)$.
3. A point is chosen with probability proportional to $D(x)^2$ that decides which point should be decided as center of the chromosomes.
4. Steps 2 and 3 are repeated until the completion of choice of all the chromosome centers.
5. After the choice of the initial centers, standard K-means clustering is employed as follows:

   i. $C \leftarrow$ A sample point is randomly chosen from X
   ii. While $|C| < K$ do
      a. Sample $x \in X$ with probability $P(x \in C) \propto D(x)^2$
      b. $C \leftarrow C \bigcup \{x\}$
   iii. End while.

4 Memetic Algorithm

The term "Memetic Algorithm" was first coined by Pablo Moscato in his technical report in 1989, inspired from both Darwinian principles of natural evolution and Dawkins' idea of a meme, where an MA is observed to be quite similar to population-based hybrid Genetic Algorithm (GA) tied with an individual learning capacity to perform local refinements [6]. This relatively new terminology is used to cover a wide category of metaheuristics where the mechanisms fulfilling general-purpose processes are aimed to direct an underlying heuristic. This algorithm is an extension of the classical GA where a local search technique is employed to minimize the probability of premature convergence [23]. The metaphorical comparison to Darwinian evolution, on one side, as well as amongst memes and domain-specific (local search) heuristics on another side, are captured within MA thereby leading to design an approach that efficiently maintains balances between generality and problem specificity.

The foundation of this algorithm lies on a population of agents [6] and has become quite popular in a number of problem domains especially for the approximate solution of optimization problems. In contrast to classical evolutionary computation approaches, MA is intrinsically designed to utilize all the existing knowledge about the problem under consideration. The assimilation of domain knowledge about the concerned problem is not a voluntary mechanism, but a fundamental feature that typifies the MA. The term "memetic" is clearly denoted by this Philosophy of functioning.

In recent times, MAs have found its application in several domains like automated industrial technologies, bioinformatics, computer vision tasks, etc. These domains in many cases require the training of machine learning models that can serve the purpose of automation efficiently. This requires an optimal level of feature selection [24] which is achieved by MAs that improve the choice at every iteration until the best solution is achieved. Other real-time applications of MAs include instruction detection, pattern classification problem, prototype selection stopping of premature classification convergence, and prediction problem [25].

In view of the above context, it becomes feasible to ensure a general model for a simple MA. A general procedure for the implementation of MA in this work, is elaborated stepwise as follows:

1. First, the population is initialized.
2. Until the convergence criterion is not reached, the following steps are implemented:
   i. The fitness function is evaluated.
   ii. Parent selection is done for suitable offspring.
   iii. Recombination is done to produce offspring.
   iv. Mutation is done within offspring.
   v. Offsprings are improved via local search.
   vi. Individuals are replaced by the new ones.

First of all, the process to generate the initial population is accountable for producing the initial set of candidate solutions. This is implemented by either simply generating a population of random configurations or by employing a more sophisticated seeding technique say, for example, some constructive heuristic. By means of this mechanism, instances of high quality are transmitted into the initial population. This can be achieved in another way using a local search scheme [26]. In case of determining a function for deciding the criterion of terminating the algorithm, it can be constructed in a way quite close to local search. Until the termination criterion is fulfilled, the steps for a hill-climbing local search method are repeated as follows:

1. New population is selected from the neighborhood of current population.
2. If fitness of the new population is better than the current population then:
   i. Current population = New population.

There are certain instances of possibilities that enable the MA to reach termination like putting a cap on the total number of times the MA procedure shall be iterating, reaching a pre-defined maximum count of iterations if the MA has not reached any improvement or completing a certain number of restarts in the population, and so on. The process for the generation of the initial population forms the core of the MA. Basically, this algorithm can be viewed as a pipelined process comprising stages, each of which consists of procuring configurations from the earlier stage and producing new configurations by applying an operator.

Next, the process of updating the population is implemented to reassign the current population by replacing the old population with the newly generated population. Taking que from the idea of the strategy of evolution, two main possibilities are given rise to, for continuing on this reconstruction: (1) the plus strategy, where the current population is constructed by taking the best size among the population configurations, and (2) the comma strategy where the best size among the population configurations is taken from the newly updated population. In this work, the comma strategy is chosen.

5 Proposed System

A stepwise modular approach is adopted for the development of an efficient multi-lingual scene text detecting system. The modules involved in the system, as shown in the flowchart in Fig. 2, are: (1) clustering module where intensity values from each color channel, are clustered using K-means++ algorithm. MA is then used to get an optimal set of candidate components from the color maps while eliminating

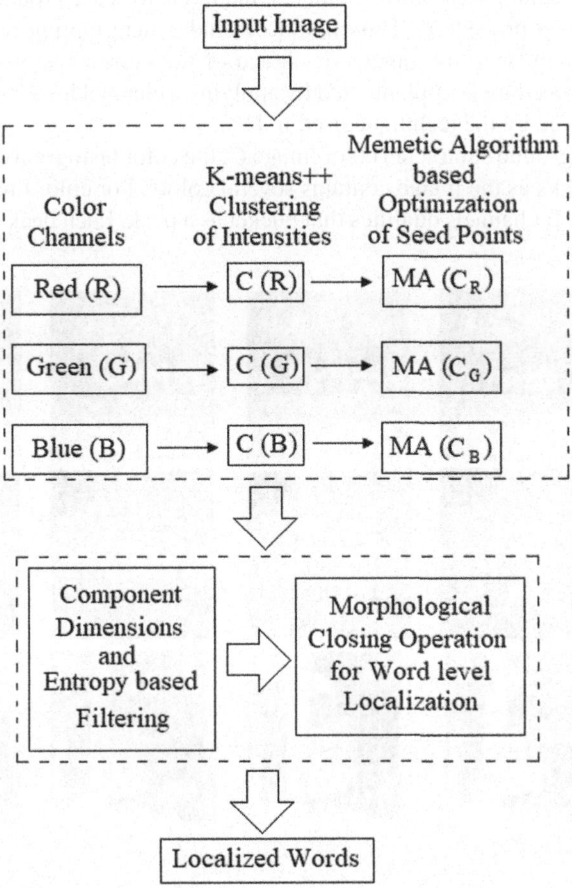

Fig. 2 Flowchart of the proposed system for detecting multi-lingual text regions in any natural scene image. Here, the color channels, i.e., Red, Green, and Blue are denoted as R, G, and B, respectively, C denotes the cluster function and MA denotes the function for memetic algorithm

the background, (2) filtering module where the spurious components are removed on the basis of their area, width and height that are either too large or too small to be text component. Entropy values are also used to determine the possible text regions and eliminating the spurious ones and finally, (3) word-level localization module where morphological closing operation is applied such that the text components are joined to represent word-level text regions. The techniques involved in every module of the proposed methodology, are described in detail in the following sub-sections.

5.1 Color Clustering

Initially, the input image is disintegrated into R, G, and B color channels as shown in Fig. 3, and for each of these color channels, the intensities are grouped accordingly using K-means++ procedure. This module includes determining an initial center vector set U_0 signifying the ranges of choice of the center vectors in an optimal manner. This procedure is implemented by applying a channel level histogram-based technique inspired by the technique used in [17].

For the background subtracted color image C, the color histogram can be obtained with several peaks as the image contains several colors. For color images, a peak in any of the R, G, B channels qualifies that bucket as a peak. Each peak corresponds to

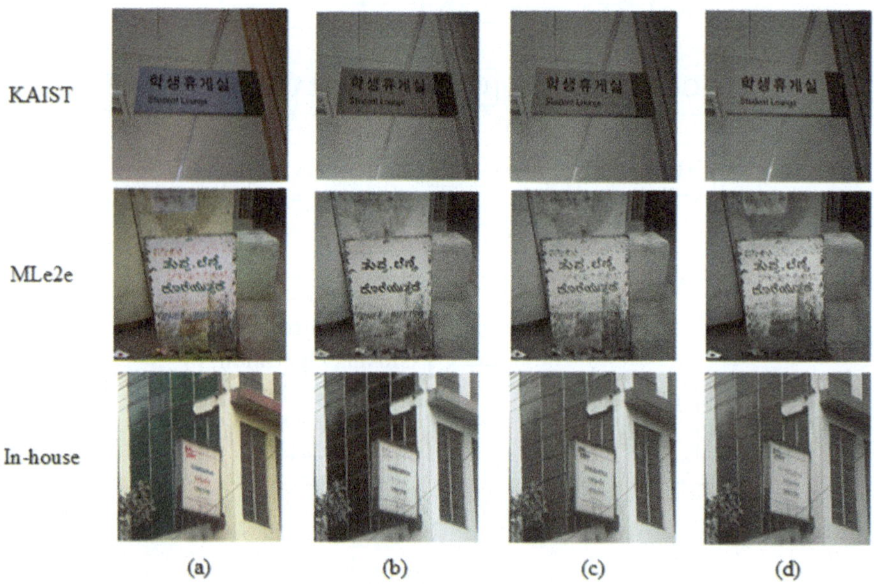

Fig. 3 Sample images containing multi-lingual texts from standard dataset KAIST and MLe2e, and in-house dataset. **a** Input image resized to 512 × 512 dimension, **b** Red component **c** Green component **d** Blue component

one object and separated from the other peaks by a valley. A peak's height signifies the count of pixels falling in the bin that corresponds to the peak's position. A given color space is split into I color bins and the color histogram of the input image with n pixels is denoted as a vector $H = [h_0, \ldots, h_{i-1}]$ where every entry h_i specifies the statistical figures of the colors in the image that have its place in the i-th bin:

$$h_i = \frac{n_i}{n}, i = 0, 1, \ldots, I - 1 \tag{1}$$

where n_i is the count of pixels having color values in the i-th color bin.

Consider, the RGB color space be traversed along the R, G, and B axes having the numbers N_R, N_G, and N_B respectively. Then, the total count of bins $N = N_R \times N_G \times N_B$ is available. These bins are coded in an arrangement from R to G and then from G to B. As per the discretization and coding scheme specified, the index of every single bin can be represented as:

$$i = R + N_G \times G + N_B^2 \times B \tag{2}$$

where $R = 0, 1, \ldots, N_R - 1$, $G = 0, 1, \ldots, N_G - 1$ and $B = 0, 1, \ldots, N_B - 1$.

Then the pixel (r_p, g_p, b_p) is supposed to be in the bin having index as i_p which is represented as:

$$i_p = \frac{r_p N_R}{256} + N_G \times \frac{g_p N_G}{256} + N_B^2 \times \frac{b_p N_B}{256} \tag{3}$$

where $\lfloor\rfloor$ is an integral operator.

After running K-means++ algorithm, the number of clusters is obtained where an intensity value in each bin is chosen for center vector optimally using MA-based procedure.

5.2 *Cluster Optimization*

Since the goal of this work is to extract text candidates from the scene image, the choice of centroids as well as the Euclidean distance based on histogram information from the color channels may result in breaking of a potential text component thereby adversely affecting the detection performance. Hence, searching the best partition is fulfilled by MA that helps to find an optimal partition matrix. Later, the optimal partitions are mapped into a single matrix where each partition is represented by an individual color, as shown in Fig. 4.

(a) (b) (c)

Fig. 4 Clustered regions of sample input images in Fig. 3a, using K-means++ and MA at different color levels taken from **a** KAIST, **b** MLe2e, **c** In-house datasets. This clustering ensures maximum text components to remain intact

The MA-based optimization procedure is implemented using the following steps:

1. **Chromosome Determination**: Given a set of m-dimensional vectors, the chromosome in a population is denoted by a string consisting of c number of m-dimensional real vectors encoding the center vectors that corresponds to c clusters in K-means++. Given a set of observations $(x_1, x_2, \ldots, x_n)$ where each of them is a d-dimensional real vector, K-means++ clustering aims to split the n observations into $k(\leq n)$ sets $S = \{S_1, S_2, \ldots, S_k\}$ so as to minimize the Within Cluster Sum of Squares (WCSS) (i.e., variance). Formally, the objective is to find:

$$\underset{S}{\operatorname{argmin}} \sum_{i=1}^{k} \sum_{x \in S_i} \| x - \mu_i^2 \| = \underset{S}{\operatorname{argmin}} \sum_{i=1}^{k} |S_i| Var S_i \tag{4}$$

Where μ_i signifies mean of the points in S_i which is equivalent to minimizing the pairwise squared deviations of points in the same cluster:

$$\underset{S}{\operatorname{argmin}} \sum_{i=1}^{k} \frac{1}{2|S_i|} \sum_{x, y \in S_i} x - y^2 \tag{5}$$

$$\sum_{x \in S_i} \| x - \mu_i^2 \| = \sum_{x \neq y \in S_i} (x - \mu_i)(\mu_i - y) \tag{6}$$

The equivalence can be construed from identity. Owing to the constant nature of the total variance it becomes equivalent to maximizing the sum of squared deviations between points in different clusters Between Cluster Sum of Squares (BCSS), which follows from the law of total variance.

2. **Population Initialization:** The center vectors in the initial population are chosen through the K-means++ method and the string consisting of the center vectors, is formed.

3. **Fitness computation**: For utilizing an MA, it becomes a necessity to define an objective function. Here, all the center vectors are considered as the vector set used in the function.
   Population undergoes some enhancements in each generation. Initially, the population is created randomly and ranked on the basis of the fitness value of the chromosomes. Then local search is performed at the start of each generation, in each chromosome until an improved chromosome string is obtained. Then, roulette wheel for parent selection is used then crossover is performed. Crossover is applied to generate two offspring after which, the offspring is considered for mutation. The mutation is performed by substituting strings in present generation with a new string from the same bin where the original strings existed. The mutation occurs with a fixed probability.
4. **Stopping Criterion**: There are two termination criteria for the algorithm:
   No visible improvements after a certain number of generations.
   The algorithm has reached a certain iteration limit.
5. **Original image retrieval:** For the retrieval of the original image, the partitioned matrix needs to be transformed into a crisp one. Let $P = [p_{ij}]$. where $i = 1, \ldots, c$ and $j = 1, \ldots, n$. be the c-partition matrix. It is quite acknowledged that p_{ij} denotes the membership grade for pixel j belonging to cluster i. A partition matrix, P_p, is represented as:

$$P_{p_{ij}} = \frac{p_{ij}}{\sum_{j=1}^{n} p_{ij}} \tag{7}$$

As per the present partition, the crisp partition matrix, $P_c = [p_{c_{ij}}]$., is represented as:

$$p_{c_{ij}} = \begin{cases} 1, P_{p_{ij}} = \max\limits_{i=1}^{c} \left((P_{p_{ij}}) \right) \\ 0, \text{ otherwise} \end{cases} \tag{8}$$

Thus, it becomes evident that in the crisp-partition matrix, each pixel falls under a certain cluster.
After the required clusters are obtained, the components in each of them undergo through filtering procedure to eradicate the spurious components, such that only the text components are retained.

5.3 Component Filtering

Although the background clusters are discarded in the previous module, certain spurious components stay behind along with the text components. These unwanted components tend to have random dimensions, i.e., the width, height or the whole area they occupy, are either too small in order to be recognized as text or too large such that they supersede the significance of any recognizable text component. In other words, text components are usually observed to have certain dimensions that allow any viewer or observer to recognize them as "something written". Hence, in this module, lower and upper thresholds are set such that the components beyond these limits get discarded, leaving behind any recognizable text component(s).

It is also observed that filtering solely on the basis of component dimensions may not guarantee maximum elimination of spurious regions that may have similar dimensions as that of text regions. These regions are in many cases observed to cover a relatively homogenous area, noticeable in a grayscale or color image. To tackle such situations, entropy of all the candidate regions are calculated to make note of the amount of disturbance within a bounded region. A region with textual content is said to have high disturbance due to the presence of background and foreground (text) with contrasting intensities. This observation prompts the filtering out the spurious regions since they have less entropy compared to the text regions, based on a specific threshold value. The residual components are combined morphologically to represent words as shown in Fig. 5.

(a)

(b)

(c)

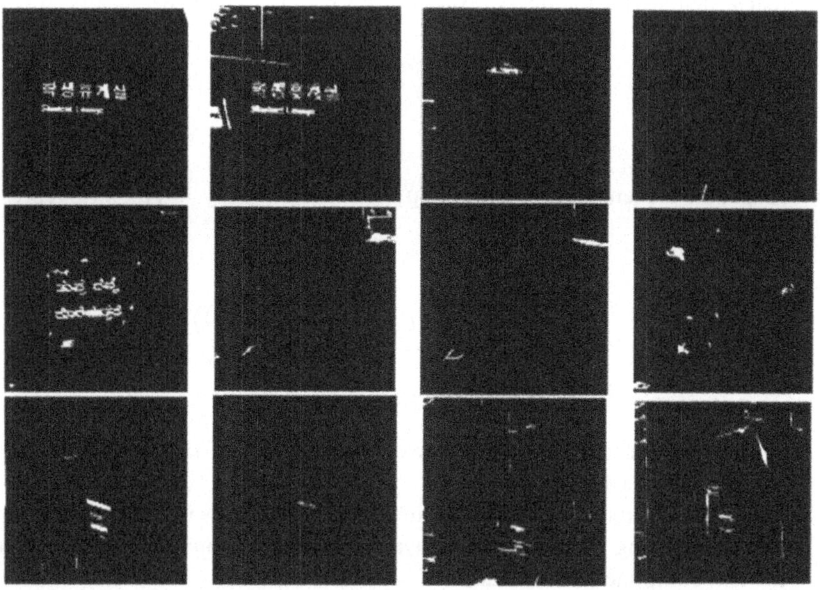

Fig. 5 Sample clusters of images from datasets **a** KAIST, **b** MLe2e, **c** In-house, from Fig. 4, represented as white components in binary masks. Text candidates are obtained from these white components after background elimination. This is done by filtering on the basis of their dimensional properties like width, height, and area. Also, entropy values of the candidate regions are calculated to further filter out the spurious regions having values less than a certain threshold. Morphological closing operation is performed to keep the characters of a word attached to each other

(a) (b) (c)

Fig. 6 Detected text regions are bounded in green rectangular boxes, in images from **a** KAIST, **b** MLe2e, **c** In-house datasets

5.4 Word-Level Localization

The resultant text components that stay back after running the filtering procedure, are combined using the morphological closing operation with ‘line’ structuring element. This ensures that the detected text regions are localized at word level as shown in Fig. 6.

6 Experimental Setup and Analysis

Experimental procedures of the overall system are elaborated in the following sub-sections where brief descriptions about the datasets used, are provided along with the choice of parametric values for the procedures applied and their justification.

6.1 Dataset Preparation and Evaluation

The proposed system is evaluated on images of standard publicly available multi-lingual datasets like KAIST and MLe2e. The system is also experimented on an in-house multi-lingual dataset having 400 images with texts in Bangla, English, and Hindi. The KAIST dataset has images with texts in English and Korean, and MLe2e has images with texts in English, Kannada, Korean, and Chinese.

Evaluation of the proposed system is done by following the standard metrics of Precision (P), Recall (R) and F-measure (F). Word level text bounding for the in-house dataset is done at word level. The pixels within a bounding box of the ground truth are taken to be True Positive (TP) and those outside the bounding box considered as True Negative (TN) for the datasets considered. If any spurious region is wrongly localized by the system, then the pixels within this localization is considered as False Positive (FP) and if the system is unable to detect a text region, then the pixels within the undetected text region is taken to be False Negative (FN).

In order to get a quantitative insight into the performance of the proposed system, standard formulation calculating the P, R and F for the proposed system, is followed. Following are the equations used for calculating the values of P, R and F:

$$P = \frac{TP}{TP + FP} \tag{9}$$

$$R = \frac{TP}{TP + FN} \tag{10}$$

$$F = \frac{2 \times P \times R}{P + R} \tag{11}$$

From the Eqs. (9) and (10), a vivid idea is realized about P and R of any scene text detection system. Precision of any such system signifies how many pixels actually represent the text regions out of the total number of pixels identified by the system. Recall ,on the other hand, gives an insight into the number of pixels identified by the system as representing text regions out of the total number of pixels actually representing the text regions. In this work, for a given dataset, precision and recall are calculated for every image and their averages are recorded. The average values of precision and recall are then used to calculate the F-measure. These measurements are used for comparing the proposed method's performance with that of some of the state-of-the-art methods in the next sub-section.

6.2 Parameter Setting

By experimenting on a trial-and-error basis, upon 100 images from each dataset, optimal values are decided for the parameters involved in the system. For K-means++ procedure, the number of clusters is set to 3 since the intensities are observed to fall under three distinct bins, dark, gray, and bright. Hence it is expected that the candidate components appear more prominent. In order to realize an optimal balance between achieving better clustering output and feasible execution time of the MA-based procedure, the population size is fixed at 50 and the number of generations is also fixed at 50. Probabilities of crossover and mutation are set to 0.8 and 0.08 respectively. The original size of the input image is resized in order to further reduce the execution time of the system.

Table 1 Comparison of performance of the proposed method with some state-of-the-art methods for different multi-lingual scene text datasets

Dataset	Method	Precision	Recall	F-measure
KAIST	Özgen et al. [14]	0.57	0.54	0.55
	Agrawal et al. [15]	**0.92**	0.34	0.48
	Proposed	0.74	**0.86**	**0.79**
MLe2e	Gomes and Karatzas [9]	0.48	0.55	0.51
	Gomes et al. [10]	0.51	0.62	0.56
	Proposed	**0.72**	**0.88**	**0.79**
In-house	Mukhopadhyay et al. [4]	**0.98**	0.41	0.58
	Dutta et al. [16]	0.48	0.84	0.61
	Panda et al. [13]	0.62	**0.90**	0.73
	Chakraborty et al. [12]	0.59	0.84	0.68
	Proposed	0.63	0.89	**0.74**

6.3 Results and Discussion

The performance of the proposed system is determined for in-house and, benchmark multi-lingual datasets like KAIST and MLe2e, and compared with some state-of-the-art techniques. The overall performance of the proposed system in comparison with some standard methods is depicted in Table 1.

From Table 1, it can be seen that the proposed framework has better precision (P) and recall (R) scores that are quite satisfactory in case of MLe2e dataset when compared with some recent methods. However, in case of KAIST dataset, the precision value is not at par with that of the benchmark score. Having said that, the proposed method scores high on recall indicating that the loss in text components, is successfully minimized. In case of the in-house dataset, the recall falls short of 1% of the benchmark score, but bettering the F-measure by 1%. Overall, in all the cases, it can be inferred that certain amount of non-text components do stay back thereby affecting the precision, however, the high recall scores suggest that the proposed method succeeded in detecting and localizing maximum scene text regions.

7 Conclusion

In the present work, a novel approach for detecting multi-lingual texts in natural scene images is proposed using K-means++ algorithms and MA. Here, the intensities in every color channel R, G, and B are clustered in a way such that text components stay intact and the background as well as other spurious regions are eliminated. The filtering mechanism significantly diminishes spurious regions with minimal loss of the text components. The proposed method proves to be quite robust and betters the performance of some standard methods. The significant improvement in recall measure indicates that the proposed mechanism is able to detect most of the scene texts. This improvement in recall is achieved by compromising precision up to a certain optimal level. From the performance data in Table 1, it can be observed that some of the standard methods have very good precision scores. But these methods achieve recall scores that fall short of being deemed satisfactory. This is because these methods employed filtering mechanisms that included too many refinement criteria that becomes unmanageable thereby losing many text regions. The framework proposed in this work overcomes this drawback and includes a manageable number of filtering criteria like a condition based on the area of the component and a condition based on the entropy present within a candidate region. This moderate level of filtering helps in retaining maximum possible text regions and reducing the number of spurious regions to an optimal level. The limitation, however, lies in the formulation of thresholds for component area and entropy. Certain images with complex intensity distribution posed challenges to the formulations and hence there is either a certain amount of loss in text regions or some spurious regions get past through the filtering mechanism or both. Hence, in future, these issues shall be aimed at to be addressed and further improvements shall be made for achieving the desire for dynamic thresholding.

Acknowledgements This work is partially supported by the CMATER research laboratory of the Computer Science and Engineering Department, Jadavpur University, India, PURSE-II and UPE-II, project. SB is partially funded by DBT grant (BT/PR16356/BID/7/596/2016). RS, SB and AFM are partially funded by DST grant (EMR/2016/007213).

References

1. Y. Zhu, C. Yao, X. Bai, Scene text detection and recognition: recent advances and future trends. Frontiers Comput. Sci. **10**(1), 19–36 (2016)
2. H. Chen, S.S. Tsai, G. Schroth, D.M. Chen, R. Grzeszczuk, B. Girod, Robust text detection in natural images with edge-enhanced maximally stable extremal regions, in *2011 18th IEEE International Conference on Image Processing* (IEEE, 2011, September), pp. 2609–2612
3. B. Epshtein, E. Ofek, Y. Wexler, Detecting text in natural scenes with stroke width transform, in *2010 IEEE Computer Society Conference on Computer Vision and Pattern Recognition* (IEEE, 2010, June), pp. 2963–2970
4. A. Mukhopadhyay, S. Kumar, S.R. Chowdhury, N. Chakraborty, A.F. Mollah, S. Basu, R. Sarkar, Multi-lingual scene text detection using one-class classifier. Int. J. Comput. Vis. Image Process. (IJCVIP) **9**(2), 48–65 (2019)
5. D. Arthur, S. Vassilvitskii, *k-means++: The Advantages of Careful Seeding* (Stanford, 2006)
6. P. Moscato, C. Cotta, A modern introduction to memetic algorithms, in *Handbook of Metaheuristics* (Springer, Boston, MA, 2010), pp. 141–183
7. K. Fan, S.J. Baek, A robust proposal generation method for text lines in natural scene images. Neurocomputing **304**, 47–63 (2018)
8. L. Li, S. Yu, L. Zhong, X. Li, Multilingual text detection with nonlinear neural network, in *Mathematical Problems in Engineering* (2015)
9. L. Gomez, D. Karatzas, A fine-grained approach to scene text script identification, in *2016 12th IAPR Workshop on Document Analysis Systems (DAS)* (IEEE, 2016, April), pp. 192–197

10. L. Gomez, A. Nicolaou, D. Karatzas, Improving patch-based scene text script identification with ensembles of conjoined networks. Pattern Recogn. **67**, 85–96 (2017)
11. J. Matas, O. Chum, M. Urban, T. Pajdla, Robust wide-baseline stereo from maximally stable extremal regions. Image Vis. Comput. **22**(10), 761–767 (2004)
12. N. Chakraborty, S. Biswas, A.F. Mollah, S. Basu, R. Sarkar, Multi-lingual scene text detection by local histogram analysis and selection of optimal area for MSER, in *International Conference on Computational Intelligence, Communications, and Business Analytics* (Springer, Singapore, 2018, July), pp. 234–242
13. S. Panda, S. Ash, N. Chakraborty, A.F. Mollah, S. Basu, R. Sarkar, Parameter tuning in MSER for text localization in multi-lingual camera-captured scene text images, in *International Conference on Computational Intelligence in Pattern Recognition* (Springer, Singapore, 2020), pp. 999–1009
14. A.C. Özgen, M. Fasounaki, H.K. Ekenel, Text detection in natural and computer-generated images, in *2018 26th Signal Processing and Communications Applications Conference (SIU)* (IEEE, 2018, May), pp. 1–4
15. A. Agrawal, P. Mukherjee, S. Srivastava, B. Lall, Enhanced characterness for text detection in the wild, in *Proceedings of 2nd International Conference on Computer Vision & Image Processing* (Springer, Singapore, 2018), pp. 359–369
16. I.N. Dutta, N. Chakraborty, A.F. Mollah, S. Basu, R. Sarkar, Multi-lingual text localization from camera captured images based on foreground homogenity analysis, in *Recent Developments in Machine Learning and Data Analytics* (Springer, Singapore, 2019), pp. 149–158
17. Y. Li, Vehicle extraction using histogram and genetic algorithm based fuzzy image segmentation from high resolution UAV aerial imagery, in *ISPRS08* (2008,) p. B3b, 529
18. S. Saha, N. Chakraborty, S. Kundu, S. Paul, A.F. Mollah, S. Basu, R. Sarkar, Multi-lingual scene text detection and language identification, in *Pattern Recognition Letters* (2020)
19. T. Khan, A.F. Mollah, Text non-text classification based on area occupancy of equidistant pixels. Proc. Comput. Sci. **167**, 1889–1900 (2020)
20. R. Bagi, T. Dutta, H.P. Gupta, Cluttered textspotter: an end-to-end trainable light-weight scene text spotter for cluttered environment, in *IEEE Access* (2020)
21. J. Han, J. Pei, M. Kamber, *Data Mining: Concepts and Techniques*. Elsevier (2011)
22. P. Jana, S. Ghosh, S.K. Bera, R. Sarkar, Handwritten document image binarization: an adaptive K-means based approach, in *2017 IEEE Calcutta Conference (CALCON)* (IEEE, 2017, December), pp. 226–230
23. M. Ghosh, T. Kundu, D. Ghosh, R. Sarkar, Feature selection for facial emotion recognition using late hill-climbing based memetic algorithm. Multimedia Tools Appl. **78**(18), 25753–25779 (2019)
24. M. Ghosh, S. Begum, R. Sarkar, D. Chakraborty, U. Maulik, Recursive memetic algorithm for gene selection in microarray data. Expert Syst. Appl. **116**, 172–185 (2019)
25. M. Ghosh, S. Malakar, S. Bhowmik, R. Sarkar, M. Nasipuri, Feature selection for handwritten word recognition using memetic algorithm, in *Advances in Intelligent Computing* (Springer, Singapore, 2019), pp. 103–124
26. M. Ghosh, S. Malakar, S. Bhowmik, R. Sarkar, M. Nasipuri, Memetic algorithm based feature selection for handwritten city name recognition, in *International Conference on Computational Intelligence, Communications, and Business Analytics* (Springer, Singapore, 2017, March), pp. 599–613

Recent Advancements in Medical Imaging: A Machine Learning Approach

Nitin Dang, Shailendra Tiwari, Manju Khurana, and K. V. Arya

Abstract This chapter discusses the importance of machine learning in the field of medical imaging for reconstructing medical images from the measured raw data. Besides discussing the use of machine learning for medical image reconstruction, a general overview is also provided on all the existing techniques of medical imaging. Mathematical models are provided in order to understand better the use of machine learning for reconstruction purposes. We discuss both unsupervised techniques like dictionary learning, auto-encoders, and supervised techniques which include learning of hyperparameters and various regularization methods used in deep learning models that replace various steps in iterative algorithms used for image reconstruction such as for image enhancement.

1 Introduction

Medical imaging is one of the most invaluable resources in today's computational world as it deals with the clinical analysis and visual representation of the human body. It provides researchers and physicians with an abundance of essential knowledge for the understanding, modeling, diagnosis, and treatment of the diseases. With the availability of abundant health data in the form of Electronic Health Records (EHRs), it has become really important to process such big data and extract valuable knowledge out of it. One of the most prominent approaches to analyze such data is by using machine learning in order to provide quality healthcare support to the patients besides reducing medical negligence to correctly identify the problem.

N. Dang · S. Tiwari (✉) · M. Khurana
Computer Science and Engineering Department, Thapar Institute of Engineering and Technology (TIET), Patiala 147004, India
e-mail: shailendra@thapar.edu

K. V. Arya
Multimedia and Information Security Research Group, ABV-Indian Institute of Information Technology and Management, Gwalior 474015, India

P. Kumar et al. (eds.), *Machine Learning for Intelligent Multimedia Analytics*, Studies in Big Data 82,
https://doi.org/10.1007/978-981-15-9492-2_10

More recent advancements in the field of medical imaging have been done over the past decades in terms of multi-modalities, high quality temporal and spatial resolution, cost optimization, etc. Still, several improvements can be anticipated such as minimizing the radiation exposure of patients while increasing the reconstructed image, artifacts reduction, etc. Machine Learning (ML) presents a promising approach to extract the features, interpretation, and decision-making in an autonomous or semi-autonomous fashion. Hence, ML has a great interest in speed-up the reconstruction process due to its ability to "learn from experience" and the fast-growing field in this domain.

This chapter has been structured as follows: It starts with the introduction of Recent Advancements in Medical Imaging. After that in Sect. 2 Inverse problems of imaging have been discussed. In Sect. 3, we will discuss how Artificial intelligence will be used in medical imaging. In Sect. 4 the role of Supervised Learning in image reconstruction is described. After that how the images are enhanced using deep learning that has been discussed in Sect. 5. Direct mapping approach has been discussed in Sect. 6. Then in Sect. 7, Generative Adversarial Networks (GANs) have been discussed which train the loss function from data, thus improving the perceptual image quality. In Sect. 8, we discuss unsupervised learning techniques like dictionary learning (DL) and K-sparse autoencoders (KSAE) for reconstructing medical images. Then finally Sect. 9 discusses the Conclusion and Future Perspective following with the references at the end.

2 The Problem of Inversion in Imaging

The inverse problems are the most promising problems that give an idea about what types of parameter values cannot directly observe. It arises in various applications of science and mathematics such as computer vision, image and signal processing, remote sensing, optics, communication theory, medical imaging, and all the non-destructive testing [1]. Some of the Medical Imaging Techniques have been shown in Fig. 1.

Medical Image reconstruction is an inverse problem and is ill-posed in the sense of Hadamard [2] and hence finding their solution requires special attention and care. The main purpose of medical image reconstruction is to reconstruct the raw data obtained from various scanning methods like Computed Tomography (CT), Magnetic Resonance Imaging (MRI), Positron Emission Tomography (PET), Single Positron Emission Computed Tomography (SPECT), etc. into a useful medical image of tissues, bones, and other internal parts of the human body. So, in order to reconstruct the medical image that can be represented as $u \in N^{K_u}$ from the measured projection data $v \in N^{K_v}$ given the following mathematical equation:

$$v = A(u) + z, \tag{1}$$

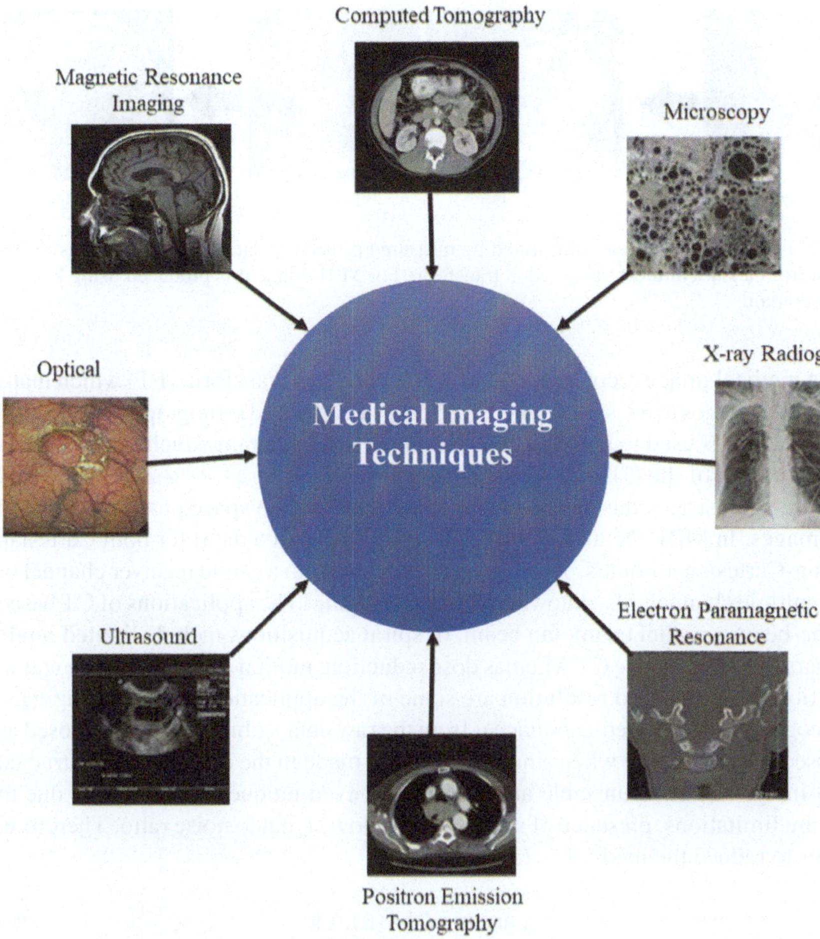

Fig. 1 Various application of medical imaging

where the field $N = \Re, C$ denotes real and complex numbers respectively, $z \in N^{K_v}$ is a random noise added and u, v are vector spaces whose dimensions are denoted by K_u and K_v respectively. A is the forward operator showing the mapping between measured data and the reconstructed images and is denoted as $N^{K_u} \rightarrow N^{K_v}$. For a K-dimensional field i.e. N^K, l_p-norm for $p >= 1$ is used for a vector t belonging to this field such that:

$$\|t\|_p = \left(\sum_{i=1}^{K} |t_i|^p \right)^{1/p}, \quad (2)$$

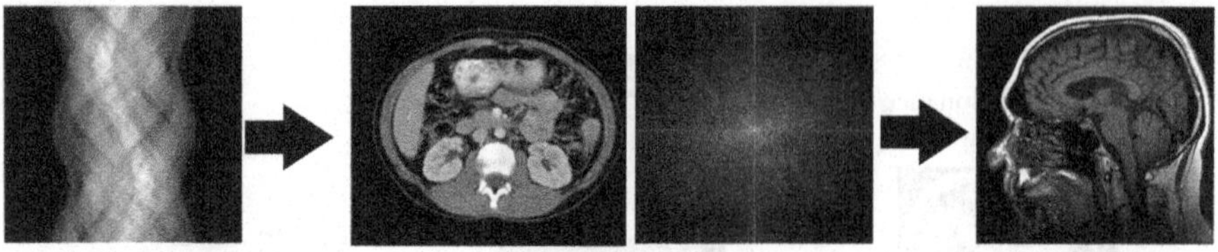

(a) Sinogram (b) CT image (c) K-space (d) MRI Image

Fig. 2 CT and MRI images are illustrated by measured projection data. A CT image **b** is reconstructed from a sinogram (**a**) using radon transform. For MRI image **d** is produced using k-space Fourier domain

A in medical image reconstruction of MRI is Fourier Transform (FT) which maps the medical images to a *k*-space, whereas in case of computed tomography (CT) radon transformation is used to map the medical image to a sinogram which is obtained as raw data output of the CT scan of the patient [3].

Figure 2 illustrates the measurement spaces and image spaces for both CT and MRI images. In MRI, the undersampled *k*-space (measured data) for both Cartesian and Non-Cartesian accounts. This data can be either from a single receiver channel or from multiple channels also known as Parallel imaging. The applications of CT based on cone-beam, parallel beam, fan beam, or spiral acquisitions include limited-angle CT, sparse view, low dose CT whereas dose reduction, motion correction, attenuation correction, and improved resolution are some of the applications of PET imaging.

Reconstructing the medical image u from the raw data v obtained is an ill-posed as discussed above because when small changes are made in the data, the reconstructed output image might be unstable and may not be even unique. This happens due to hardware limitations, presence of sparse data, or low signal to noise ratio. Therefore, in order to reduce the misfit, Eq. (3) is used:

$$u^* \in \underset{u \in N^{K_u}}{\operatorname{argmin}}\, D[A(u), v], \tag{3}$$

where $D: N^{K_v} \times N^{K_u} \rightarrow \Re$ represents the mapping of the statistical properties of the data. The least-square solution is calculated in the case of normally distributed noise in case of k-space data of MRI

$$u^* \in \underset{u \in N^{K_u}}{\operatorname{argmin}}\, 1/2\|A(u) - v\|_2^2, \tag{4}$$

Similarly, Kullback-Leiber divergence [4] is commonly used in the case of Poisson distributed samples in PET or even in CT.

$$u^* \in \underset{u \in N^{K_u}}{\operatorname{arg\,min}} \sum_{j=1}^{K_v} \left(A(u)_j - v_j \log\left(A(u)_j\right)\right), \tag{5}$$

where j denotes discrete sampling locations. The solution space should be restricted and an approximated solution should be calculated in order to avoid noise in the solution as minimizing the above problems often leads to noisy data due to overfitting. There are two possibilities that can be performed. One of them is to stop the solution early which acts as a regularization and the other one being is, by adding a regularization term which adds additional constraints to our reconstructed image that is, u leading to a variational model [5]:

$$u^* \in \underset{u \in N^{K_u}}{\operatorname{argmin}} \lambda D[A(u), v] + R[u], \tag{6}$$

where the effect on regularization R and data consistency term D is controlled by weight parameter $\lambda > 0$. Total Variation (TV) [6] semi-norm is used as a regularization term in medical imaging in contrast to the use of priors of the image which is more of a traditional method

$$\begin{aligned} R[u] = \|Du\|_{2,1} \sum_{i=1}^{K_u} \sqrt{\sum_d \left(|Du|_i^{(d)}\right)^2} \|t\|_p \\ = \left(\sum_{i=1}^{K} |t_i|^p \right)^{1/p}, \end{aligned} \tag{7}$$

where the gradient of the image is approximated using $D: N^{K_u} \to N^{K_{u,d}}$,$i$ denotes the discrete pixel locations and d denotes the dimensions of the image space. Sharp discontinuities are observed in the reconstructed image due to the use of TV semi-norm since it approximates the sparsity in the edges of the image. Besides this, the piecewise solution given by this method in order to describe the reconstructed image is not an appropriate regularization measure. The regularization part and weight parameter accelerate the image acquisition process which has a major effect on the solution and since TV semi-norm is convex in nature global optimization is achieved in the model [6].

Compressed Sensing (CS) [7] is also another technique used for reconstructing medical images in which a few measurements are taken into account and sampled the value by the Nyquist rate but it needs certain conditions to be fulfilled before applying the reconstruction approach. Firstly, incoherency among the artifacts which arises when the under-sampled measurement data v is encoded and the second one is based on the representation of the image that is, it needs to be done in a sparse manner which can be done in a transform domain using TV semi-norm and can be further extended to Total Generalized Variation (TGV) [8] as shown in the Eq. (7). Efficient global optimization is achieved by the use of l_1-norm in order to estimate the sparse nature of the modified image. Therefore, consistency of the data measured and sparsity in the transform domain combine to give the third condition for CS in non-linear reconstruction.

Even though the CS approach shows promising results in medical imaging but it is insignificant to use it for clinical purposes. MRI is majorly used for clinical purposes even though it doesn't satisfy the first condition discussed above, i.e., the artifacts are coherent here, but due to the use of sparsity in the transform domain, it is not able to capture complex image data and often leads to an unnatural or blocky image besides being computationally expensive. Therefore, adjusting the hyperparameters in the CS-based approaches is a major task to avoid under and overregularization and proper details are shown in the output images. The sensitivity of the hyperparameters for the reconstruction of the image makes it difficult to use for clinical purposes where the imaging situations vary.

While considering reconstruction methods that are iterative in nature each subtask is considered to be a new problem that needs to be optimized by the use of precedent knowledge and this is quite opposite of the way how physicians analyze patterns and look past the redundant artifacts of the image. Considering this observation of physicians, a machine learning model is prepared in which the parameters of the models are set taking both the raw data and the clean images into consideration in order to provide an optimized reconstructed image. Now in the following sections, we will discuss the role of AI in medical imaging and also how machine learning strategies like supervised and unsupervised improve the regularizer and overcome the challenges of the CS-based approaches.

3 Artificial Intelligence (AI) in Medical Imaging

Artificial Intelligence (AI) is the most promising field in medical imaging [9]. Since, machine learning ML is a subset of it and it has been widely used in the current research application of deep learning such as used: computer-aided design and detection (CAD) [10]. There have been major breakthroughs in data acquisition, segmentation, classification, and image reconstruction by the use of deep learning models. Deep learning has shown tremendous growth and is constantly propelling the field of medical imaging at a very fast pace. Earlier in the radiology, manual assessment of the medical images was done by the trained physicians for the detection and treatment of the diseases but with the use of AI, we are able to automatically recognize complex patterns in the collected patient data and are able to provide quantitative assessments of the medical images.

The recent advancements of the AI in perceiving data have allowed better representation and interpretation of data even if its nature is complex in turn leading to advancement in all the tasks which were once done better by humans ranging from searching the web and natural language processing to computer vision. Deep learning is similar to the human brain in a way it functions as it uses neural networks in order to learn patterns from the data and use that to approximate complex non-linear relationships which even excels human performance. The schematic model of how artificial intelligence methods are applying the medical imaging is shown in Fig. 3. Combining these algorithms with large amounts of available data and processing power helps

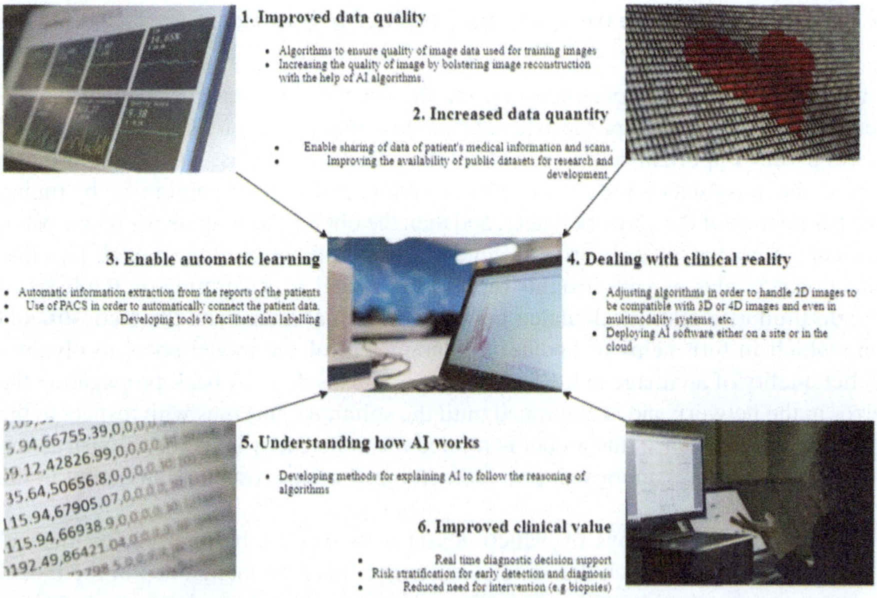

Fig. 3 The schematic model of artificial intelligence in medical imaging

the model to navigate the data space and learn directly from it giving them higher problem-solving capabilities.

3.1 *Challenges of AI in Medical Imaging*

A common problem with AI-based tasks is that they are based on narrow intelligence, that is they are unable to address more than one task at a time. Fully automated newer deep learning methods take years or even decades to be implemented for clinical radiology tasks. Data is the most central and major component in the design of AI-based models. Besides the availability of a moderate amount of data, it is not useful as this data is not curated which is a major issue in the path of learning an AI model. Curation [11] helps to remove unwanted variances in data since the data is collected from various sources and unwanted artefacts affect the quality of the training data. Curation is an expensive process since it is important for it to be done by a trained professional and the data which needs to be curated is huge and is different for different AI problems. In the case of mathematical based models, it is possible to question every solution obtained from the model but it is not possible in deep learning models since their inner working is opaque, and shifting through thousands of nodes in the neural network along with the connections is an unattainable task.

4 Supervised Learning in Image Reconstruction

In the case of image reconstruction using the supervised techniques network architecture [12] along with appropriate data and loss function in order to train the model plays a very important role in reconstructing medical images. During the training phase, the noisy data is fed into a network system and is in turn optimized by tuning the parameters of the network like Θ, and then the obtained output image is compared with the reference image in order to generate the loss function of the model. This loss function provides us with insights on how the model is performing on the basis of the dissimilarity matrix calculated between the actual image and the reconstructed one which in turn helps to update the parameters of the model so as to obtain a better quality of an image in the next iteration. This is done by back-propagating the error in the network and is continued until the solution converges with respect to the reference image. Once this model is trained it can be easily used for reconstructing other testing images and it will provide optimized image results as an output to the raw data.

There are various ways by which neural networks are being used for medical image reconstruction such that data consistency. It can be maintained in the model to give optimized results and image enhancement can be achieved in order to reduce the reconstruction errors in the final output image and is as close as to the reference image. In order to maintain data consistency, a regularizer is learned using the corrupted and the reference images, and also another approach is by learning the weight parameters which ensures the consistency of the measured data by the use of bi-level optimization. Thus, we can say that it is an unrolled iterative reconstruction technique that maintains consistency in every single unrolled step by the application of certain operations. Now we see that in order to enhance the image, the measured data is improved in the measurement domain rather than training the model on the initial measured data itself to obtain the reconstructed image.

4.1 Learning an Improved Regularization Function

While using a model the most important question is what regularization function R must be selected along with its parameters like the weight λ in order to reconstruct the image. The underlying handcrafted regularizer in the case of bi-level optimization [13] does not perform well in complex reconstruction tasks even though it addresses the problem of weight parameter. Therefore, there is a huge focus on techniques like bi-level optimization in recent times since they have non-convex regularizers encased in it. Recently, Convolution Neural Networks (CNN's) are trained and used as an additional regularizer in the model which in turn helps to reduce error, thus increasing performance.

4.1.1 Nonconvex Regularization

In cases where a distinct solution exists, convex regularizers achieve global optimization and TV-seminorm is one of them. Even though l_1-norm does not effectively describe the distribution that exists in the medical images, it is still used over l_0-norm which produces a sparse output. l_q norm properly describes the pre-distribution better in turn leading to superior results of the model. Therefore, on moving from convex regularizers to the non-convex ones, highly sophisticated algorithms are required in order to find a good solution that overcomes spurious local minima and also is efficient in computation.

Field of Experts (FOE) model [14] is a non-convex regularization method which is an extended version of TV-seminorm:

$$R[x] = \sum_{i=1}^{N_k} (l_i(G_i u), 1), \tag{8}$$

The above equation depicts the extended version of TV-seminorm consisting of $l_i(z)$ potential functions such that $l_i(z) = \left(l_{i,1}(z_1)\ldots l_{i,K}(z_K)\right) \in N^{K_u}$ and G_i as the convolution operator belonging to $N^{K_u} \to N^{K_u}$. Here in the above Eq. (8) depicts a vector which consists of all ones. The non-linear functions along with the convolution operators need to parameterized in the FOE model which is effectively learned with the help of data. The potential functions which are selected as shown above are directly related statistically to the clean images.

4.1.2 Bi-level Optimization

Selecting the weight criteria λ is a natural question that arises when working on the variational model is an important question. Reinforcement learning [15] and bi-level optimization [16] are two approaches to solve this problem. In case of bi-level optimization, the method is divided into two subtasks or problems in order to achieve the optimization, one is to minimize the mean square (MSE) which is the higher-level problem and the outcome of which depends on the second i.e. the lower level problem to solve the equation in the image reconstruction process where a regularizer is added in order to smoothen the process of image reconstruction. Bi-level optimization can be mathematically understood as follows:

$$\min_{\theta>=0} \zeta\left(z^*(\theta), z_{\text{ref}}\right) \quad s.t\ z^*(\theta) \in \arg\min_{z \in N^{K_x}} (E(z,\theta)) \tag{9}$$

Here, z^* is the output of the second subtask above, and $\arg\min_{z\in N^{K_x}}(E(z,\theta))$ depends on the training parameters θ and ζ is the loss function which we need to minimize, hence representing the first subtask which compares z^* and z_{ref}.

In order to find the solution of bi-level optimization, the second subtask should be completed with high veracity and since it calculates the gradient of the first subtask it should support the second derivative i.e.:

$$\frac{\partial \zeta}{\partial \theta} = -\frac{\partial^2 E}{\partial u^* \partial \theta}\left(\frac{\partial^2 E}{(\partial u^*)^2}\right)^{-1} \frac{\partial \zeta}{\partial u^*} \tag{10}$$

where, $\beta, \lambda > 0$. Hence, we can say that proposals for non-smooth functions to be used as the second subtask in this type of optimization are considered. This type of optimization technique refines the output of the reconstruction by learning of appropriate weight parameters. But still, the underlying architectures are too simple to be used for medical imaging, and focus has shifted to learn plenty of parameters of both function as well as the filters rather than learning just a unique parameter. But bi-level optimization has drawbacks in terms of the computational requirement and to achieve accuracy in the second subtask which is a critical task in medical image reconstruction in order to achieve an output that is as close as to the referenced image.

4.1.3 Convolution Neural Networks as Regularization

There exist various methods by which convolution neural networks can be used in the existing models. Use of CNN in Magnetic Resonance (MR) for reconstructing an image is visualized as a model where the noisy data $A^*(y)$ is mapped to the noiseless clean image x_{ref} which is further used to formulate compressed sensing. The generated output of the convolution model acts as clean output image for further problems.

$$x^* \in \arg\min_{x \in K^{N_x}} \frac{\beta}{2}\left\| f_\theta\left(A^*(\gamma) - x\right)\right\|_2^2 + \frac{\lambda}{2}\|(A(x) - \gamma)\|_2^2 + \mathcal{R}[x] \tag{11}$$

The parameters $\beta, \lambda > 0$ are parameters of the model, A signifies the forward operator, $R[x]$ is a regularizer which is fixed and A^* is the adjoint operator. Also, convolutional neural networks enhance the performance of the model in terms of the generated output image by replacing the existential mapping of the iterative procedures. In iterative techniques like Papoulis–Gerchberg (PG) method [17], the regularization constraint is fulfilled by employing the proximal mapping $\text{prox}_{\tau R}$ while the gradient is concluded with respect to the D which ensures the consistency of data in the model:

$$x^{t+1} = \text{prox}_{\tau\mathcal{R}}\left(x^t - \tau\nabla_x D\left[A\left(x^t\right), \gamma\right]\right) \tag{12}$$

And the proximity term can be depicted as:

$$\text{prox}_{\tau\mathcal{R}}(\hat{x}) = \arg\min_{x \in K^{N_x}} \frac{1}{2\tau}\left\|(x - \hat{x})\right\|_2^2 + \mathcal{R}[x] \tag{13}$$

Heide et al. [18] and Venkatakrishnan et al. [19] found that the existential mapping is coincident with the Maximum a Posteriori estimate of the gaussian denoiser. Therefore, the proximity term was replaced with algorithms in order to act as a denoiser for the model. Some of the algorithms used are non-local means (NLM) [20] and Block Matching with 3-D filtering (BM3D) [21]. The denoising steps BM3D and NLM have been adopted for both computed tomography and magnetic resonance images. The plug and play framework and the PD algorithm motivated Meinhardt et al. [22]. to replace the existing proximity operator with a convolutional neural network capable of removing the noise:

$$x^{t+1} = f_\theta\left(x^t - \tau\nabla_x D\left[A\left(x^t\right), \gamma\right]\right) \tag{14}$$

The CNN model is trained based on both corrupted and clean reference images and along with it, the intermediary results are also fed into the model to increase its performance. Promising results have been obtained for both computed tomography and positron emission tomography because it has an edge over other methods because the model can be easily trained even if it is disconnected and medical images can be depicted easily in the convolutional network. The learned proximity term does not vary throughout the entire problem but can be used with other problems by varying D.

4.2 Learning an Iterative Reconstruction Model

The Landweber method [23] motivated to learn the iterative reconstruction methods rather than optimizing one problem at a time. Gradient descent [24] is applied on the method of least squares, in turn, giving an iterative algorithm where x^0 depicts the primary solution and τ^t depicts the step size which varies with the iterations:

$$x^{t+1} = x^t - \tau^t\nabla_x D\left[A\left(x^t\right), \gamma\right] \tag{15}$$

This procedure should be put to a halt after T iterations are concluded in order to ensure that the model does not overfit the raw data y. After the inclusion of a regularizer, this methodology can be expanded to the proximity model for T iterations which use the regularizer and the consistent term for data in an alternative fashion.

$$x^{t+\frac{1}{2}} = f_1\left(x^t\right) \tag{16}$$

$$x^{t+1} = f_2\left(x^{t+\frac{1}{2}}\right) \tag{17}$$

The main task includes learning a proper optimized technique for a total of T iterations which include weight parameters, previous model, and the used step sizes besides ensuring the data is consistent. Gregor and LeCun proposed [24] the Learned

Iterative Shrinkage and Thresholding algorithm (LISTA) in the year 2005. This methodology is used in order to train an encoder network that approximates the codes in the Iterative Shrinkage and Thresholding Algorithms (ISTA) [25, 26] which are sparse in nature. This scheme can be extended to various optimization algorithms and is not just restricted to the gradient method.

On using the above equations and solving them with the proximity gradient in which the function f_1 evaluates the derivative with respect to R and the function f_2 evaluates the proximity term D resulting in the following result:

$$x^{t+1} = x^t - \tau^t \nabla_x \mathcal{R}\left[x^t\right], \tag{18}$$

$$\begin{aligned} x^{t+1} &= \mathrm{prox}_{\tau^t \lambda D[A(.),\gamma]}\left(x^{t+\frac{1}{2}}\right) \\ &= \arg\min_{x \in K^{N_x}} \frac{1}{2\tau^t} \left\| \left(x - x^{t+\frac{1}{2}}\right) \right\|_2^2 + \lambda D\left[A(x), \gamma\right], \end{aligned} \tag{19}$$

Another approach was suggested by Diamond et al. [27] and Schlemper et al. [28] to replace the stochastic gradient descent with a function f_θ that can be taught and represents a convolutional neural network. This function can be varied in each iteration t and even sharing of weights θ can take place in between different iterations forming a residuary neural network from the above equation. The step size in the above equation is not taken into consideration since its effect is already covered by λ and the convolutional neural network and is varied after each loop.

5 Image Enhancement Using Deep Learning

In contrast to the iterative reconstruction model wherein every reconstruction step, the measured data is accounted for the physical modeling of internal view of the objects. It is not accounted for enhancing the reconstructed images which can be achieved using two techniques that are as preprocessing and postprocessing. In the data domain, pre-processed is achieved using $v^* = f_\theta(v)$ whereas in the image space post-processed as $u^* = f_\theta\left(u^0\right)$. The enhanced data v^* obtained after post-processing is used for reconstruction whereas it is not the case for pre-processing where the value of u^0 is used for reconstruction purposes and then further enhanced to obtain u^*. As the physical model is not used, training is highly efficient as the physical model is heavily time-consuming to evaluate numerically, therefore we can say that it serves as an advantage but it does have a disadvantage that the used network architecture only works on the information extracted from the training data and on the initial guess. Also, data consistency is not maintained for post-processing methods and since the algorithms work in the image domain various post-processing methods like artifact removal, denoising, and super resolution fall into this class of problems. These algorithms are applied to the top of other correction methods or after the reconstruction of x^0.

Now, we will discuss some of the proposed methodologies in this domain. In image reconstruction using magnetic resonance combined images are used as input for the model. In other technique, a fully convolutional network was used in order to reconstruct MR images, and also similar approaches were extended to 3D and dynamic MRI [29]. Also, additionally in order to suppress the image modality-specific noise, Jin et al. [29] proposed an approach to use wavelet transformed images of both the input and as well as the reference. Residual learning is an approach in which rather than providing the mapping between the noisy input image and also the clean output image (reference), it gives a method to learn the unwanted artifacts present in the input image which after subtracting from the input image give a resultant output.

Kwon et al. [30] proposed a Multi-Layer Perceptron (MLP) which is a subset of feedforward artificial networks and unravels the raw data one line at a time in single-coil images identical to encoding technique. SENSE [31] requires the network architecture to be remodeled for various sizes of matrix alongside different sampling techniques. Also, deep learning offers a way to learn the mapping between the low quality and higher quality images which leads to higher performance in comparison to the standard interpolation methods like cubic interpolation used to achieve super resolution in images.

Image denoising is one of the applications of CNN for medical image reconstruction which uses the data of low-dose computed tomography. Deep convolutional frameworks are also used to study this data and a mathematical relationship is drawn between the convolution networks and the representation of the framelets. Besides these variational networks can be used as an alternative for both limited angle and low-cost computed tomography. Also, VNs provide another alternative for both low-dose and limited-angle CT. Even though noise appears in the reconstruction techniques where a low amount of radiation is used still various algorithms like residual learning using a fully convolutional network, deep convolutions, and even various other neural network combinations are used in order to obtain output in sparse view computed tomography where the radiations dosage and measurement time are reduced in significant portions. Metal Artifact Correction (MAR) [32, 33] is another image enhancement method that uses convolutional neural networks in which it produces an image with less noise from the images that were priory corrected and this generated image is further used to make the projections error-free, hence generating a noise-free image as an output.

In the case of image enhancement of PET, a multilayered perceptron along with various weights for regularization is used in order to obtain an enhanced output. Xu et al. [34] proposed an approach for PET using a fully convolutional network which mapped the low dose images to the images where a proper dose of radiation was used. In another approach to producing a positron emission tomography image, Xiang et al. [35] provided a solution using convolutional neural networks generated using PET images that use a lesser dose of radiation along with weighted magnetic resonance images.

Antholzer et al. [36] proposed a convolution network-based approach for photoacoustic tomography (PAT) in order to produce a quality of reconstructed images even when the raw data is sparse in nature. Another method was proposed by Schwab et al. [37] in which a fully convolutional network was applied to the reconstructed image produced by the raw data which is spare in nature after its length is dynamically corrected using a proper algorithm. Rather than executing the post-processing method in the image domain to enhance the quality of reconstructions another approach of preprocessing is used to correct the raw data which includes the detection as well as the removal of the noise. Park et al. [38] used a fully convolutional network architecture in order to make the sinogram error-free which was earlier affected due to the metals. Claus et al. [39] proposed a method that used multi-layer perceptron in order to improve the quality of data even when some values are missing in it.

6 Direct Mapping Approach

Instead of training a neural network in order to reconstruct an image, partial or complete mapping can be learned between the raw data obtained from the medical machines and the output image for reconstruction purposes. In order to learn the complete mapping, no detail of the internal architecture is used for reconstruction purposes that is:

$$u^* = f_\theta(v) \tag{20}$$

Besides this, in order to achieve efficient results, a large amount of training data along with parameters of the model are required alongside huge memory requirements. But even after fulfillment of all the requirements, it does not guarantee the quality of the reconstructed image. Even though efficient implementations exist it is a challenging question that whether neural networks should be implemented in the model instead of Fourier transform. Automated Transform by Manifold Approximation (AUTOMAP) [40, 41] is an example of direct mapping where CNN is used on the connected layers of the model in order to form a network. Since it deals with a direct mapping between the input and the output which leads to an increased memory requirement and in turn has a direct impact on the scalability of the model for larger problems.

U-net architecture [42] is used to obtain a computed tomography image as an output when images acquired at various angles are fed into the system as an input. CNN architecture is used to directly map the sinogram which is the raw measurement data with the image domain. Instead of using direct mapping in the model which has a huge memory requirement, neural networks are used to replace steps of restoration which enhances the performance of the system besides reducing the memory load and is shown below where A^* denotes the transform function:

$$u^* = A^*(f_\theta(v)) \tag{21}$$

The concept of back projection is used in order to train the network model to produce optimal results where a back-projection layer is added to the network and it optimizes the hyperparameters after a number of iterations thus, enhancing the performance of the model in terms of quality of the resultant image produced.

6.1 *Availability of Training Data*

ImageNet database [43] is considered to be a great source to train the deep neural networks which has images present in various categories and as of now, there are 14 million images as reported in 2019 [42]. Besides providing access to the data it also serves as a testing benchmark to the new development in comparison to the earlier presented methodologies. Multiple projects are constantly being initiated in order to allow access to people to large image datasets for research purposes since training data plays a vital role in the performance of the model. Hence large clean datasets serve the purpose to train the model without compromising the performance. Some of the projects are the UK-Biobank [44], the human connectome project that stores the reconstructed images, and not the data in the raw form obtained by the machines used for medical imaging. Depending on the domain of research different datasets are available and can be used for further researches. Therefore, it can be concluded that lack of proper clean datasets has a direct impact on the training and validation of the new research models in this domain.

6.2 *Transfer Learning as a Learning Approach*

Present network architectures involve detailed computing specifications, a wide range of free parameters, training time, and robust datasets. The concept of transfer learning is to provide fair insights into the outcomes of approaches that are learning-based even when there is a lack of availability of large datasets. The idea behind transfer learning is that large repositories of datasets like ImageNet and even data generated using simulations are used in order to pretrain the network model and depending on the area of research the network is trained on that particular data. Recently, neurological data was analyzed by the use of transfer learning approach combined with convolution architecture for MR based image reconstruction.

6.3 Image Quality Analysis and Loss Functions

Loss functions play a vital role in supervised learning approaches and have a direct impact on the performance of the model apart from the training data and the architecture of the network since a proper loss function helps in reevaluating the hyperparameters of the model which in turn helps in increasing its performance. The loss function works by computing a dissimilarity matrix between the output image x obtained after reconstruction and the clean image x_{ref} which is used as a reference. This loss function has a direct impact on the hyperparameters of the model which in turn affects the output image. Mean square error (MSE) [45] is a common approach in order to evaluate the loss function and is shown as follows:

$$\mathcal{L}_{\text{MSE}}(x, x_{\text{ref}}) = \frac{1}{2N}\left\|x^T(\theta) - x_{\text{ref}}\right\|_2^2, \tag{22}$$

where, N determines the number of pixels. Even though MSE can be easily optimized it is not robust against the outliers due to its averaging behavior which results in blurred reconstructions and when the data contains noise the output quality of the image is affected even more. Various evaluation measures related to MSE are used in order to evaluate the performances of the model like Normalized root mean square error (NRMSE) [46] in which data comparison happens at a variety of scales

$$\mathcal{L}_{\text{NRMSE}}(x, x_{\text{ref}}) = \frac{\left\|x^T(\theta) - x_{\text{ref}}\right\|_2}{\|x_{\text{ref}}\|_2} \tag{23}$$

Peak Signal to Noise Ratio (PSNR) [47] is another method for the quantitative evaluation of the model which involves maximum intensity values of the image $I_{\max}$ into consideration:

$$\mathcal{L}_{\text{PSNR}}(x, x_{\text{ref}}) = 20\log_{10}\frac{I_{\max}\sqrt{N}}{\left\|x^T(\theta) - x_{\text{ref}}\right\|_2} \tag{24}$$

All these methods above evaluate the model by comparing the output image with the referenced image pixel by pixel and are known to represent the human perceptual system poorly.

Therefore, another evaluation parameter Structural Similarity Index (SSIM) [48, 49] is used which instead of focusing on pixels takes local patch statistics into consideration:

$$\begin{aligned}\text{SSIM}(x, x_{\text{ref}}) &= \frac{1}{N}\sum_{i\in\Omega} l\big(x(i), x_{\text{ref}}(i)^{\alpha} c\big(x(i), x_{\text{ref}}(i)^{\beta}\big)\big)\\ &= \frac{1}{N}\sum_{i\in\Omega}\left(\frac{2\mu_{x(i)}\mu_{x_{\text{ref}}(i)} + C_1}{\mu_{x(i)}^2 + \mu_{x_{\text{ref}}(i)}^2 + C_1}\right)^{\alpha}\left(\frac{2\sigma_{x(i)}\sigma_{x_{\text{ref}}(i)} + C_2}{\sigma_{x(i)}^2 + \sigma_{x_{\text{ref}}(i)}^2 + C_2}\right)^{\beta},\end{aligned} \tag{25}$$

where l is the luminance term, c is a constant, Ω is the domain, i determines the location of the pixel, c_1 and c_2 depicts the constants defined by the user, and $\mu_{(i)}$ and $\sigma_{(i)}$ is the value for each image patch of mean and standard deviation whose value is determined by the Gaussian filter σ_G when convolved over the entire image. This filter has a direct impact on the performance of the model in terms of the generated output. The loss equation considering SSIM whose value lies in the range $[-1, 1]$ is given below:

$$\mathcal{L}_{\text{SSIM}}\big(x^t(\theta), x_{\text{ref}}\big) = 1 - \text{SSIM}\big(x^t(\theta), x_{\text{ref}}\big), \tag{26}$$

Another approach was proposed by Wang et al. [50] which assesses and integrates different resolution values of SSIM as the Multi-Scale Structural Similarity Index (MS-SSIM). For similarity measures like Feature Similarity Index (FSIM) which are complex in their nature cannot be used to adjust the parameters of the model when used as a loss function. Hence, we can say that not every loss function is applicable to train a model.

7 Generative Adversarial Networks

Generative Adversarial Networks (GANs) train the loss function from data, thus improving the perceptual image quality. They have two networks competing against each other. One being the generator network that gives rise to an output reconstructed image obtained from the raw data and the other being the discriminator network which differentiates between the reconstructed and the reference image but training such networks is a tedious task on its own due to the presence of cross-entropy losses when the problem is being formulated initially.

So, in order to stabilize the training, various other strategies like Wasserstein Generative Adversarial Networks (wGANs) [51, 52] come into the picture. Thus, further increase of stability can be achieved by using techniques like gradient penalty along with wGANs or by the use of least-squares GANs. In case of lack of data, GANs introduce artificial structure also known as hallucinations but should be avoided in medical imaging since they affect the reconstructed image by sharpening it [51]. Therefore, l_1 and l_2 which are pixel-based content losses are added in order to guide GAN for efficient reconstruction. The use of perceptual losses, which is based on feature responses of the VGG network, further expanded this notion. To reduce noise in low-dose CT, Wolterink et al. [53] used Generative Adversarial Network alongside l_2 loss. Thaler et al. [54] proposed the use of GAN with cyclic loss in which the second generator was used to ensure that the generated images and the reference images are similar. Least squares GANs approach was studied in combination with l_2 loss for accelerated MR. Seitzer et al. [55] used GAN architecture along with VGG and matching loss to learn the residual image in order to refine the reconstructions trained with MSE. Another method was suggested to minimize picture noise by integrating wGANs with gradient penalties with MSE as a function. Hence, researching the implementation capacity of GANs for therapeutic purposes and also restricting sparse data acquisitions is an important research path.

8 Unsupervised Learning in Image Reconstruction

In this Section, we discuss unsupervised learning techniques [56] like dictionary learning (DL) [57, 58] and K-sparse autoencoders (KSAE) [59, 60] for reconstructing medical images. Optimal sparse signal transformations produced during the training process are checked in dictionary learning and is a drastically different method to signal modeling in comparison handcrafted signal models such as wavelet. The models used in unsupervised learning may either be pre-trained from a collection of skewed training results or can be learned simultaneously during the process of reconstruction called blind compressed sensing. Compressed Sensing (CS) is a novel theory in the acquiring and processing of information that restores signals accurately. This is done by resolving an optimal problem where the signals are sparse and interpreted using a base matrix and transforming higher dimensionality space into a lower number of dimensions. Deep Learning is a standard compressed sensing image reconstruction process in which sample data is compressed into a particular transformation domain and the coefficients of transformation are mapped into a lower dimensionality while retaining the important information of the image, hence converting a complex reconstruction problem to an optimization problem. Here an over completed dictionary is designed in order to represent the signal sparsely. The accuracy of the model depends a lot on the designed dictionary. It should be both self-adaptive as well as flexible. Hence, the undersampled image reconstruction issue is essentially solving the undetermined linear equation method $F_x u = v$ by decreasing the l_o quasi norm in such a way that u has a sparse nature after transformation $\Psi \in \Re^{M \times N}$ is applied. Hence our optimization problem is:

$$\min_{u} \|\Psi u\|_o \quad s.t \; F_x u = v, \tag{27}$$

This is a problem of sparse coding which in turn has a NP-hard nature and hence cannot be explicitly solved therefore approaches like greedy algorithms are used to solve it by methods like orthogonal matching pursuit (OMP) [61, 62]. Here u is the final reconstructed image and F_x contains the estimations for v in order to achieve reconstruction. On replacing the norm to l_1 norm instead of the existing l_o norm, linear programming is used to solve the problems lying in the real domain whereas, in the complex domain, second-order cone programming does the task. z is the image of dimensions $n \times n$ which is decayed into tiny patches of $n' \times n'$ size on the basis of slide distance and patchsize.

There are mainly four dictionary learning algorithms currently to solve the dictionary learning problem as follows:

- Direct Method (DM): Due to the direct extraction procedure in DM all information on the sample images is retained. Using appropriate patches, we can fully recover the target image which is high in resolution [63].
- K-means single value decomposition (K-SVD): It is an iterative procedure that updates the atoms of the dictionary to suit the data better. Besides updating the current dictionary atoms and coefficients simultaneously and also removes the error using SVD. It is the most widely used method with excellent convergence and sparsity [64].
- Generalized Principal Component Analysis (GCPA): GCPA is a procedure that uses subspaces in order to segment and model the mixed data [65]. This method introduces a broad range of the spectrum for data modeling and clustering since it introduces algebraic models into data clustering.
- Method of optimal directions (MOD): The coefficient corresponding to the dictionary vectors is set in this process, and then the atoms which are the dictionary columns are modified by reducing residuals between the training vector and its representations.MOD is of benefit as it enables the dictionary vectors to be optimally modified in each iteration [66].

Therefore, we can summarizing the dictionary learning techniques for image reconstruction by block diagram in Fig. 4, we can say that dictionary learning [67–69] is a method in which an initial dictionary u_o is initiated from the input measured data and sparse representation is achieved as output after proper training and then transforms are used in order to update u resulting in output after several iterations on repeating the above steps.

Ravishankar et al. [57] proposed an approach where the sparsifying transform is learned instead of the dictionary giving a generalized model for analysis. On applying

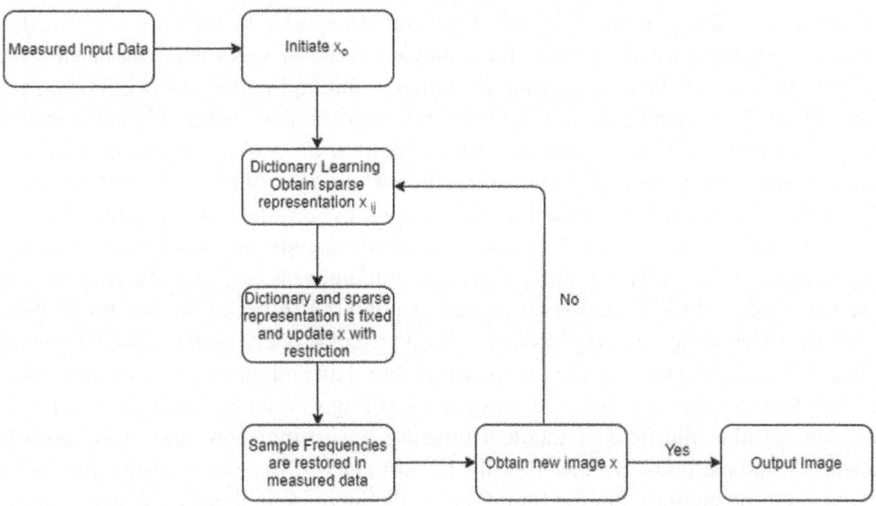

Fig. 4 The block diagram of image reconstruction dictionary learning techniques

to blind compressed sensing, we get the model as shown below:

$$\min_{\substack{x \in K^{N_x} \\ W \in K^{N_p \times N_p} \\ \alpha \in K^{N_\alpha \times N_s}}} \sum_{i=1}^{N_s} \left(\frac{1}{2} \| W P_i x - \alpha_i \|_2^2 + \beta \| \alpha_i \|_0 \right) + \frac{\lambda}{2} \| A x - \gamma \|_2^2, \quad s.t., W * W = I, \tag{28}$$

where $\beta, \lambda > 0$, W represents the transform matrix and α denotes a matrix consisting of vectors α_i which are sparse in nature. Learning W is not as expensive as it was in the case of dictionary learning due to the presence of solutions which are closed in nature on the basis of the threshold values for the sparse codes. The patches are approximated by the use of dictionary atoms in a sparse combination in both DL and transform learning. Also, an optimization of the above method is proposed by Wu et al. [70] by the use of the Separable Quadratic Surrogate (SQS) algorithm.

9 Conclusion and Future Perspective

Recent advances in medical imaging serve as pillars to provide medical assistance in today's computational world. The availability of quality hardware in terms of resolution and sensitivity has led to a more detailed analysis of the medical data which cannot be analyzed even with a trained eye, now can be done easily with the help of AI. In contrast to traditional methods that are based on the predefined feature set, deep learning algorithms improve the performance of the trained model in terms of accuracy and reduction in time and effort with the increase in the scaled data. AI produces quality results for a narrow range of tasks whereas in the case of multiple tasks it lacks performance which is unlike humans. Hence advantages that AI brings to medical imaging must not be overstated since advances in this field fall in the narrow AI category which focuses on one and only one task at a time even though it exceeded human intelligence by a substantial amount. Rational thinking must be used in AI field in order to possess top-down knowledge which is how the human brain works. As AI advances and exceeds the human performances, the human operators will be able to oversee outcomes and are also able to interpret the reasoning behind it besides discovering potential information that might have been overlooked by trained physicians. Rather than processing the medical images directly, the focus is to acquire the measurement data and then process it in order to extract features out of it since curating and labeling of data is a tedious task, hence we can say that the field of medical imaging is shifting more and more towards unsupervised methods. Besides this, the performance of machine learning models is expected to be increased in the future due to the increase in the availability of large, clean, and diverse public datasets.

The use of machine learning with modern computational technology and tools such as bio-sensors, bioelectronics, and the wearable devices, etc. It helps for better diagnosis and easily monitoring the lifestyle of patients and provides a new paradigm shift in the area of medical sciences using P4 models Predictive, Prevention, Personalized, and Participating. Lastly, we conclude that there are unlimited opportunities to improve the healthcare system using machine learning, AI, and other recent technologies.

References

1. J. Blitz, G. Simpson, *Ultrasonic Methods of Non-destructive Testing*, vol. 2 (Springer Science & Business Media, Berlin, 1995)
2. P. Gong, P. Song, S. Chen, Ultrafast synthetic transmit aperture imaging using Hadamard-encoded virtual sources with overlapping sub-apertures. IEEE Trans. Med. Imaging **36**(6), 1372–1381 (2017)
3. A. Averbuch, I. Sedelnikov, Y. Shkolnisky, CT reconstruction from parallel and fan-beam projections by a 2-D discrete Radon transform. IEEE Trans. Image Process. **21**(2), 733–741 (2011)
4. P.P.B. Eggermont, Multiplicative iterative algorithms for convex programming. Linear Algebra Appl. **130**, 25–42 (1990)
5. C. Samson, L. Blanc-Féraud, G. Aubert, J. Zerubia, A variational model for image classification and restoration. IEEE Trans. Pattern Anal. Mach. Intell. **22**(5), 460–472 (2000)
6. L.I. Rudin, S. Osher, E. Fatemi, Nonlinear total variation-based noise removal algorithms. Physica D **60**(1–4), 259–268 (1992)
7. M. Lustig, D.L. Donoho, J.M. Santos, J.M. Pauly, Compressed sensing MRI. IEEE Signal Process. Mag. **25**(2), 72–82 (2008)
8. K. Bredies, T. Valkonen, Inverse problems with second-order total generalized variation constraints, in *Proceedings of SampTA 2011* (2011)
9. F. Pesapane, M. Codari, F. Sardanelli, Artificial intelligence in medical imaging: threat or opportunity? Radiologists again at the forefront of innovation in medicine. Eur. Radiol. Exp. **2**(1), 35 (2018)
10. K. Suzuki (ed.), *Machine Learning in Computer-Aided Diagnosis: Medical Imaging Intelligence and Analysis* (IGI Global, USA, 2012)
11. P.M.A. Ooijen, Quality and curation of medical images and data, in *Artificial Intelligence in Medical Imaging* (Springer, Cham, 2019), pp 247–255
12. L.C. Jain, M. Seera, C.P. Lim, P. Balasubramaniam, A review of online learning in supervised neural networks. Neural Comput. Appl. **25**(3–4), 491–509 (2014)
13. B. Colson, P. Marcotte, G. Savard, An overview of bilevel optimization. Ann. Oper. Res. **153**(1), 235–256 (2007)
14. S. Roth, M.J. Black, Fields of experts: a framework for learning image priors, in *2005 IEEE Computer Society Conference on Computer Vision and Pattern Recognition (CVPR'05)*, vol. 2 (IEEE, 2005), pp. 860–867
15. Z. Yang, Z. Fu, K. Zhang, Z. Wang, Convergent reinforcement learning with function approximation: a bilevel optimization perspective (2018)
16. J.Z. Liang, R. Miikkulainen, Evolutionary bilevel optimization for complex control tasks, in *Proceedings of the 2015 Annual Conference on Genetic and Evolutionary Computation* (2015), pp. 871–878
17. P. Chatterjee, S. Mukherjee, S. Chaudhuri, G. Seetharaman, Application of Papoulis-Gerchberg method in image super-resolution and inpainting. Comput. J. **52**(1), 80–89 (2009)

18. F. Heide, M. Steinberger, Y.-T. Tsai, M. Rouf, D. Pająk, D. Reddy, O. Gallo et al., Flexisp: a flexible camera image processing framework. ACM Trans. Graph. (TOG) **33**(6), 1–13 (2014)
19. S.V. Venkatakrishnan, C.A. Bouman, B. Wohlberg, Plug-and-play priors for model-based reconstruction, in *2013 IEEE Global Conference on Signal and Information Processing* (IEEE, 2013), pp. 945–948
20. A. Buades, B. Coll, J.-M. Morel, Non-local means denoising. Image Process. OnLine **1**, 208–212 (2011)
21. K. Dabov, A. Foi, V. Katkovnik, K. Egiazarian, Image denoising by sparse 3-D transform-domain collaborative filtering. IEEE Trans. Image Process. **16**(8), 2080–2095 (2007)
22. T. Meinhardt, M. Moller, C. Hazirbas, D. Cremers, Learning proximal operators: using denoising networks for regularizing inverse imaging problems, in *Proceedings of the IEEE International Conference on Computer Vision* (2017), pp. 1781–1790
23. L. Landweber, An iteration formula for Fredholm integral equations of the first kind. Am. J. Math. **73**(3), 615–624 (1951)
24. H. Bruder, R. Raupach, J. Sunnegardh, M. Sedlmair, K. Stierstorfer, T. Flohr, Adaptive iterative reconstruction, in *Medical Imaging 2011: Physics of Medical Imaging*, vol. 7961 (International Society for Optics and Photonics, 2011), p. 79610J
25. A. Beck, M. Teboulle, A fast-iterative shrinkage-thresholding algorithm for linear inverse problems. SIAM J. Imaging Sci. **2**(1), 183–202 (2009)
26. I. Daubechies, M. Defrise, C. De Mol, An iterative thresholding algorithm for linear inverse problems with a sparsity constraint. Commun. Pure Appl. Math. J. Issued Courant Inst. Math. Sci. **57**(11), 1413–1457 (2004)
27. S. Diamond, V. Sitzmann, F. Heide, G. Wetzstein, Unrolled optimization with deep priors.arXiv preprint arXiv:1705.08041 (2017)
28. J. Schlemper, J. Caballero, J.V. Hajnal, A. Price, D. Rueckert, A deep cascade of convolutional neural networks for MR image reconstruction, in *International Conference on Information Processing in Medical Imaging* (Springer, Cham, 2017), pp. 647–658
29. K.H. Jin, M. Unser, 3D BBPConvNet to reconstruct parallel MRI, in *2018 IEEE 15th International Symposium on Biomedical Imaging (ISBI 2018)* (IEEE, 2018), pp. 361–364
30. K. Kwon, D. Kim, H.W. Park, A parallel MR imaging method using multilayer perceptron. Med. Phys. **44**(12), 6209–6224 (2017)
31. K.P. Pruessmann, M. Weiger, M.B. Scheidegger, P. Boesiger, SENSE: sensitivity encoding for fast MRI. Magn. Reson. Med. Official J. Int. Soc. Magn. Reson. Med. **42**(5), 952–962 (1999)
32. J. Timmer, Metal artifact correction in computed tomography. U.S. Patent 7,340,027, issued March 4, 2008
33. W. Lu, K.B. Pauly, G.E. Gold, J.M. Pauly, B.A. Hargreaves, SEMAC: slice encoding for metal artifact correction in MRI. Magn. Reson. Med. Official J. Int. Soc. Magn. Reson. Med. **62**(1), 66–76 (2009)
34. J. Xu, E. Gong, J. Pauly, G. Zaharchuk, 200x low-dose PET reconstruction using deep learning. arXiv preprint arXiv:1712.04119 (2017)
35. L. Xiang, Y. Qiao, D. Nie, L. An, W. Lin, Q. Wang, D. Shen, Deep auto-context convolutional neural networks for standard-dose PET image estimation from low-dose PET/MRI. Neurocomputing **267**, 406–416 (2017)
36. S. Antholzer, M. Haltmeier, J. Schwab, Deep learning for photoacoustic tomography from sparse data. Inverse Probl. Sci. Eng. **27**(7), 987–1005 (2019)
37. J. Schwab, S. Antholzer, R. Nuster, M. Haltmeier, DALNet: high-resolution photoacoustic projection imaging using deep learning. arXiv preprint arXiv:1801.06693 (2018), pp. 1–18
38. H.S. Park, S.M. Lee, H.P. Kim, J.K. Seo, Y.E. Chung, CT sinogram—consistency learning for metal-induced beam hardening correction. Med. Phys. **45**(12), 5376–5384 (2018)
39. B.E.H. Claus, Y. Jin, L.A. Gjesteby, G. Wang, B. De Man, Metal-artifact reduction using deep-learning based sinogram completion: initial results, in *Proceedings of 14th International Meeting Fully Three-Dimensional Image Reconstruction Radiology and Nuclear Medicine* (2017), pp. 631–634

40. Bo. Zhu, J.Z. Liu, S.F. Cauley, B.R. Rosen, M.S. Rosen, Image reconstruction by domain-transform manifold learning. Nature **555**(7697), 487–492 (2018)
41. M.S. Rosen, B. Zhu, B.R. Rosen, System and method for automated transform by manifold approximation. U.S. Patent Application 16/326,910, filed July 11, 2019
42. J. Deng, W. Dong, R. Socher, L.-J. Li, K. Li, L. Fei-Fei, Imagenet: a large-scale hierarchical image database, in *2009 IEEE Conference on Computer Vision and Pattern Recognition* (IEEE, 2009), pp. 248–255
43. Li. Fei-Fei, J. Deng, K. Li, ImageNet: constructing a large-scale image database. J. Vis. **9**(8), 1037 (2009)
44. C. Bycroft, C. Freeman, D. Petkova, G. Band, L.T. Elliott, K. Sharp, A. Motyer et al., The UK biobank resource with deep phenotyping and genomic data. Nature **562**(7726), 203–209 (2018)
45. L. Gjesteby, H. Shan, Q. Yang, Y. Xi, B. Claus, Y. Jin, B. De Man, G. Wang, Deep neural network for CT metal artifact reduction with a perceptual loss function, in *Proceedings of the Fifth International Conference on Image Formation in X-ray Computed Tomography*, vol. 1 (2018)
46. C. Quinsac, A. Basarab, D. Kouamé, J.-M. Grégoire. 3D compressed sensing ultrasound imaging, in *2010 IEEE International Ultrasonics Symposium* (IEEE, 2010), pp. 363–366
47. W. Yuanji, L. Jianhua, L. Yi, F. Yao, J. Qinzhong, Image quality evaluation based on image weighted separating block peak signal to noise ratio, in *International Conference on Neural Networks and Signal Processing, 2003*, vol. 2 (IEEE, 2003), pp. 994–997
48. R. Dosselmann, X.D. Yang, A comprehensive assessment of the structural similarity index. Signal Image Video Process. **5**(1), 81–91 (2011)
49. D. Brunet, E.R. Vrscay, Z. Wang, On the mathematical properties of the structural similarity index. IEEE Trans. Image Process. **21**(4), 1488–1499 (2011)
50. Z. Wang, E.P. Simoncelli, A.C. Bovik, Multiscale structural similarity for image quality assessment, in *The Thrity-Seventh Asilomar Conference on Signals, Systems & Computers*, vol. 2 (IEEE, 2003), pp. 1398–1402
51. Q. Yang, P. Yan, Y. Zhang, H. Yu, Y. Shi, X. Mou, M.K. Kalra, Y. Zhang, L. Sun, G. Wang, Low-dose CT image denoising using a generative adversarial network with Wasserstein distance and perceptual loss. IEEE Trans. Med. Imaging **37**(6), 1348–1357 (2018)
52. M. Ran, J. Hu, Y. Chen, H. Chen, H. Sun, J. Zhou, Y. Zhang, Denoising of 3D magnetic resonance images using a residual encoder–decoder Wasserstein generative adversarial network. Med. Image Anal. **55**, 165–180 (2019)
53. J.M. Wolterink, T. Leiner, M.A. Viergever, I. Išgum, Generative adversarial networks for noise reduction in low-dose CT. IEEE Trans. Med. Imaging **36**(12), 2536–2545 (2017)
54. F. Thaler, K. Hammernik, C. Payer, M. Urschler, D. Štern, Sparse-view CT reconstruction using wassersteinGANs, in *International Workshop on Machine Learning for Medical Image Reconstruction* (Springer, Cham, 2018), pp. 75–82
55. M. Seitzer, G. Yang, J. Schlemper, O. Oktay, T. Würfl, V. Christlein, T. Wong et al.,Adversarial and perceptual refinement for compressed sensing MRI reconstruction, in *International Conference on Medical Image Computing and Computer-Assisted Intervention* (Springer, Cham, 2018), pp. 232–240
56. D. Shen, Wu. Guorong, H.-I. Suk, Deep learning in medical image analysis. Annu. Rev. Biomed. Eng. **19**, 221–248 (2017)
57. S. Ravishankar, Y. Bresler, MR image reconstruction from highly undersampled k-space data by dictionary learning. IEEE Trans. Med. Imaging **30**(5), 1028–1041 (2010)
58. Q. Xu, H. Yu, X. Mou, L. Zhang, J. Hsieh, G. Wang, Low-dose X-ray CT reconstruction via dictionary learning. IEEE Trans. Med. Imaging **31**(9), 1682–1697 (2012)
59. A. Makhzani, B. Frey, K-sparse autoencoders. arXiv preprint arXiv:1312.5663 (2013)
60. L. Gondara, Medical image denoising using convolutional denoisingautoencoders, in *2016 IEEE 16th International Conference on Data Mining Workshops (ICDMW)* (IEEE, 2016), pp. 241–246

61. B. Yang, S. Li, Pixel-level image fusion with simultaneous orthogonal matching pursuit. Inf. Fusion **13**(1), 10–19 (2012)
62. M. Sajjad, I. Mehmood, S. WookBaik, Sparse coded image super-resolution using K-SVD trained dictionary based on regularized orthogonal matching pursuit. Bio-Med. Mater. Eng. **26**(s1), S1399–S1407 (2015)
63. S. Li, H. Yin, L. Fang, Group-sparse representation with dictionary learning for medical image denoising and fusion. IEEE Trans. Biomed. Eng. **59**(12), 3450–3459 (2012)
64. L. Wang, Lu. Ke, P. Liu, R. Ranjan, L. Chen, IK-SVD: dictionary learning for spatial big data via incremental atom update. Comput. Sci. Eng. **16**(4), 41–52 (2014)
65. K. Kreutz-Delgado, J.F. Murray, B.D. Rao, K. Engan, T.-W. Lee, T.J. Sejnowski, Dictionary learning algorithms for sparse representation. Neural Comput. **15**(2), 349–396 (2003)
66. I. Tosic, P. Frossard, Dictionary learning. IEEE Signal Process. Mag. **28**(2), 27–38 (2011)
67. K. Hammernik, F. Knoll, Machine learning for image reconstruction, in *Handbook of Medical Image Computing and Computer Assisted Intervention* (Academic Press, Cambridge, 2020), pp. 25–64
68. J. Wang, J. Liang, J. Cheng, Y. Guo, Li. Zeng, Deep learning based image reconstruction algorithm for limited-angle translational computed tomography. PLoS ONE **15**(1), e0226963 (2020)
69. H.-M. Zhang, B. Dong, A review on deep learning in medical image reconstruction. J. Oper. Res. Soc. China 1–30 (2020)
70. D. Wu, K. Kim, G. El Fakhri, Q. Li, Iterative low-dose CT reconstruction with priors trained by artificial neural network.IEEE Trans. Med. Imaging **36**(12), 2479–2486 (2017)

Solving Image Processing Critical Problems Using Machine Learning

Ajay Sharma, Ankit Gupta, and Varun Jaiswal

Abstract The history of vision can go back around 543 BC. From fossil studies, researchers found out that no. of animal species floats out and soon develops the first animal with eyes. As evolution in bigger species around fifty percent of neurons in our body are involved in the visual processing. Vision allows human to perceive, comprehend, and understand surrounding them while computer vision aims to duplicate the impact of human vision electronically in the form of images. The history of making artificial vison starts around 1545, Gemma Frisius, and Encylopedie in eighteenth century. The foundation period of the concept lies from 1943 to 1956, during this period research is cantered towards understanding the concept of machine learning intelligence, artificial neural system, and automata theory computation. The first work in the field of Artificial intelligence in computer vision and image processing was done in MIT called "Summer Vision Project". The rise of artificial intelligence starts and it is used widely in every field of technology. As the technology advances machine learning is used in almost every field. This book chapter provides you some details how the machine learning is used in image processing and to solve the current problem and challenge in computer vison like: Denoising, Image compression, Image color enhancement, segmentation, etc. Machine learning algorithm like Neural Network, Support Vector Machine, Genetic Algorithm, Convolutional Neural Network, etc. were discussed in the book chapter.

A. Sharma (✉)
Department of Biotechnology and Bioinformatics, Jaypee University of Information Technology JUIT, Solan, Himachal Pradesh 173215, India
e-mail: ajaysharmadharmani@gmail.com

A. Gupta
Interactive Technologies Institute (ITI/LARSyS), Madeira-ITI, University of Madeira, Caminho da Penteada, 9020-105 Funchal, Portugal
e-mail: gupta.ankit894@gmail.com

V. Jaiswal
National Centre for Disease Control NCDC, New Delhi, India
e-mail: computationalvarun@gmail.com

Gachon University, 1342 Seongnam-daero, Bokjeong-dong, Sujeong-gu, Seongnam-si, Gyeonggi-do, South Korea

P. Kumar et al. (eds.), *Machine Learning for Intelligent Multimedia Analytics*, Studies in Big Data 82,
https://doi.org/10.1007/978-981-15-9492-2_11

Keywords Image processing · Machine learning · Neural network · Genetic algorithm · Support vector machine · CNN

1 Introduction of Machine Learning

Machine Learning is one of the powerful and influential technology in today's world. In 1959 Arthur Samuel an American computer scientist coined this term with the statement that "It gives computer the ability to learn without being explicitly programmed". Artificial Intelligence and Machine Learning researches the assessment, examination and advancement of calculations that can gain up from and make prediction on given data sets. Machine learning is an interdisciplinary field that utilizes algorithmic mathematical, procedures, and a use of artificial intelligence which enable computer systems and other electronic devices to learn and get it. Machine learning investigates the examination and development of calculations that can gain from and make prediction on given data sets. Among the diverse kinds of Machine Learning methods, an urgent refinement is drawn among is supervised and unsupervised learning [1]. In Supervised machine learning the computer program is "prepared" on a pre-characterized set of "preparing models", which at that point encourage its capacity to achieve an exact end when given new information while in unsupervised machine learning the program is given a pack of information and must discover examples and connections in that. Machine learning is divided into different classes as follows.

Classification of machine learning as follows:

1. Supervised Learning
2. Unsupervised Learning
3. Semi-supervised Learning
4. Reinforcement Learning

1. **Supervised Learning**: It is also called as the inductive or predictive learning model. In the supervised learning, the prediction of the future outcome is based on the historical labeled data. The machine learning models are trained on the basis of labeled data for the outcome which consists of the input parameter and we know the expected outcome form the given data set. The model can then be used to predict the samples with unknown labels by calculating the features for that sample, similar used for training the model [2]. One of the issues with the usage of these models to the data is their selection. A solution to this problem is Meta-Learning which selects the attributes from the feature set and look for correlations between those attributes and performance of algorithms [3]. For example, Classification of persons whether the person is male of female based on the given data set, machine learning algorithm will learn based on the features and some patterns in the data and classify based on the training data.
   Supervised learning algorithm classified into two types of category, classification, and regression algorithm.

(a) **Classification Algorithms**: This type of algorithm classifies the data into some classes of some labels within the dataset. For the classification, we have algorithm like K-nearest Neighbor (KNN) [4, 5].
(b) **Regression Algorithm**: Algorithm like Linear regression and Logistic regression determines the mathematical relationship between the different data points within the dataset [6]. These algorithms are used in the predicting the output dependency and two or more variables. Regression algorithm can be classified into the linear regression and logistic regression. Logistic regression includes Support Vector Machine (SVM) [7], Random Forests, and decision tree algorithm [8].

2. **Unsupervised Algorithm**: Unsupervised algorithm is known as the descriptive models where the output is unknown and we don't have any idea that what the algorithm will spit out from the input data. The input data for the training is unlabeled for unsupervised machine learning. The algorithms belonging to this category are used to cluster the data based on the similarity between the data points. However, these algorithms differ in their functionality for data clustering. For example, K-means clustering, the data is clustered into the different group based on some properties between the different input variables [9, 10].

   (a) **K-means Clustering**: In the k-means clustering, the collected data is grouped together in a similar cluster based on their similar properties in the data points. It is a combination of two words k and means, K refers to no. of cluster of centroids' in the group and means refers centroid centre point with respect to that centroid or cluster formed with in the given data set [11]. The commonly used data clustering method is k-means clustering where k is the number of clusters to which data is divided [12–14]. Deciding the number of clusters is tricky and data-dependent, and is often decided by elbow method, which works on heuristics.

3. **Reinforcement Learning**: It is a new type of learning technique where the algorithm learns and behaves based on the previous action within a problem. for every decision algorithm will work on the score whether it is positive of negative based on that machine will took the decision that what to do next. This type of learning uses the concept of agent-environment. The agent takes the input and current state of the environment for changing the state by an action. The-changed state is further shared using reinforced signal. The action selected by the agent is based on the score whose value should be as large as possible. Reinforcement model works on trial and error and hence follows heuristics For the algorithm to take decision is depend upon the process called as the Markov Decision Process [15]. In recent year's reinforcement learning algorithm works in so many areas, e.g., in 2016 AlphaGo game use the reinforcement learning for and beat the world champion [16].
4. **Semi-supervised learning**: Is a technique that uses the small amount of labeled data for the training purpose within the large amount of unlabeled data sets. By using the labeled data will allow the algorithm to reduce the pattern and helps in

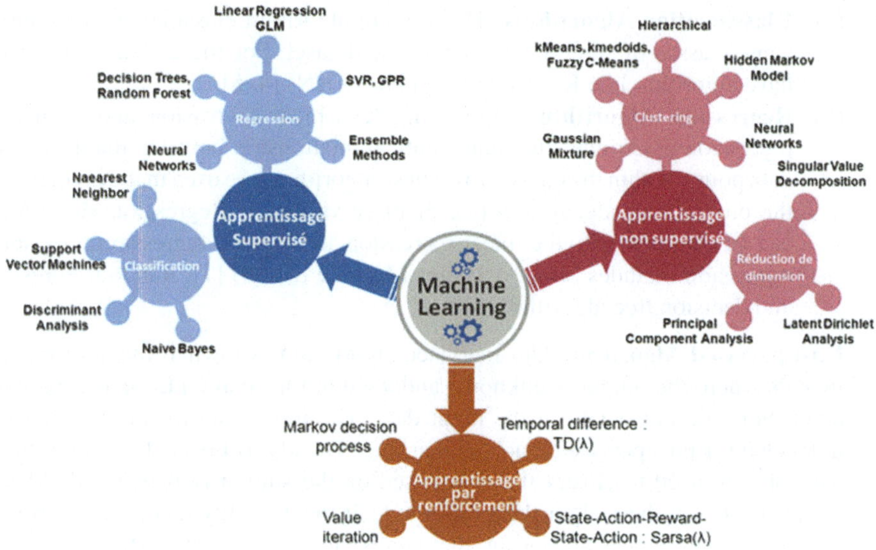

Fig. 1 Different machine learning classification and their algorithm

the identifying the relationships with in your target and the information which your dataset contains already has [17], e.g., if you have a dataset of records where the thousands of record are known and rest of the data is unknowns then the semi-supervised learning model will automatically classify that based on the behavior on the previous learn model and the algorithm will build the quick model on the basis of labeled data and applies it on the unlabeled data set in case of credit card fraud detection of bank loan approval task in the banking system, web page classification on the web, speech recognition and some in bioinformatics for genetic sequencing [18] (Fig. 1).

2 Image Processing Challenges

Image is a multidimensional signal, which requires certain operations for the extraction of useful information. Information extraction from the various image modalities are not straightforward and have certain issues or challenges associated with it. To list few, noise associated with the images, large amount of data for transferring images from one place to other digitally. Few image processing challenges in the computer vision and machine learning are as follows: Image compression, enhancement, Recognition, and visualization [19]. With the help of using machine learning lot of work is going to development of appropriate solutions using image processing algorithms. Here are some of the image processing challenges in which the machine leaning work is going on are as follows:

2.1 Image Compression

Modern information and communication technologies include transferring images from one place to another e.g., television signal digitization. Image compression enables the sending and receiving of image frames using the available bandwidth [19]. Compression deals with reducing the size of images in order to make them transferrable at a faster rate, depending on the application. Compressing an image is tricky and requires good precision so that compressed and decompressed image should represent the same information. Designing a compression algorithm is a two-step process consisting of encoding the information while compressing and decoding while decompressing.

2.2 Image Enhancement

Image enhancement is a technique of enhancing the quality of image. Every image acquisition system has a noise content associated with it. The noise makes it hard to acquire a clean image. Image with higher noise content makes it difficult to extract plateful information. Hence, image denoising is one of the predominant challenges in image processing and requires high level of expertise to deal with it. Machine learning algorithm enhances the quality of images, below there is discussion of some of the algorithm used for the image enhancements.

2.3 Image Recognition

Image recognition deals with the identification and classification of objects in the image. This is a hard problem for the computer systems to process as these systems recognize images by just their pixel intensity values. A typical image recognition system tries to identify the pattern by looking into the texture color shapes of the objects in an image and classify the objects. Building a generic image recognition system is difficult to identify large number of objects. Moreover, it is easy to figure out patterns if the objects are similar but relatively harder for alike objects with different color, texture, or shapes. Artificial intelligence algorithm finds out the patterns within the image and classify accordingly.

2.4 Visualization

It is one of the subdomains of Computer Graphics and deals with generation of models and objects. The challenging part of the visualization is to model the dynamic objects.

Dynamic objects are the objects that changes as a function of time like moving hairs, curtains, trees, etc. Secondly, the generated models should be genuine which is computationally expensive and requires high computational configuration.

3 Image Processing Task, Noise Removing/Denoising of Images

Before understanding the technique of denoising of image, we have an idea about the noise and what are its cause and how it effect in the digital images. Let us understand the concept of noise. In the digital image processing noise is a topic, which makes lot of confusion in the students, research scholars or any other person who is hearing this term noise. Just in case of sound noise, refer to the audible distortion, in case of image it is visual distortion in the image. For example: if you took photograph with a closed cap on the lens, the resulting photo isn't black totally it is approximate to black there must be some hinge, random brightness or some kind of distorted pixels in the image. In digital sensor noise cause by the sensors when attempting to capture tiny white light. In the real word scenario our camera is trying to focus on capturing the dark matter on the image as a result the electrical signal windup and as a fleck or noise [20]. Machine Learning algorithms are capable of removing the noise with in the image and helps in the denoising in the images. There are different type of noise found in the image while capturing the image and this can be removed with the help of using machine learning.

3.1 Types of Noise in Digital Image

In every image, there is always some kind of noise due to physical properties of noise, so technically we can say that every image contains some kind of noise.

- **Fixed Pattern Noise**: This sort of noise incorporates what is called as "hot pixels," which are characterized as such when a pixel's intensity far outperforms suppresses that of random noise arbitrary variances/fluctuations. Fixed pattern noise generally appears for the most part and shows up in long exposures and is exacerbated by higher temperatures. Fixed example clamor is special in that it will show nearly a similar circulation of hot pixels whenever taken under similar conditions (temperature, length of exposure, ISO speed). The fixed pattern noise is more offensive and is easier to remove from the captured image due to its repetitive nature. Camera internal electronic circuits has to just reveal the repetitive pattern of noise and just applying some algorithm subtract this noise reveal out the true color image [21].
- **Random Noise**: Random noise is also called shot noise/photon noise, which is categorized by intensity and color fluctuations above and below the actual

image intensity. The light composed of photons, which are discrete, random in nature. There will consistently be some random, irregular noise at any exposure introduction length and it is most impacted by ISO speed in camera. The random pattern noise commotion changes regardless of whether the exposure camera settings are similar or identical. The light emits and reflects everything you see the image. However, this thing is not fixed every time there is always graininess in the image. E.g. a very dim light bulb may emit on an average of 500 photons of light per second, but every time individual second will be a bit different—486 photons, 528 photons, 466 photons, 481 photons, 560 photons, and so on. On the off chance that, if you are taking a one-second long image of this light, you won't get the very same outcome each time. This is what photographers call "shot noise/random noise/photon noise" in an image. Random noise is less pragmatic than the fixed pattern noise in the latest imaging devices, but a slight amount of noise could create more errors than the fixed pattern noise and difficult to remove [22].

- **Bending Noise**: This type of noise is also called as the digital noise and it is depended on the digital image device, internal electronics of cameras in which image is being captured. This noise is also visible in the high ISO speed and shadows in an image, when an image is extremely brightened or white balance on an image depending upon the camera model [23].

Different type of noise reduction model in image processing are

Unwanted of unnecessary information within an image like as blurred objects, corners, unseen lines, edges, artifacts, disturbed background sense contribute the noise. This noise is produced form the CCD (Charged Coupled Device), CMOS (Complementary Metal Oxide Semi-Conductor) sensors with in the digital image devices. To reduce this redundant noise, prior dectection and learning of model needs to be develop. In some cases mathematical function like Modulation transfer function (MTF), point spreading function (PSF) are used for quantitative and qualitative analysis of mode of noise model [23].

4 Color Enhancement

For visual perception of human being color image provide the more and accurate information as compare to the grayscale images. Color enhancement is the key issue in the high-quality pictures and some other HDTV cameras. The quality of images is easily affected by the light and weather condition which leads to suffer from the loss of information. The are some color enhancement technique such as gamma correction technique, histogram equalization, contrast stretching, Contrast-limited adaptive histogram equalization which are older one and give the mean performance in the Root Mean Square deviation (RMSE), Peak signal to noise ratio (PSNR) and mean absolute error. The newly developed algorithms are used with the machine learning model give us the better accuracy for the color enhancement. These are the algorithm such as Retinex, Homomorphic and Wavelet Multiscale technique are popular for the image enhancement [24].

With the help of machine learning color enhancement is done in various domains like underwater animal detection and classification [25], Iris Recognition [26], Smart laser surgery [27], Brain image enhancement [28, 29], Ultrasound image enhancement [30] etc.

Colored image provides and hold a vast amount of information regarding object in image. It is difficult for the human eye to recognize and analyze each and every little potion of image such as color intensity and texture with in an image [31]. In colored images we need an algorithm which segments out all the parts with in an image and gives us the better results especially in case of the satellite image because they contain lot of noise with It and it is a trivial task of computer. Color enhancement is more difficult in case of color image rather than the gray scale image due to the following reasons.

1. In case of color we need to consider vector instead of scalar image due to presence of color.
2. Complexity of image is also a problem in case of color image.

Color or contrast enhancement is another technique in which the intensity of pixel is changes to maximize or enhance the density of pixel with in image. The term "contrast" used for the separation of the dark and bright areas within an image. Due to contrast enhancement we are able to remove the dark areas from the image and we are able understand the content within an image [32]:

$$s = (r - c)\left(\frac{b - a}{d - c}\right) + a$$

Mathematical equation: a and b are the upper limits for the 8-bit gray scale picture, c and d are lower and upper value of histogram.

Image Enhancement Techniques: This provide the information for human to understand the image by interpretation and perception of viewer against image via providing better input for automated image processing techniques. Image enhancement also classified into the two type of category:

- **Spatial and Frequency Domain**: In the special domain we directly deal with the image pixels. People do the manipulation with the pixels with for the enhancement in the image. And in case of the frequency domain we change the frequency of pixels and change the image into the frequency. So, this states that we have need to apply the Fourier transform within the image computed first and after that we have to apply the inverse Fourier transform in order the get the resultant image [33].
- **Histogram Equalization**: is a technique which normalize the contrast with in the image by using the histogram equalization. It is based on the idea of near-uniform

probability distribution function. It redistributes the probability distribution function within an image. For example, if in an image there are peaks and valley, after applying the histogram it will normalize the peaks and valleys will be shifted to normal curve. This results in improving the image contrast. There are types of the Histogram equalization namely Global Histogram Equalization (GHE), Adaptive Histogram Equalization (AHE), Block based Histogram Equalization (BHE) [34, 35]. In GHE each pixel with in an image assigned a new value based on the cumulative distribution function. To perform Global Histogram Equalization (GHE), the first original histogram of the grayscale picture should be adjusted or equalized. The aggregate/cumulative histogram from the input picture should be evened or equalized out to 255 by making a new force an incentive by applying [34].

$$s = T(r) = \int p_r(w)\mathrm{d}w$$

$\Pr(r)$ = is probability distribution function, r = gray level of input image

$$p_s(s) = \begin{cases} 1 \text{ for } 0 \le s \le 1 \\ 0 \quad \text{otherwise} \end{cases}$$

$$S_k = T(r_k) = \sum_{j=0}^{K} \frac{n_j}{N^2} = \sum_{j=0}^{k} P_s(r_j), \quad 0 \le r_k \le 1, k = 0, 1, \ldots, L-1$$

n_j = count of jth pixel, N is the no of pixels in horizontal and vertical direction in an image.

- **Homomorphic Filter**: Homomorphic filter work in the principle of normalization of brightness and increase the contrast across the image. This will help the user to see the dark area within the image. For the removal of multiplicative noise, Illumination and reflectance we use the homomorphic filter. Since illumination and reflectance are combined multiplicatively, the components that are made added substance by taking the logarithm of the picture intensity, so that therein frequency domain linearity is formed. Illumination or brightening varieties can be thought of as a multiplicative noise and can be decreased by separating in the log domain area.
- **Retinex**: In 1964 this theory was first proposed by Edwin Land. Multi-Scale Retinex (MSR) and Multi-Scale Retinex with modified color restoration are the other technique in the Retinex [36].

$$L = R/E$$

L is the incident of light, R is the object reflection, E is the reflected light.

There is another technique called the single-scale Retinex method proposed by the Jobson, in which estimate the illumination by using the low pass filter for an input color image.

Wavelet Multi-scale Transform: In recent year, wavelet transform becomes a famous tool for the image processing. Wavelet transform (WT) is a powerful mathematical tool for the processing of images of multiple resolutions. It provides a septal and frequency characteristic with in an image. WT transform for an image the high-frequency part stores the original image details and the low frequency or imaginary part is stored with in the low-frequency part in an image [37]. The imaginary and low frequency part determine the dynamic image with in the image, but in case of the low frequency of the annotation there is the loss of information, so that why some of the information is stored in the high-frequency domain that is why the image reconstructed by the Inverse Wavelet Transform has the more details and widely used for the image processing [37, 38].

$$\psi(t) = 2\sin c(2t) - \sin c(t) = \frac{\sin(2\pi t) - \sin(\pi t)}{\pi t}$$

Equation for continuous wavelet transform

$$\psi_{m,n}(t) = \frac{1}{\sqrt{a^m}}\psi\left(\frac{t - nb}{a^m}\right)$$

Discreate wavelet Transform equation

5 Sharpening of Images in Machine Learning

Image sharpening belongs to the category of spatial filtering in the image processing deals with the enhancement of detailed information within the images [39]. Sharpening of the images increases the brightness within the image. This could be done via using the high pass filter within the image. On the spatial regions there is small image features and image features correspond to the detailed information about high-frequency components of an image. Image sharpening is dependent on the two factors which are resolution and acutance. The resolution of the image deals with the no of pixel within the image more the pixel in the image more will celerity or sharpness within the image so this doesn't concern with the subjective nature of image. Acutance is little bit more complicated to understand and is linked with the subjective measure of image dealing with the edge with in the image. Edge on the image has more contrast appear more accurately in the image and human visual system. So, to increase the acutance within the image there is only one way to increase the acutance within the image at the edge contrast [40, 41].

Technique developed by Google like RAISER: Rapid and Accurate Image Super-Resolution "use machine learning to produce high quality of images from low resolution of images from 10 to 100 times faster than the currently available methods". SISR (Single Image Super Resolution) is the technique which converts the low-resolution images into the high resolution [42]. This method is based in the upscaling the image in the linear interpolator including the nearest neighbor in the image [43, 44] (Figs. 2, 3 and 4).

Binary reconstruction of gray scale and nonlinear processing images: Grayscale images contain only black and with color. It permits the removal of small regions that are disjoint forms the larger objects without distorting the small features within the large objects. Neural networks have been proposed and a variety of works in binary reconstruction of grayscale images [45, 46]. This technique is mainly used in X-ray images and MRI images and helps in the detection of cancer. The interpretation of these gray scale images is difficult task and depends on the expertise and experienced radiologist. Thus, extracted features were classified with the help of Artificial Neural Network. In medical imaging diagnostic the detection of presence

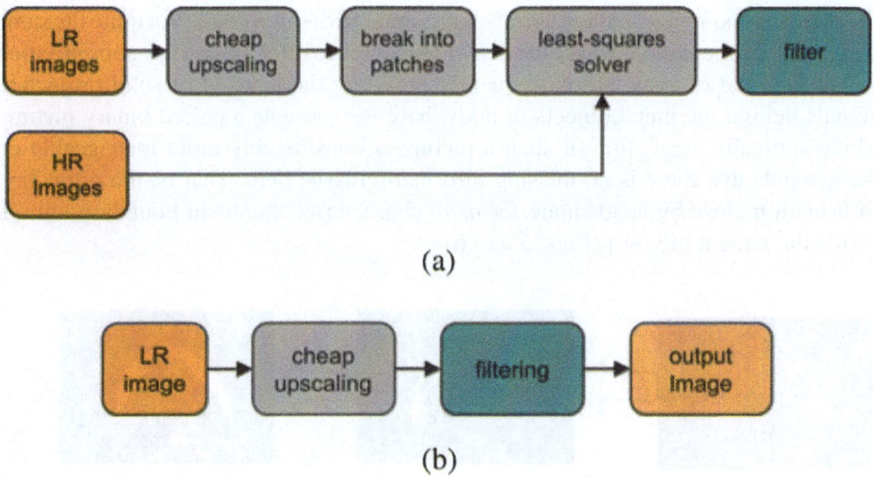

Fig. 2 Basic learning with global filters, low resolution to high resolution. **a** learning stage, **b** upscaling stage

$$\begin{bmatrix} -1/9 & -1/9 & -1/9 \\ -1/9 & 1 & -1/9 \\ -1/9 & -1/9 & -1/9 \end{bmatrix}$$

Fig. 3 Array of kernel in high pass filter for sharpening of image

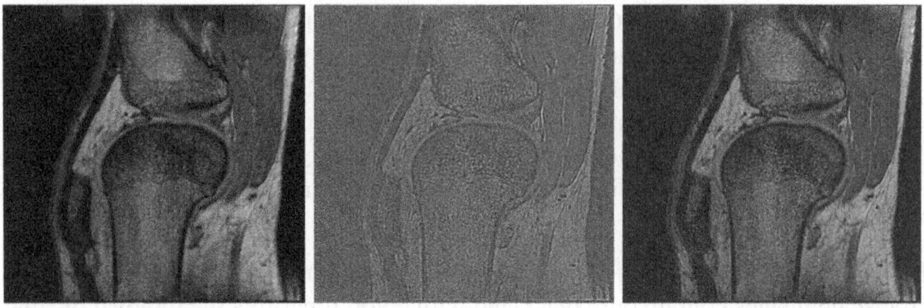

Fig. 4 Sharpening of MRI (Magnetic resonance image) images. **a** MRI image of knee, **b** knee MRI with sharpened filter sharpened knee MRI [44]

of metal within the body which makes it important to diagnose of detect the metal object with in the body with the help of x-ray image or MRI CT images [47]. The CT images are classified into two categories one is high density metal and low-density tissues with in the human body. Within the sight of a metal object, linear attenuation coefficient at the limit of the metal changes out of suddenly, and the identification of the metal boundary is identical to transfer a signal with sharp edge. Not quite the same as regular CT recreation, where the center focuses is to infer the density appropriation inside the field of view and is frequently tested by the missing of solid projection signals behind the metal objects in body, here we estimate a paired binary picture. Mathematically, recreation of such a picture is considerably more manageable on the grounds that there is no missing information issue here. That is, the projection information close by is adequate for us to characterize the metal boundary with in inside the patient [48, 49] (Figs. 5 and 6).

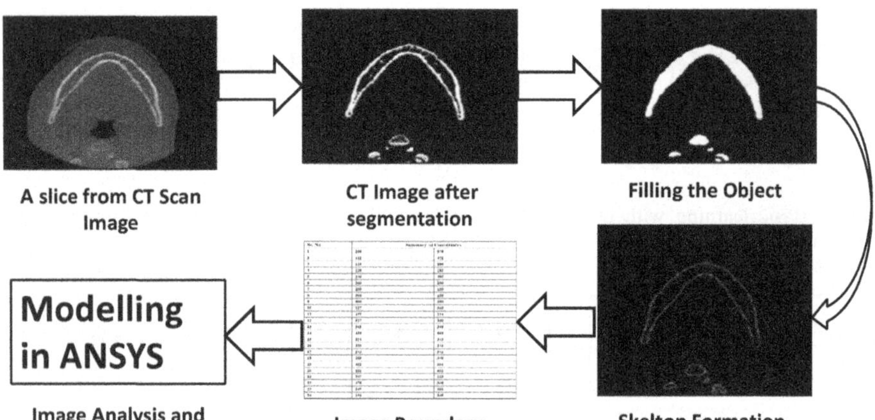

Fig. 5 Binary image reconstruction

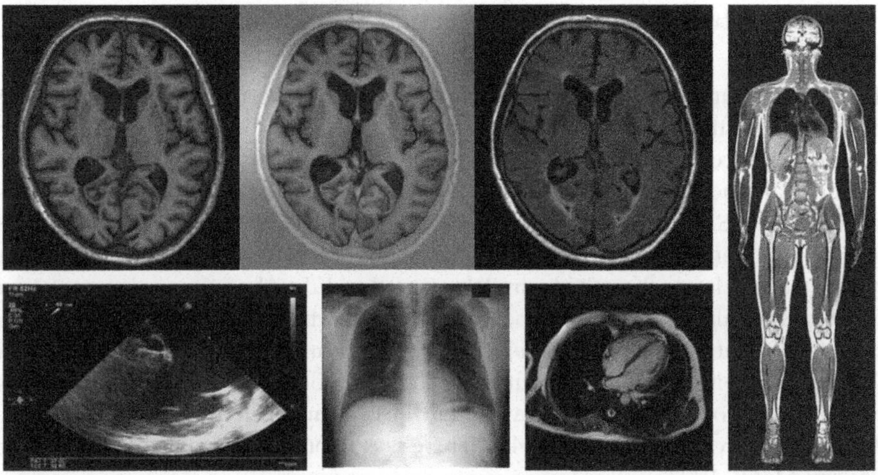

Fig. 6 Reconstruction of gray-scale images have some advantage as well as the disadvantage also

Advantages

- Easy to acquire: straightforward computerized digital cameras can be utilized along with basic frame stores, or minimal low-cost scanners or thresholding might be applied to gray level pictures or images.
- Low storage: close to 1 bit/pixel, frequently this can be diminished as such pictures are entirely amiable to compression in the image (for example run-length coding).
- Simple handling and processing: the calculations by the algorithm are by and large a lot more straightforward than those applied to gray level pictures/images obtained through cameras.

Disadvantages

- Limited application: as the representation is just an outline silhouette, application is restricted to undertakings where inside detail isn't required as a distinctive characteristic.
- Does not stretch out to 3D: The 3D idea of items can once in a while be represented to by outlines silhouettes (What could be compared to binary processing preparing utilizes voxels, spatial inhabitance occupancy of little small cubes in 3D space).
- Specialized lighting is required for outlines: it is hard to get solid, reliable binary image without limiting, restricting the environment. The simplest example among is model is an overhead projector or light box.

6 Segmentation and Edge Detection Using Neural Network

The development of technology in our society we can expect that the image sensing device detects the things around you. Image segmentation is an important step for any image processing and computer vision algorithm. Analyzing and extracting and object form and image is an important for building intelligent machines for our surroundings. Segmentation will help in the retrieval of images form a large-scale database for content-based image retrieval system software. In a region, a group of connected pixels are there with similar properties. Edge detection is segmentation by finding similar pixels on their boundary. This can be termed as the binary image classification at pixel level. To solve this problem use of fixed and adaptive and feature selection in concurrence with the use of Support vector machine [49]. In this case the first work was done by David Marr in 1970 and then first success full work done in the David Lowe in 1987 [49]. There was improvement in the Algorithm by Shi and Malik, 1997.

The image segmentation is the process of partitioning the image into different segments. This is done through the different patterns based on the pixel's similarity within the image. Segmentation is the way toward recognizing, distinguishing objects in the dataset from their environmental surroundings factors in order to encourage the creation of geometric models. For instance, in clinical biomedical imaging it is regularly important to measure the shape, surface area, or volume of tissues in the body. Once the dataset is segmented, those quantities are easily measured. The image segmentation is needed to identify the content within the image. Image segmentation is used to classify the data within the image. Image segmentation is used in almost every field ranging from the automated driving, medical cancer diagnosis, Handwritten digit recognition, underwater object classification/detection, food industries, etc. It has a vital role and more important among other problems in image processing. Due to introduction of machine learning algorithm computer researchers developed new algorithm to automated segmentation and edge detection within the image. Algorithm like neural network and other clustering algorithm will classify the segmented part within the image. Hence there lot of work is needed to be done for image segmentation and edge detection.

The objective of the edge detection is to find the pixels in the picture that relate to the edges of the object found in the picture captured through the cameras. This is normally finished with a first and second derivative estimation following by a test which denotes the pixel as either having a belong to the place with which an edge is there or not. The outcome is a binary image which contains just the recognized or detected object edge pixels. Edge detection is the fundamental tool for the image segmentation. There are several tools and software that are used for the image segmentation and for the edge detection also. For the image segmentation, every method is not appropriate so there are several techniques which are used for the image segmentation are as follows [50].

- Region-based Segmentation
- Edge-based Segmentation

- Threshold
- Feature-based Clustering

- **Region-Based Segmentation**: It is based on the process that the group of the pixels are based on the similar properties. Partition of the image takes place based on similar region. Region in an image may be consist of the object or some entity within the image. This type of classification is suitable for the noisy image where the border with the object are difficult to detect or border are blur. Fair level of accuracy is achieved in the care of region-based segmentation [51].
- **Edge-Based Segmentation**: Segmentation within the image based on the discontinuous similar intensity values in an image. Edge detection can be found where there are abrupt changes within the brightness and the intensity in the pixel and a problem in the primary value in the image analysis. Hence by detection the edge with in the image we do the image segmentation. Edge-based Segmentation is further classified into sub-category: Filtering and Enhancement, Detection of edge points, and edge localization [51].
- **Filtering and Enhancement**: So as to encourage the identification of edges, it is fundamental to repress as much noise as could reasonably possible be expected and decide changes in power in the area of a point, without destroying the true edges.
- **Detection of Edge Points**: Figure out which edge pixels ought to be disposed as of noise and which ought to be retained (as a rule, thresholding gives the criterion used for the identification or detection).
- **Edge Localization**: Not all of the points in an image are edges for a particular application. Edge localization determines the exact location of an edge. Edge thinning and linking are usually required in this step.
- **Threshold**: Thresholding is simple and powerful tool for image segmentation. This is beneficial where the light object in the shady background. It will convert the multi-level image into the grayscale image. The separation of the object forms the background is done by selecting the values threshold value of T. Local thresholding and Global thresholding. When the T is not constant it is called as the local thresholding and when T is constant global thresholding.
- **Feature-based Clustering**: Clustering the process of grouping the individual together who have similar properties and dissimilar properties into another cluster [51]. Good quality clustering produces high intracluster and low inter clustering properties. Issues arise in the feature-based clustering while dealing with the image are as follows:

  - How to represent the image.
  - How to organize the data from the image.
  - How to classify the image into a certain cluster (Table 1)

  Clustering is further classified into the Fuzzy C-means clustering and K-means clustering. K-means is the fast, robust, and simple unsupervised learning method for the clustering. Fuzzy clustering allows the object which form different cluster with membership. It is an effective method for the pattern recognition.

Table 1 Comparison between the different algorithm used for the segmenting and edge detection

Parameter	Threshold-based method	Region-based method	Edge-based segmentation	Cluster-based method	Fuzzy C-means method
Spatial information	Ignored	Considered	Ignored	Considered	Considered
Region continuity	Fast	Good	Reasonable	Reasonable	Good
Speed	Reasonable	Slow	Moderate	Fast	Moderate
Computation complexity	Less	Rapid	Moderate	Rapid	Moderate
Automaticity	Semiauto	Semiauto	Inactive	Automatic	Automatic
Noise resistance	Less	Less	Less	Moderate	Moderate
Multiple object detection	Poor	Fair	Poor	Fast	Fair
Accuracy	Moderate	Fine	Moderate	Moderate	Moderate

Automatic Edge Detection Using Deep Learning:
Edge detection, identification is an image processing method used to find and distinguish the sharp discontinuities with in a picture captured from camera. Edge detection is as often as possible the initial phase in recouping data of information from pictures. One of the eldest classical examining studies in computer vision is image processing in which further most of research work in done in edge detection or identification. There are many edge detection operations are available, which recognizes vertical, level, horizontal, corner, and step edges in an image. The quality, nature of edges identified by these operators is profoundly subject to, noise, clamor, lighting conditions, objects of similar intensities and the density and the thickness of edges in the scene [52]. There are two alternatives which are used for the detection of edges in the image are as follows.

1. Classification of all the pixels that satisfy the criteria of homogeneous.
2. Detection of all the pixels that are on the border in different homogenous areas (Fig. 7).

Instead of using the classical edge detection method neural network or deep convolutional neural network are used more widely for the classification of this purpose (Fig. 8).

$$L = \frac{1}{N_p}\sum_{i=1}^{N_p} - \log\left(\frac{e^{f_{i,y_i}}}{\sum_j e^{f_{i,j}}}\right) + \lambda \sum_{j,k} W_{jk}^2$$

Mathematical equation for the Edge Detection

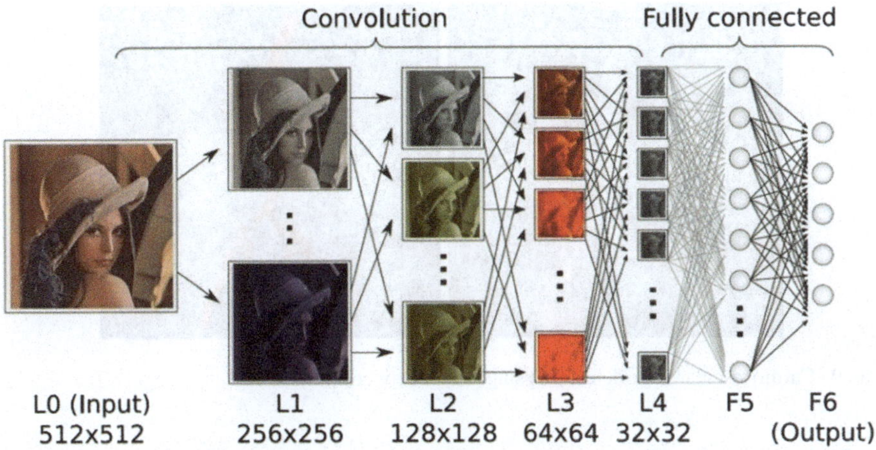

Fig. 7 Convolutional neural network for image classification [52]

Fig. 8 Image shows the changes of the edge detected output image of the proposed technique; it is obvious that the best

Cardio vascular edge detection using the neural networks to perform the flow of blood flow simulation. In this method is used to identify the blood vessels in the entire region of image for image segmentation. This method used a fully connected convolutional neural network for classification of image segmentation of cardiovascular disease within the lung nodules [53, 54] (Fig. 9).

$$L = \frac{1}{N_p} \sum_{i=1}^{N_p} - \log\left(\frac{e^{f_{i,y_i}}}{\sum_j e^{f_{i,j}}}\right) + \lambda \sum_{j,k} W_{jk}^2$$

Mathematical equation for the loss functions in the deep neural network

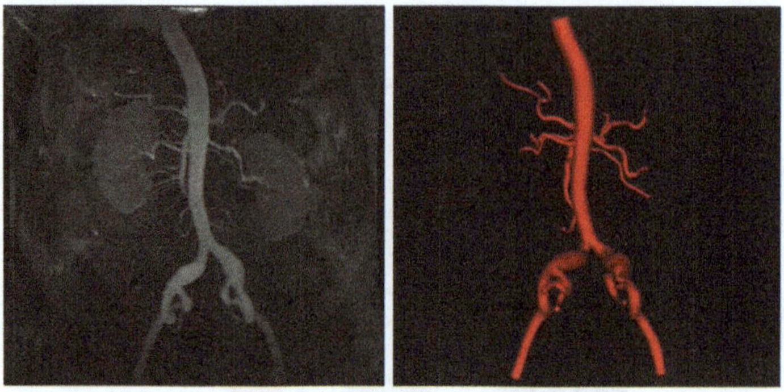

Fig. 9 Cardiovascular edge detection using neural network [55]

Brain tumor classification with the help of deep learning is possible. Convolutional neural network is used for the training the model on the brain tumor data set, testing is performed on the unlabeled data set and the accuracy is measured. Automated segmentation and detection is also possible with the help of these Deep learning based algorithms [56].

7 Image Compression

Image compression is the method of minimizing the size of the image by minimizing the sizes in bytes of graphics file with reducing the quality of the image. This will create more space and allow storing more images in an available memory size. Today on the internet ninety percent of the data is in the form of image and videos. Thus, most of the companies will use the data compression technique in order to run out of space. Machine Learning algorithms like support vector machine combines with the DCT discrete cosine transform achieve the level as compared to the RBF and multilayer preceptor [54, 57]. DCT is based on the JPEG compression algorithm one of the most widely used image compression algorithms. In the resulting value of the matrix it divides the whole image into pixel of squares and apply a discrete cosine function [57].

$$F(u, v) = \left(\frac{2}{N}\right)^{\frac{1}{2}} \left(\frac{2}{m}\right)^{\frac{1}{2}} \sum_{i=0}^{N-1} \sum_{j=0}^{M-1} \left(\Lambda(i)\Lambda(j) \cos\left[\frac{\pi u}{2N}(2i+1)\right] \cos\left[\frac{\pi v}{2M}(2j+1)\right]\right)$$

Mathematical representation of Fourier transforms for discrete solution

Image compression with the help of machine learning:

1. **Neural Network**
2. **Genetic Algorithm**

- **Neural Network**: NN generates the rules to adjust and compute the complexity of image compression. They are generally composed of an input layer, hidden neurons, and an output layer. There may be one or more layer of hidden neurons. The hidden layer networks are logic functions that, based on input from one or more neurons in preceding layers will output information to one or more neurons in subsequent layers [58, 59]. Each of the connections between the neurons is designed to weight the input into the next neuron, hence if, for example, neuron A on input layer, outputs the value 2.46 to neuron B on hidden layer, there may be a weight, or bias, between both that doubles up the value, and consequently neuron B receives 4.92 as an input. In this case, the weight for the connection (synapses) was 2, doubling up the value, but this weight can be modified by giving a network training data. If a Neural Network is provided with a sufficiently large training data set, eventually the value of the weights in the synapses will be such that the results obtained from a neural network computation will be close to the expected results. While applying neural network method in image we can reconstruct and compress the image from the following provided image data set on which machine learning model is trained [60]. Fast data transmission in industrial scale on IoT devices is possible due to machine learning algorithm for the image compression [61]. In medical science for the detection of diabetic retinopathy is possible because of the image compression and machine learning image processing plays a vital role in the development of better treatment for diabetic patients [62].
- **Genetic Algorithm**: Genetic algorithms (GA's) are based on the processes of evolution and natural selection. They are a subset of evolutionary algorithms. Genetic algorithms take an iterative approach to problem-solving, generally following these steps [63, 64].
  - **Initialization**: A population of algorithms is generated (randomly or semirandomly) to solve a particular problem.
    E.g. $x = a + b$, $x = 2b - 3c$, and $x = a + b - c$ for deriving the quadratic equation.
  - **Selection**: The algorithms to be combined are selected, based on the proceeding of fitness if it is not the first iteration.
  - **Crossover**: Parent algorithms are combined to produce new children algorithms, the operation used to combine them is a factor to be determined by the supervisor, and may have a significant effect on the computational efficiency of the process.
    E.g. $x = 2b - 3c$, and $x = a + b - c$ are added to produce $2x = a + 3b - 4c$
  - **Mutation**: A random number or factor is added to the algorithm, mimicking the process of mutation in nature. For our example, this could be $2x = a + 3b - 4c + 8$. This will generate new, potentially unique, algorithms for solving the particular problem

- **Proceeding of Fitness**: The generated algorithms are tested against a training set, determining the fitness of the solutions. The solutions that are inaccurate above a certain threshold are dropped, and the rest are iterated through the process again, going to step 2: selection This process is repeated until a maximum number of iterations are taken or the solution is close enough to the training set.

Combining the machine learning algorithm and compression algorithm:

- GA and DCT
- Neural Network and DCT

Genetic Algorithm and Discrete Cosine Transform

Genetic Algorithm is one of the most common used optimization algorithms; so in image compression, this could be used for the optimizing the fractional compression and DCT (Discrete Cosine Transform). In DCT for optimal compression genetic algorithm could be a better solution for the optimization to improve the quantization with in the image.GA also applied in the fractional image compression. In fractal compression are searching the domain blocks and transforming the domain block to match the range blocks, it is these processes that are optimized by the genetic algorithm from Mtira and Morthy (1998) [64].

Neural Network and DCT

Neural networks have been used to improve image compression extensively. Both as standalone methods as well as combined with other techniques, traditional or more novel approachs.

Setiono and Lu used a standalone neural network in order to compress images, using the image as input data for the network, and the reconstructed (compressed) image as the output. The images used where black and white, and the compression ratio achieved was about 10, producing good quality, albeit noticeably lossy, images. Figure shows Lena, uncompressed on the left, and compressed with the neural network on the right. The loss of information is particularly noticeable around the hair and the band on the hat, due to this being high-frequency areas [64].

Parodi and Passaggio approached the compression problem by combining DCT and Neural Networks. The approach they take to first divide the image into low and high activity area blocks, which will be compressed at different ratios and by different neural networks. The high and low activity areas are classified as such by using DCT's; high activity areas are those with higher coefficients within the DCT transform, indicating that they contain a wider range of frequencies. Each of this image is posteriorly provided to the corresponding neural network, either high or low capacity, with a number of input neurons proportional to the size of the block to be processed. The images are then compressed by the neural network, which uses the original block as a target for the learning process. The results obtained indicate a significant improvement in image quality over standard neural network compression of images, without compromising compression ratio, they achieve higher PSNR and lower MSE.

8 Facial Expression Analysis

In recent years there is growing interest in emotional intelligence in the human-computer interaction. For the facial expressional analysis there is need for the compute there is need for the computer to interact naturally with the user via speech, body gestures to display emotions. One of the important way humans display emotions is through facial expressions via vocal, visual and other physiological means like Neutral, anger, disgust, fear, joy, sadness, surprise. Support vector machine and neural network can classify the facial expression with a speed and accuracy. Data set Cohn and Kanade's DFAT-504 have images of 100 of university students age ranging from 18 to 30 years [65] (Fig. 10).

- **Face Finding**—FaceReader finds a precise position of the face via utilizing the well-known Viola-Jones calculation.
- **Modeling**—A precise 3D modeling demonstrating of the face is made, which depicts more than 500 key points with in the face
- **Classification**— A neural network system utilizes more than 10,000 pictures to arrange the basic emotional expressions, passionate articulations, and various properties of facial expression
- **Deep face classification**—This procedure permits FaceReader software to straightforwardly order the face from picture pixels utilizing an artificial neural network system to perceive patterns with in the image. This permits the product to break down the face regardless of whether a part of it is covered up with (Fig. 11).

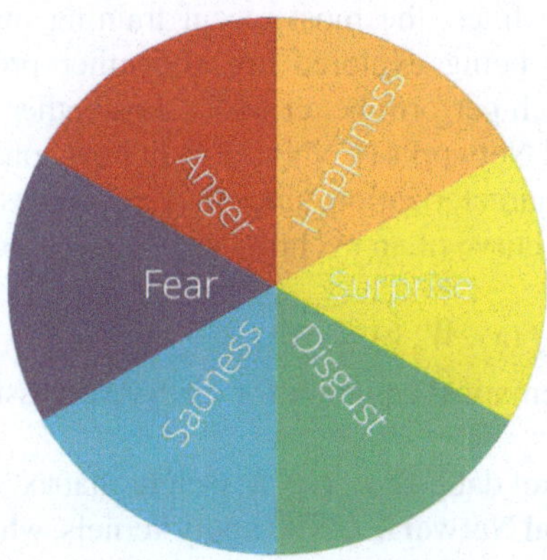

Fig. 10 The pi chart indicates the emotions of human facial expression analysis software works by following these consecutive steps

Fig. 11 Image showing the different facial expressions for classification

Deep learning algorithm for the face expression analysis:

- Multiclass Support Vector Machines (SVM)
- Convolutional Neural Networks (CNN)
- Recurrent Neural Networks (RNN)
- Convolutional Long Short-Term Memory (ConvLSTM)

- **Multiclass Support Vector Machines (SVM)**: Are supervised machine learning algorithms that investigate, analyze, group information, and classify the data, and they perform well while predicting, classifying human facial expressions. In any case, they possibly do so when the pictures are made in a controlled lab setting with steady head postures, brightening and a constant illumination [66].
- **Support Vector Machine:** SVMs perform less well when characterizing pictures are captured "in the wild," or in unconstrained, unspontaneous, uncontrolled settings. Along these lines, the most recent training machine leering or deep learning architecture being explored are altogether profound as deep neural network systems which perform better under those other available architectures. Convolutional Neural Networks (CNN) are right now considered the go-to neural systems for picture characterization, since they get on designs small patterns of a picture, for example: curve of an eyebrow or lips style, nose style etc.

$$f(x_i, W, b) = Wx_i + b$$

Mathematical equation for the SVM classification

where x_i = image pixel data, $W = C * K$ weight matrix, b = biased term.

- **Convolutional Neural Network**: CNNs apply kernels, which are matrices smaller than the image, to chunks of the input image [67]. By applying kernels to inputs, new initiation lattices, at times referred to as feature maps, are created and gone as a contribution, pass the input to the next layer of the network. Along these lines, CNNs process increasingly granular elements inside a picture and make a better via classification between the two similar at recognizing two comparable emotion classification. VGG16 is the state-of-the-art architecture for the CNN network with 16 fully connected layers, with 3*3 convolutional and 2*2 pooling layers [68, 69] (Fig. 12).

- **ResNet50**: It is one the most popular CNN architecture for the face recognition. It has the additional mapping capacity as compare to the VGG16 networks [71] (Fig. 13).
- **Recurrent Neural Networks (RNN)**: Utilize dynamic transient behavior while the classification of an image. This implies when an RNN forms an input information model, it doesn't simply take a data at the information from that example model. It additionally takes a data from past sources or layer with in the training

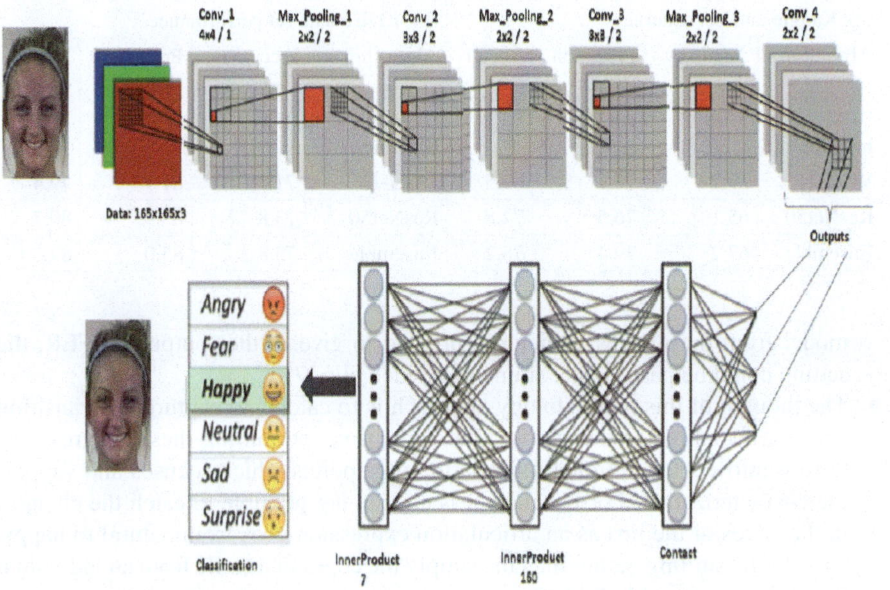

Fig. 12 CNN architecture for facial expression analysis [70]

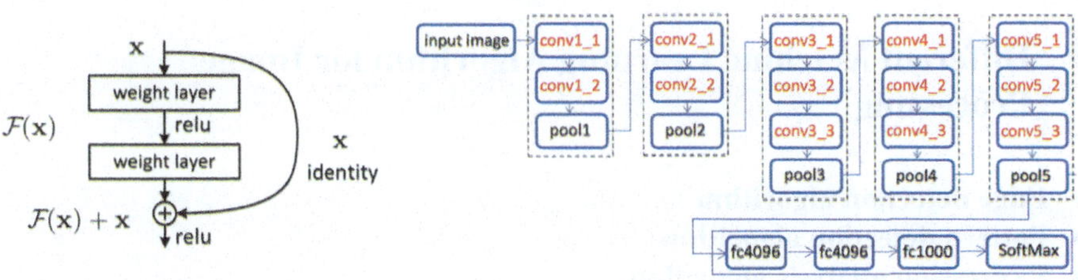

Fig. 13 ResNet block diagram with additional mapping identity

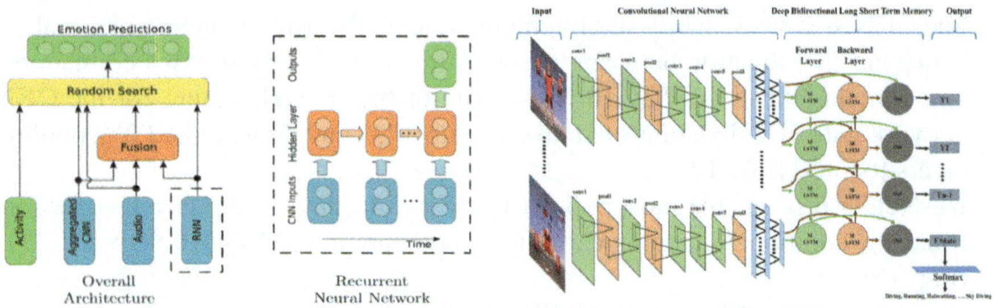

Fig. 14 VGG network for face classification [99100]

Table 2 **a** Kaggle dataset performance [73], **b** Showing the KDEF data set performance of different learning model [73]

(a) Kaggle data set accuracy				(b) KDEF data set performance			
Algorithm	Accuracy (%)	Precision (%)	Recall (%)	Algorithm	Accuracy (%)	Precision (%)	Recall (%)
SVM baseline	31.8	43.7	54.2	SVM baseline	37.9	50.1	54.9
VGG-16	59.2	70.1	69.5	VGG-16	71.4	81.9	79.4
ResNet50	65.1	76.5	74.8	ResNet50	73.8	83.3	80.7
Ensemble	67.2	79.4	78.2	Ensemble	75.8	85.0	82.3

model from the images, which are utilized to give further output. In FER, the setting could be past picture frame of a video clip [72].

- The thought of this methodology approach is to catch and capture the transition between facial patterns example after some time, permitting these progressions to turn into extra additional information data points which focuses and support the characterization. For instance, it is conceiving possible to catch the changes in the edges of the lips as an articulation expression goes from natural to happy, cheerful by smiling, rather than the simply the edges of a smile from an individual image frame (Fig. 14; Table 2).

9 Different Machine Learning Algorithm for Image Processing

- **Edge detection algorithm**
- **Pattern detection algorithm**
- **Expression analysis algorithm**

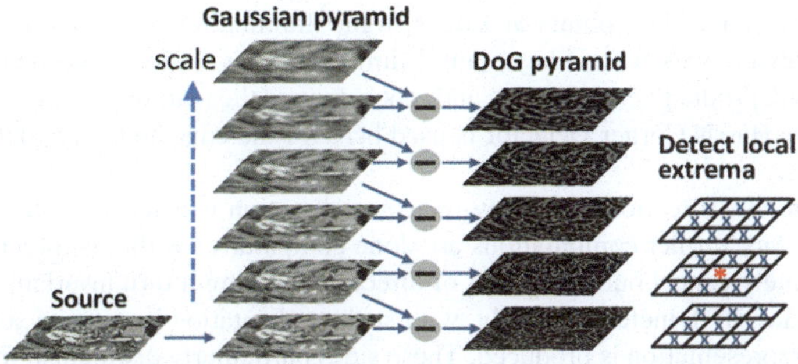

Fig. 15 Edge detection algorithm Gaussian, DoG pyramid [080]

- **Edge Detection Algorithm**
  The point at which the image brightness changes sharply and organized into the lines or curves are termed as the edges. Edge detection is the fundamental tool in the image processing especially in the computer vision and particularly playing the role in the feature extraction. Feature detector such as Scale Invariant Feature Transform SIFT (DoG), Harris, and SUSAN are one of the good methods which yield high quality of features but due to too much computationally intensive nature for the use in real-time application of any complexity [74]. Canny edge detector is an edge detection technique which uses a multi-stage algorithm which is developed by the John F. Canny in 1986 to detect a wide range of edges in an image. Machine learning can be used to derive a feature detector which can process the PAL video less than seven percent of processing time. ML can compare with Harris detector (120%) and SIFT (300%) faster at full frame rate [75, 76] (Fig. 15).

Working of SIFT algorithm: SIFT algorithm is widely used and can be divided into the following parts as follows [77]:

1. Constructing a scale space This is the underlying initial arrangement. You can create internal representation of the original image to make sure that there is scale invariance. This is finished by creating a “scale-space”.
2. LoG Approximation: The Laplacian of Gaussian is incredible for finding intriguing interesting points (or key points) in a picture image. Be that as it may, it’s computationally expensive. In this way, we cheat and surmised approximate it using the representation created earlier.
3. Finding key points with the super-fast estimate approximation, we currently attempt to discover out the key points. These are maxima and minima in the Difference of Gaussian picture we have to figure out.

4. Get rid of bad key points at Edges, light illumination and contrast and low complexity areas are bad key point. Eliminating these points make an algorithm efficient, productive, and strong and robust toward the feature learning. A method like the Harris Corner Detector is used here for the edge and curve detection in objects.
5. Assigning a direction to the key focuses a direction is determined for each key point. Any further computations are done comparative with this direction. This adequately cancel out the impact of direction, making it turn invariant.
6. Generate SIFT includes Finally, with scale and rotation invariance set up, one more representation is produced. These sides particularly distinguish features of the objects in an image. The algorithm will be implemented in the Open Cv.

- **Pattern Detection Algorithm**
  Repetitive sequence within the image called pattern. Pattern in the images can be either seen physically or can be found with the help of mathematics by applying some technique. In case or colored image, there should be more than one pattern in the image. Pattern detection is the process of classification of image data based on the supervised learning or statically information extracts out form the unlabeled image data. Pattern reorganization has done some great work in the speech recognition, Speaker identification, multimedia document recognition (MDR), and medical image analysis and could be used in the biometric identification and Fraud detection also. It could be done via supervised learning (Classification, Regression), Clustering, ANN, and reinforcement learning. In the literature, so many algorithms are available for the classification like support vector machine for multi-class pattern recognition, ANN, CNN, etc (Fig. 16).

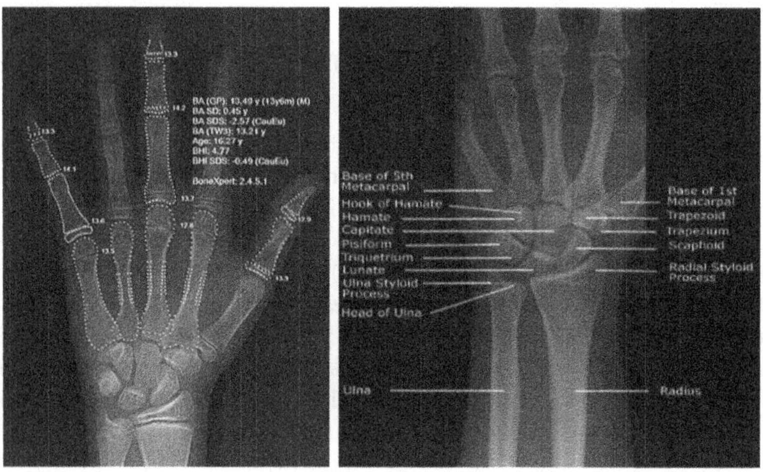

Fig. 16 Pattern detection in medical image analysis

- **Expression Analysis Algorithm**
  Face expression technology could be used in the variety of applications such as access control and surveillance and identity authentication. Human being uses the facial expression to interact with the socially. Since the rise of artificial intelligence proceeded with research interests for empowering computer system frameworks to perceive the expressions and to utilize the emotive data inserted in them in human–machine interfaces. Research activity in face recognition have been increased in the past couple of years. SVM is recently proposed and widely used algorithm for the facial expression analysis. The Support Vector Machine (SVM) classification trains, identifies the different displacement patterns features form unknown sample images and based on the learning experiences it classifies the different facial expressions like anger, fear, joy, sorrow provided during the training period (Fig. 17; Table 3).

RAISAR Sharp Image with Machine Learning:
Technique developed by the Google team that incorporates the machine learning in order to produce the high-resolution images from the low scale image. Its 10–100 times better than the currently available algorithm and can be run on the mobile devices [78] (Fig. 18).

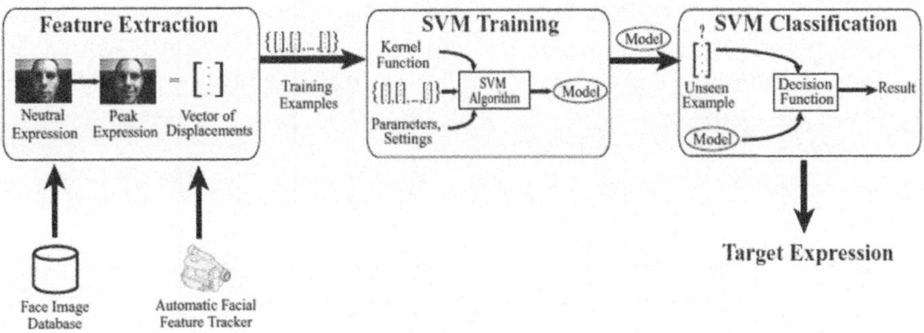

Fig. 17 Facial expression analysis pipeline using support vector machine SVM

Table 3 Showing the classification accuracy of different emotions and total accuracy of 87.9%

S. No.	Emotions	Anger (%)	Disgust (%)	Fear (%)	Joy (%)	Sorrow (%)	Surprise (%)	Overall (%)
1	Anger	–	72.5	61.8	97.1	93.8	97.1	84.1
2	Disgust	72.5	–	63.2	94.9	88.9	100.0	83.9
3	Fear	61.8	63.2	–	90.9	66.7	100.0	76.9
4	Joy	97.1	94.9	90.9	–	96.8	97.1	95.3
5	Sorrow	93.8	88.9	66.7	96.8	–	100.0	89.4
6	Surprise	97.1	100.0	100.0	100.0	100.0	–	98.8

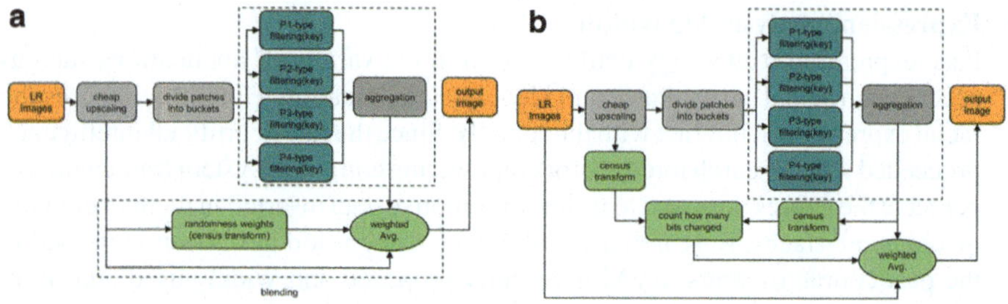

Fig. 18 **a** Image showing pre-learned filters from LR to HR. **b** Figure showing LR to HR with weighted average

Upsampling:

Is the process of producing an image of larger size with significantly more pixel and higher image quality from low quality of images. The term up sampling is associated with the process of resampling in Digital image signal processing (Fig. 19).

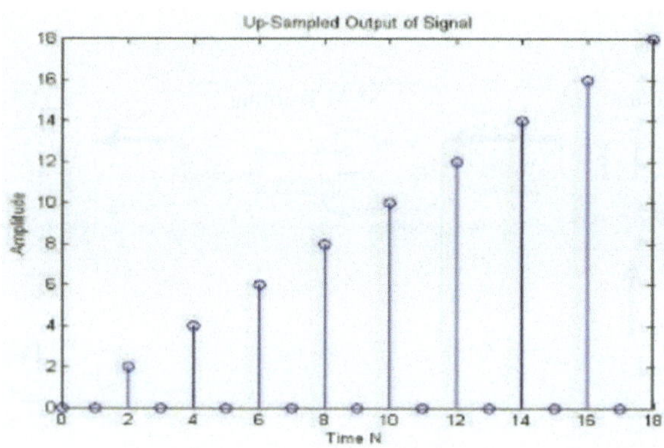

Fig. 19 Upsampling and interpolation of image signal in image

1. **Algorithms: Hashing Based Learning and Upscaling**

(i) **Algorithm: RAISER Hashing Based Learning**

INPUTS:
1: s –Upscaling factor
2: $\{X_i\}_{i=1}^{L}$ - The ground truth HR images (Optionally sharpened/enhanced).
3:$\{Z_i\}_{i=1}^{L}$ - The LR (optionally compressed) versions of $\{X_i\}_{i=1}^{L}$.

OUTPUT
1: $\mathcal{H}_t : (j) \longmapsto h_{j,t}$ – Hash table that maps each key.
$j = (\theta, \lambda_1, \mu)$ to the corresponding d^2 –dimensional filter $h_{j,t}$, where $1 \leq t \leq s^2$ is the pixel/patch type.

Process
$\boldsymbol{Q}_{j,t} \leftarrow 0$, $\boldsymbol{V}_{j,t} \leftarrow \boldsymbol{0}$ for all the hash-table keys denoted by j and pixel-type t
for$i = 1$**to**L**do**
compute y_i, an initial interpolated version of z_i
Initialize $A_{j,t}$ and $b_{j,t}$ to be empty matrices for all the possible hash-tables
Denoted by j and pixel type t
for$each pixel k$ in y_i**do**
Denote by $j = (\theta_k, \lambda_1^k, \mu k)$ the hash-table key of the pixel k
Denoted by $\boldsymbol{P}_K \in \mathbb{R}^{d^2}$ the patch extracted from y_i centred at k
Denoted by $x_i(k)$ ground truth of the pixel of x_i located at k
Denoted by t the type of the pixel k
$\boldsymbol{A}_{j,t} \leftarrow [\boldsymbol{A}_{j,t} ; \boldsymbol{p}_k^T]$,i.e., concatenate
p_k^T to the end of $\boldsymbol{A}_{j,t}$
$\boldsymbol{b}_{J,t} \leftarrow [\boldsymbol{b}_{j,t}; \boldsymbol{x}_i(k)]$, i.e., concatenate
$x_i(k)$ to the end of $\boldsymbol{b}_{j,t}$
end
for $each key j and pixel - type t$ **do**
$\boldsymbol{Q}_{j,t} \leftarrow \boldsymbol{Q}_{j,t} + \boldsymbol{A}_j^T t \boldsymbol{A}_{j,t}$
$\boldsymbol{V}_{j,t} \leftarrow \boldsymbol{V}_{j,t} + \boldsymbol{A}_j^T t \boldsymbol{b}_{j,t}$
end
end
for $each key j and pixel - type t$ **do**
$\boldsymbol{h}_{j,t} \leftarrow argmin_{\boldsymbol{h}} \parallel \boldsymbol{Q}_{j,t} h - V_{j,t} \parallel_2^2$
$\mathcal{H}_t(\text{j}) \leftarrow \text{h}_{j,t}$
end

(ii) **Algorithm: RAISER Hashing Based Upscaling**

INPUTS

1. $s-$ Upscaling factor.

2. $\mathbf{z}-$LRimage

3. $\mathcal{H}_t : (j) \longmapsto \boldsymbol{h}_{j,t}$ – Hash-table that maps each key $j = (\theta, \lambda_1, \mu)$ to the corresponding d^2 –dimensional filter$\boldsymbol{h}_{j,t}$ where $1 \leq t \leq s^2$ is the pixel/patch type.

OUTPUT

1: $\hat{X}$–**HR**estimate of **z**

PROCESS

Compute$y$, an initial interpolated version of **z**

for$eachpixelkiny$ **do**

Denote by $j = (\theta_k, \ \lambda_1^k, \mu k)$ the hash-table key of the pixel k

Denote by $\boldsymbol{p}_k \in \mathbb{R}^{d^2}$ the patch extracted from y centred at k

Denoted by t the type of the pixel k

$h_{j,t} \leftarrow \mathcal{H}_t(j)$

$\hat{X}(k) \leftarrow \boldsymbol{p}_k^T \boldsymbol{h}_t$

End

$\hat{X} \leftarrow$ CT-Blending $\hat{X}, y$

2. **Algorithm: SUR-Net: Predicting the Satisfied User Ratio Curve for Image Compression with Deep Learning** [79]

Let $I_1[0], I_2[0], \ldots, I_K[0]$ be a large training set of K pristine reference images. **For each** pristine image $I_k[0], k \in \{1, \ldots, K\}$,
we associate the N distorted images $I_k[n], n = 1, \ldots, N$ corresponding to the distortion levels $n = 1, \ldots, N$.

Let SUR_k denote the SUR function of image $I_k[0]$.

Regression model: $f\theta$ parameterized by θ such that: $f\theta(I_k[0], I_k[n]) \approx \boldsymbol{SUR}_k(n), k = 1, \ldots, K, n = 1, \ldots, N$.
Then we calculate $\Delta f_{\min}$, $\Delta f_{\max}$, and Δf_{avg}, between the distorted images $I_k[n]$ and the pristine image $I_k[0]$, e.g., $\Delta f_{\min} = f_{\min}(I_k[n]) - f_{\min}(I_k[0])$.

Given the training data: $\{(I_k[0], I_k[n], \mathrm{SUR}_k(n)) | k = 1, \ldots, K, n = 1, \ldots, N\}$,

Mean Absolute Error (MAE) loss function:

$$L = \frac{1}{KN} \sum_{k=1}^{K} \sum_{n=1}^{N} | f_\theta(I_k[0], I_k[n]) - \mathrm{SUR}_k(n) |$$

Prediction of SUR curve deep learning:
$I[0]$, together with its distorted versions $I[1], \ldots, I[n]$, a sequence of predicted satisfied user ratios SUR(1), ..., SUR(N) is obtained from the network. Assuming that the JND is normally distributed, we estimate the mean μ and variance σ^2 by least squares fitting:

$$\left(\hat{\mu}, \hat{\sigma}^2\right) = \underset{\mu,\sigma^2}{\operatorname{argmin}} \sum_{n=1}^{N} \left|\overline{\Phi}\left(n|\mu, \sigma^2\right) - \mathrm{SUR}(n)\right|^2$$

3. **Algorithm: AROC-DPNet Lightweight Convolutional Neural Network and Deep Clustering for Hyperspectral Image Semi-supervised Classification with Limited Training Samples** [80]

Input: Small number of available labeled training samples $\{X^L, y^{(m)}\}, y^{(m)} \in \{1, 2 \ldots M\}$, abundant unlabeled samples X^U and test samples X^T. The number of max-iteration is N.

Output: The labels of X^T.

1. Pre-train 3D-DPNet with dual loss by Eqs. (1) and (2) on $\{X^L, y^{(m)}\}$.
2. Obtain deep features F^L and F^U for training samples (X^L and X^U) and the primary labels $p^{(m)}$ of X^U, $y^{(m)} \in \{1, 2 \ldots M\}$ through 3D-DPNet classification.
3. Project all training samples (X^L and X^U) to the new deep feature space and implement AROC algorithm by Eq. (5) on$\{F^L, p^{(m)}\}$, and$\{F^U, p^{(m)}\}$, to obtain cluster labels$\{X^U, c^{(m)}\}, c^{(m)} \in \{1, 2 \ldots M\}$.
4. Comparison of $\{X^U, c^{(m)}\}$ and $\{X^U, p^{(m)}\}$, and those with matched labels are considered to have high confidence and their corresponding pseudo labels $\{X^{C_i}, s^{(m)}\}, X^{C_i} \subset X^U$ are determined.
5. Fine-tune the pre-trained 3D-DPNet using $\{X^{C_i}, s^{(m)}\}$ and $\{X^L, y^{(m)}\}$.
6. Repeat 2 to 5, until $i = N$.
7. The labels of X^T were obtained using the parameters of the last iterations of 3D-DPNet. Output the labels of X^T.

10 Conclusion and Discussion

- As the human species evolved human learned the things via sensing organs and learns, understand the daily or new activities day by day. As the evolution in the technology human tries to mimic the human brain behavior on computers with the help of some programmed algorithm. Machine learning is such a field which when combines with the computer vision solve most of the today's problem. Here, in this book chapter, we have discussed and tries for find out some of the problem of computer vision with the use of ML algorithms. We have discussed how image compression, image enhancement and noise removal technique, automated segmentation like problem will be solved with the help of machine learning.

- Broadly Machine learning is composed of Supervised, Unsupervised, and semi-supervised ML. Algorithms like KNN, k-means falls into this category, there will be further improvement day by day in these algorithm as we have a lot of data for the training our model on different datasets.
- Image compression is one of the important areas of research now these days as we have better quality of cameras which are capable of capturing the high-resolution images. Thus, transferring and storing these images on the storage system, transfer the data from one place to another require a development of new algorithms thus Machine learning helps a lot and we can develop a new image compression algorithm.
- Image processing use machine learning a lot. Automated image segmentation to edge detection, color enhancement, and visualization is possible due to machine learning. In Medical Science image processing use a lot in X-RAY, MRI, and CT scans also. Image Recognition, visualization is also one more task in image processing, thus visualization of underwater object, human body organs detection, cancer/lesion detection, and recognition is popular field of computer vision and use machine learning algorithms a lot. Denoising of images is done through the machine learning. Now we have such an algorithm which can do the automated de-noising of noise from the noisy image.
- Due to machine learning algorithm like Neural network, genetic algorithm is trained on the medical image data thus it is become possible to diagnose the patient automatically. As a result, more accurate and precise diagnosis and treatment are possible in the hospitals. The increase in the data and computing power makes the automated diagnosis more easily and prediction algorithms help a doctor in diagnosis.
- Images captures form the satellite, mobile phone, or digital cameras are large in size, due to ML algorithm it is able to compress the size of image automatically without any loss of information is possible. Applying ML algorithm in image processing enhances the colors of the image.
- Facial expression analysis is done with the help of Deep learning classification. On the web there are facial expression datasets are available and via using these data set we can train our models or we can develop a new algorithm also. Facial expression algorithms read our facial feature based on these they predict the emotion or mood of human body.
- Thus, in this book chapter author has tried to provide the new advancement in the field of computer vision and artificial intelligence/machine learning. As the technology is advancing day by day there is explosion in the data also and we need some algorithms to compress, classify, label this huge amount of data having knowledge and a skill of machine learning and image processing one can gain develop new algorithm also for the better understanding of content between the images or videos.

Acknowledgements During this period of COVID-19 everyone life's get disturbed , everyone is at their home or may be in safe place God bless them all. In this pandemic era I would like to thank Mr. Rohit Shukla for helping hand in writing the English literature (Ph.D. scholar Dept. of Bioinformatics JUIT) , Mr. Sameer for helping in drawing the Mathematical Equations (Ph.D. scholar Dept. of Mathematics JUIT) , Ms. Shashi Kala (Asst Prof Dept of Bioinformatics KMV) for providing some resources for reading, and my Grandfather for providing me Internet Resources during this COVID-19 pandemic.

Author Contribution In the era of pandemic of COVID-19, Mr. Ajay Sharma[1] (Ph.D. Research Scholar Dept of Bioinformatics JUIT) has prepared all the Manuscript of the book Chapter Most of the data, paper collection, and written part is done by the Ajay Sharma. Mr. Ankit Gupta[2] (Ph.D. Research Scholar Interactive Technologies Institute (ITI/LARSyS Madaria) has written the introduction part of the book chapter and helps in mining latest research articles for this book chapter. Dr.Varun Jaiswal (Ex. Assist Director, National Centre for Disease Control NCDC New Delhi, Current Assist prof Gachon University Dept of Biotechnology) has revised the prepared manuscript and find out the correction where it will be needed.

The book chapter contains some of the novel work. While planning and designing the book chapter entitled "Solving Image Processing Critical problems using Machine Learning" author should contain in mind that this book chapter helps those readers who don't have any or little knowledge about image processing to those who are sound in imaging. Thus, keep in mind author has divided the book into several sections like easy, intermediate, and modest. In easy part, reader can understand the theory behind the image processing algorithm and machine learning and intermediate part contains the working and the modest part contains the mathematics algorithm that could be understood by those who have sound knowledge in image processing.

Conflict of Interest The author has declared that there is no conflict of interest. The book chapter entitled "Solving Image Processing Critical problems using Machine Learning" has authentic work. The literature were mined from all the internet resources and while writing the book chapter author took the keen observation that proper credit is given to all the work via referencing and citation of research paper or book chapter.

References

1. D. Michie, D.J. Spiegelhalter, C. Taylor, Machine learning. Neural Stat. Classif. **1994**(13), 1–298 (1994)
2. S.B. Kotsiantis, I. Zaharakis, P. Pintelas, Supervised machine learning: a review of classification techniques. Emerg. Artif. Intell. Appl. Comput. Eng. **160**, 3–24 (2007)
3. A. Kalousis, J. Gama, M. Hilario, On data and algorithms: understanding inductive performance. Mach. Learn. **54**(3), 275–312 (2004)
4. M.-L. Zhang, Z.-H. Zhou, A k-nearest neighbor based algorithm for multi-label classification, in *2005 IEEE International Conference on Granular Computing* (IEEE, 2005)
5. Y. Liao, V.R. Vemuri, Use of k-nearest neighbor classifier for intrusion detection. Comput. Secur. **21**(5), 439–448 (2002)
6. D. Böhning, Multinomial logistic regression algorithm. Ann. Inst. Stat. Math. **44**(1), 197–200 (1992)
7. B. Schölkopf et al., Shrinking the tube: a new support vector regression algorithm, in *Advances in Neural Information Processing Systems* (1999)
8. J. Ali et al., Random forests and decision trees. Int. J. Comput. Sci. Issues (IJCSI) **9**(5), 272 (2012)
9. D. Yarowsky, Unsupervised word sense disambiguation rivaling supervised methods, in *33rd Annual Meeting of the Association for Computational Linguistics* (1995)
10. G.E. Hinton et al., The "wake-sleep" algorithm for unsupervised neural networks. Science **268**(5214), 1158–1161 (1995)

11. J.A. Hartigan, M.A. Wong, Algorithm AS 136: a k-means clustering algorithm. J. Roy. Stat. Soc. Ser. C (Appl. Stat.) **28**(1), 100–108 (1979)
12. H. Li et al., Illumination compensation and image denoising for low-light images based on deep learning, in *Proceedings of the 3rd International Conference on Vision, Image and Signal Processing* (2019)
13. M. Poostchi et al., Image analysis and machine learning for detecting malaria. Transl. Res. **194**, 36–55 (2018)
14. A. Kan, Machine learning applications in cell image analysis. Immunol. Cell Biol. **95**(6), 525–530 (2017)
15. R.S. Sutton, A.G. Barto, *Reinforcement Learning: An Introduction* (MIT Press, Cambridge, 2018)
16. J.X. Chen, The evolution of computing: AlphaGo. Comput. Sci. Eng. **18**(4), 4–7 (2016)
17. O. Chapelle, B. Scholkopf, A. Zien, Semi-supervised learning (chapelle, o. et al., eds.; 2006) [book reviews]. IEEE Trans. Neural Netw. **20**(3), 542–542 (2009)
18. X.J. Zhu, *Semi-Supervised Learning Literature Survey* (University of Wisconsin-Madison Department of Computer Sciences, 2005)
19. T.S. Huang, K. Aizawa, Image processing: some challenging problems. Proc. Natl. Acad. Sci. **90**(21), 9766–9769 (1993)
20. P. Patidar et al., Image de-noising by various filters for different noise. Int. J. Comput. Appl. **9**(4), 45–50 (2010)
21. B.T. Hannigan, B.A. Bradley, G.B. Rhoads, Removal of fixed pattern noise and other fixed patterns from media signals. Google Patents (2003)
22. S.O. Rice, Mathematical analysis of random noise. Bell Syst. Tech. J. **23**(3), 282–332 (1944)
23. A.K. Boyat, B.K. Joshi, A review paper: noise models in digital image processing. arXiv preprint arXiv:1505.03489 (2015)
24. A.K. Vishwakarma, A. Mishra, Color image enhancement techniques: a critical review. Indian J. Comput. Sci. Eng. **3**(1), 39–45 (2012)
25. V. Lopez-Vazquez et al., Video image enhancement and machine learning pipeline for underwater animal detection and classification at cabled observatories. Sensors **20**(3), 726 (2020)
26. M. Liu et al., Fuzzified image enhancement for deep learning in iris recognition. IEEE Trans. Fuzzy Syst. **28**(1), 92–99 (2019)
27. Y.A. Bayhaqi et al., Fast optical coherence tomography image enhancement using deep learning for smart laser surgery: preliminary study in bone tissue, in *Fourth International Conference on Applications of Optics and Photonics* (International Society for Optics and Photonics, 2019)
28. J. Fenshia Singh, V. Magudeeswaran, A machine learning approach for brain image enhancement and segmentation. Int. J. Imaging Syst. Technol. **27**(4), 311–316
29. K.J. Halupka et al., "One-size fits all" OCT image enhancement via deep learning. Invest. Ophthalmol. Vis. Sci. **60**(9), 1513–1513 (2019)
30. M. Abdel-Nasser, O.A. Omer, Ultrasound image enhancement using a deep learning architecture, in *International Conference on Advanced Intelligent Systems and Informatics* (Springer, Berlin, 2016)
31. L. Ruihua, Z. Quan, X. Huachao, An image compression processing method based on deep learning, in *2019 IEEE 2nd International Conference on Information Communication and Signal Processing (ICICSP)* (IEEE, 2019)
32. T. Celik, Spatial entropy-based global and local image contrast enhancement. IEEE Trans. Image Process. **23**(12), 5298–5308 (2014)
33. A.M. Ali, G.C. Karmakar, L.S. Dooley, Review on fuzzy clustering algorithms. J. Adv. Computations **2**(3), 169–181 (2008)
34. S.M. Pizer et al., Adaptive histogram equalization and its variations. Comput. Vis. Graph. Image Process. **39**(3), 355–368 (1987)
35. Y.-T. Kim, Contrast enhancement using brightness preserving bi-histogram equalization. IEEE Trans. Consum. Electron. **43**(1), 1–8 (1997)
36. E.H. Land, J.J. McCann, Lightness and retinex theory. Josa **61**(1), 1–11 (1971)

37. E.P. Simoncelli et al., Shiftable multiscale transforms. IEEE Trans. Inf. Theory **38**(2), 587–607 (1992)
38. G. Beylkin, R. Coifman, V. Rokhlin, Fast wavelet transforms and numerical algorithms I. Commun. Pure Appl. Math. **44**(2), 141–183 (1991)
39. D. Williams, P.D. Burns, Measuring and managing digital image sharpening, in *Archiving Conference* (Society for Imaging Science and Technology, 2008)
40. F. Russo, An image enhancement technique combining sharpening and noise reduction. IEEE Trans. Instrum. Meas. **51**(4), 824–828 (2002)
41. J.D. Buhr et al., Digital photofinishing system including scene balance, contrast normalization, and image sharpening digital image processing. Google Patents (2000)
42. W. Xu et al., Single image super-resolution based on global dense feature fusion convolutional network. Sensors **19**(2), 316 (2019)
43. Q. Zhang et al., Super-resolution of single multi-color image with guided filter. J. Vis. Commun. Image Represent. **58**, 277–284 (2019)
44. C. Xing et al., Gradation measurement of asphalt mixture by X-ray CT images and digital image processing methods. Measurement **132**, 377–386 (2019)
45. R. Bostanabad, Reconstruction of 3D microstructures from 2D images via transfer learning. Comput. Aided Des. 102906 (2020)
46. M. Iliadis, L. Spinoulas, A.K. Katsaggelos, DeepBinaryMask: learning a binary mask for video compressive sensing. Digital Signal Process. **96**, 102591 (2020)
47. C.D. Bahadir, A.V. Dalca, M.R. Sabuncu, Learning-based optimization of the under-sampling pattern in MRI, in *International Conference on Information Processing in Medical Imaging* (Springer, Berlin, 2019)
48. M. Van Dael et al., Combination of shape and X-ray inspection for apple internal quality control: in silico analysis of the methodology based on X-ray computed tomography. Postharvest Biol. Technol. **148**, 218–227 (2019)
49. G. Fang et al., Identification of corrosion products and 3D distribution in reinforced concrete using X-ray micro computed tomography. Constr. Build. Mater. **207**, 304–315 (2019)
50. N.R. Pal, S.K. Pal, A review on image segmentation techniques. Pattern Recogn. **26**(9), 1277–1294 (1993)
51. K. Sun, J. Wang, Robust local region-based active contour for inhomogeneous image segmentation, in *Recent Developments in Intelligent Computing, Communication and Devices* (Springer, Berlin, 2019), pp. 333–340
52. B.A. El-Rahiem et al., An efficient deep convolutional neural network for visual image classification, in *International Conference on Advanced Machine Learning Technologies and Applications* (Springer, Berlin, 2019)
53. C.L. Wong, Automated septal defect detection in cardiac computed tomography images. Google Patents (2019)
54. C. Jácome et al., Convolutional neural network for breathing phase detection in lung sounds. Sensors **19**(8), 1798 (2019)
55. Z. Li, Rapid methods for 2D and 3D T1 estimation using highly undersampled radial MRI data (2019)
56. K.U. Devi, R. Gomathi, Brain tumour classification using saliency driven nonlinear diffusion and deep learning with convolutional neural networks (CNN). J. Ambient Intell. Humanized Comput. 1–11 (2020)
57. M. Zhang, Q. Zhou, Y. Hu, A lossless data hiding scheme in JPEG images with segment coding. arXiv preprint arXiv:1901.11203 (2019)
58. T.L. Saaty, *The Brain: Unraveling the Mystery of How It Works: The Neural Network Process* (Rws Publications, Pittsburgh, 2019)
59. D.S. Levine, Theory of the brain and mind: visions and history, in *Artificial Intelligence in the Age of Neural Networks and Brain Computing* (Elsevier, Amsterdam, 2019), pp. 191–203
60. P. Kubal, S. Doyle, Deep learning method for pathology image compression and reconstruction, in *Laboratory Investigation* (Nature Publishing Group, New York, 2020)

61. B. Sujitha et al., Optimal deep learning based image compression technique for data transmission on industrial Internet of things applications. Trans. Emerg. Telecommun. Technol. e3976 (2020)
62. M.Y. Yip et al., Effect of image compression and number of fields on a deep learning system for detection of diabetic retinopathy. Invest. Ophthalmol. Vis. Sci. **60**(9), 1438–1438 (2019)
63. A. Mosavi, A. Bahmani, Energy consumption prediction using machine learning: A review. Preprints (2019)
64. B.S. Neysiani et al., Improve performance of association rule-based collaborative filtering recommendation systems using genetic algorithm. Int. J. Inf. Technol. Comput. Sci. **2**, 48–55 (2019)
65. M. Pantic, L.J.M. Rothkrantz, Automatic analysis of facial expressions: the state of the art. IEEE Trans. Pattern Anal. Mach. Intell. **22**(12), 1424–1445 (2000)
66. J. Weston, C. Watkins, Multi-class support vector machines. Citeseer (1998)
67. A. Raghuvanshi, V. Choksi, Facial expression recognition with convolutional neural networks. CS231n Course Projects, 362 (2016)
68. S. Haykin, *Neural Networks: A Comprehensive Foundation* (Prentice Hall, USA, 1994)
69. A. Krizhevsky, I. Sutskever, G.E. Hinton, Imagenet classification with deep convolutional neural networks, in *Advances in Neural Information Processing Systems* (2012)
70. X. Zhao, X. Shi, S. Zhang, Facial expression recognition via deep learning. IETE Techn. Rev. **32**(5), 347–355 (2015)
71. S. Hershey et al., CNN architectures for large-scale audio classification, in *2017 IEEE International Conference on Acoustics, Speech and Signal Processing (ICASSP)* (IEEE, 2017)
72. A. Graves et al., Facial expression recognition with recurrent neural networks, in *Proceedings of the International Workshop on Cognition for Technical Systems* (2008)
73. A. Savoiu, J. Wong, Recognizing facial expressions using deep learning (2017)
74. Y. Bengio, Learning deep architectures for AI. Found. Trends Mach. Learn. **2**(1), 1–127 (2009)
75. W. Shen et al., Deepcontour: A deep convolutional feature learned by positive-sharing loss for contour detection, in *Proceedings of the IEEE Conference on Computer Vision and Pattern Recognition* (2015)
76. T. Liu, G. Wang, Q. Yang, Real-time part-based visual tracking via adaptive correlation filters, in *Proceedings of the IEEE Conference on Computer Vision and Pattern Recognition* (2015)
77. P. Kumar, S. Henikoff, P.C. Ng, Predicting the effects of coding non-synonymous variants on protein function using the SIFT algorithm. Nat. Protoc. **4**(7), 1073 (2009)
78. Y. Romano, J. Isidoro, P. Milanfar, RAISR: rapid and accurate image super resolution. IEEE Trans. Comput. Imaging **3**(1), 110–125 (2016)
79. C. Fan et al., SUR-Net: Predicting the satisfied user ratio curve for image compression with deep learning, in *2019 Eleventh International Conference on Quality of Multimedia Experience (QoMEX)* (IEEE, 2019)
80. B. Fang et al., Collaborative learning of lightweight convolutional neural network and deep clustering for hyperspectral image semi-supervised classification with limited training samples. ISPRS J. Photogrammetry Remote Sens. **161**, 164–178 (2020)

Spoken Language Identification of Indian Languages Using MFCC Features

Mainak Biswas, Saif Rahaman, Satwik Kundu, Pawan Kumar Singh, and Ram Sarkar

Abstract Spoken language has been the most natural mode of communication since a long time. In today's world, the advancement of telecommunication technology has taken communication to the next level. This acts as motivation for an automatic speech recognition (ASR) process. As research in ASR progresses, the development of language identification would be necessary for any multi-lingual ASR system. This problem is highly pertinent to India due to the existence of a wide diversity of languages, each having hundreds of dialects. In this chapter, we propose a spoken language identification framework for the recognition of the six most widely used spoken languages in India *namely*, English, Hindi, Bangla, Marathi, Tamil and Telugu. Firstly, the pre-processing is performed on an input audio clip, which focuses on removing its silent frames. This, in turn, increases its probability of recognition as redundant and non-informative information gets eliminated. The silence removal parameters are optimized in order to achieve the best performance. Secondly, the Mel Frequency Cepstral Coefficient (MFCC) features are extracted from the speech signals. Using these features, a Support Vector Machine (SVM) classifier is trained, which has been established as a superior classifier, due to its large-margin classification property, and its ability to classify complex and non-linear data. The proposed framework is tested on a standard IndicTTS speech database. SVM is trained for 13

M. Biswas · S. Rahaman · S. Kundu · P. K. Singh (✉)
Department of Information Technology, Jadavpur University, Kolkata 700106, West Bengal, India
e-mail: pawansingh.ju@gmail.com

M. Biswas
e-mail: mainak.biswas.dbl@gmail.com

S. Rahaman
e-mail: saifrahaman500@gmail.com

S. Kundu
e-mail: satwikkundu25@gmail.com

R. Sarkar
Department of Computer Science and Engineering, Jadavpur University, Kolkata 700032, West Bengal, India
e-mail: raamsarkar@gmail.com

P. Kumar et al. (eds.), *Machine Learning for Intelligent Multimedia Analytics*,
Studies in Big Data 82,
https://doi.org/10.1007/978-981-15-9492-2_12

(static), 26 (static + delta) and 39 (static + delta + delta-delta) MFCC features. It is found that the best results are achieved using only 13 static features, and addition of delta and delta-delta features decreases the performance. Maximum accuracy obtained is 60.33% for the above-mentioned 6 languages, and 89.33% for only Bangla, English and Hindi.

Keywords Spoken language identification · Indian languages · Mel frequency cepstral coefficients · Support vector machine

1 Introduction

Spoken language is a system of arbitrary vocal symbols by means of which human beings communicate and co-operate with one another. The legend of the Babel's tower tells that at the very beginning of the world, humans had only one unique language [9]. Now there are over 5000 languages in the world. India alone has 22 languages recognized officially under the eighth schedule of the Indian Constitution. Language is a means of conveying ideas, expression and emotion which have led to the growth of civilization and globalization. Thus, in order to communicate effectively, a language which is mutually understandable by both participants is an important requirement. Language identification offers a solution to this problem [24].

Spoken Language Identification (SLID) is the process of detecting the language of an utterance by an anonymous speaker, irrespective of gender, accent and pronunciation. There are different aspects incorporated in speech which can be employed to represent the characteristics of a language. It is undertaken by performing computational linguistic approaches and applying many contexts. It is a challenging task because of the variations in the type of speech input. It involves understanding how humans process and interpret speech in adverse conditions and trying to model it [2].

Language Identification (LID) is important for most NLP applications to work accurately, since models are usually trained using data from a single language. If a model is trained on English text and is used for prediction on French text, we usually see a significant decrease in accuracy. Some specific applications where LID is used are as follows:

1. **Multilingual speech recognition**—Automatic Speech Recognition systems which have the ability to recognize multiple languages, need to identify the language that is being spoken first.
2. **Machine translation**—This is the use of software for translation of a text from one language to another. Here, too, LID is used as the first step, and based on the detected language, different trained models are used for the prediction step.
3. **Switchboard routing for incoming calls**—In call centres, LID can be used to first detect the language and then route the call to operators fluent in the required language [12].

The major task here is to identify what features have to be extracted in order to discriminate between languages. Some of the widely used feature extraction techniques are as follows: Linear Predictive Analysis (LPC), Linear Predictive Cepstral Coefficients (LPCC), Perceptual Linear Predictive Coefficients (PLPC), Mel-Frequency Cepstral Coefficients (MFCC), Power Spectral Analysis (FFT), Mel Scale Cepstral Analysis (Mel), Relative Spectra Filtering of Log Domain Coefficients (RASTA), First Order Derivative (DELTA), etc. [26]. Feature extraction reduces the magnitude of the speech signal devoid of causing any damage to the power of speech signal [3].

Speech features are broadly classified into low-level features and high-level features. At low-level, the most commonly used features for LID are acoustics, phonetics, phonotactics and prosodic information while at high-level, LID can be established based on the morphology and sentence syntax. The acoustic features usually modelled by MFCCs, are the compact representation of the input speech signal fulfilling a compression of the data contained in the audio waveform. Phonotactics are rules that govern permissible sequence of phonemes in speech signals. Phonotactics define permissible syllable structure, consonant clusters, and vowel sequences by means of phonotactical constraints. Phonetic features are features that are associated with the sounds of human speech and their production.

The low-level features are easier to obtain but are very volatile and are easily affected by noise and speaker variations, whereas high-level features contain more information regarding language discrimination. However, high-level features rely on large vocabulary recognition, and as a result, more training data is needed which ultimately leads to a greater level of complexity in terms of computation cost. However, before the features are extracted, there are sequences of pre-processing phases that are first carried out in order to avoid the effects of noisy and unreliable/unwanted data on the system.

A major problem in LID is how to design a specific language to represent speech utterances. It is challenging due to the significant variations introduced relating to different speech patterns, speakers, channels and background noises. Due to technological advances, data is being generated at an ever-increasing pace, and the size and dimensionality of the datasets continue to grow each day. Another key problem in SLID is to design effective representations which are specific to language information. Although advances in machine learning have led to significant improvements, performance of LID system is still lacking, especially for short duration speech utterances [10].

1.1 Contributions

1. In this study, we have developed an acoustic SLID model by making use of the static, delta and delta-delta values of MFCC methodology to extract features from speech samples across different languages. These features are then used to train an SVM classifier. We have discussed how the traditional SVM (binary classifier) classifier is modified to be applicable in a multi-class system. We also discuss a scheme for classifying a clip consisting of multiple samples.

2. We propose an effective silence removal preprocessing step and also show its contribution in ameliorating the final accuracy.
3. However, the most important contribution of this study is the comparison of three sets of MFCC features and the finding that including delta and delta-delta features worsened the performance of the classifier. This is rather unexpected, since our intuitions say that capturing the dynamic characteristics of audio ought to increase accuracy of classification. The results lead us to believe that these features are not good for the identification of the targeted set of languages.

The language set considered in our study includes six popular Indian languages namely English, Hindi, Bangla, Tamil, Telugu and Marathi.

The remainder of the paper is arranged as follows: Sect. 2 provides information about the database used in the present work. A brief overview of characteristics of the six popular Indian languages used is given in Sect. 3. Section 4 lists the literature that we have referred to, and the motivation behind our work. Section 5 contains information about the pre-processing and feature extraction methods used. Section 6 has a brief description of the SVM classifier, and the methodology used in training and testing. Section 7 lists the metrics we have used to measure performance of our model. Section 8 contains results obtained, and Sect. 9 concludes the work, giving a brief account of future plans.

2 Database Used

In a multilingual country like India, systematic efforts have been taken in developing speech corpora in major languages. Depending on the specific application, the speech corpora which are needed are very diverse. For example, speech synthesis usually requires a large amount of speech data from one or two speakers, whereas speech recognition often requires a smaller amount of speech data from many speakers.

Speech and Music Technology Lab IIT Madras has developed a database for speech synthesis and recognition, music information retrieval and music processing. This database consists of 13 Indian languages (duration of approximately 40 h each) and a unified parser (a single parser for parsing 13 different Indian languages). It contains a total number of 14 speakers, where each language is spoken by 2 speakers (male and female) belonging to different age category [29].

3 Characteristics of Selected Indian Languages

All the Indian languages are phonetic in nature. Every Indian language has its distinctive phone (smallest segment of sound in a speech stream) sets. Speech sounds or phonemes of all languages are classified into vowels and consonants. Most popularly, they are represented with some specific symbols generally called as '*aksharas*', which are an audible speech sound of that language. Most of the Indian languages have at least one distinct trill phoneme. The trills of Indian languages are 'typical trills'; i.e. there is a periodic vibration (roughly around 30 Hz) at the stricture usually by the tip of the tongue [14].

The following are some distinctive characteristics of the languages referred to in our study [1]:

Hindi is a syllable-timed language, meaning words are not distinguished based on stress alone. Default stress is given on the last syllable. Each content word except the final one has rising contour.

In Bangla, stress is at the word initial; the first syllable of the word carries the maximum stress, the third syllable carries somewhat weaker stress and all units with odd number of syllables carry very weaker stress. Bangla has uniform pattern of pitch contour for focus intonation and it has identical phonological form.

Tamil is syllable-timed; it has lexical stress with a complex vowel quantity. Tamil intonation is rising contour that occurs on each lexical word except the last in a phrase.

In Marathi, stress is at the word initial and is weight sensitive. Words with open syllables with a short vowel have light stress while closed syllables and open syllables with a long vowel have heavy stress.

Telugu is a mora-timed language. Default stress is on the first syllable in Telugu. Telugu intonation is classified into falling (F), rising–falling (RF), rising (R) and falling–Rising (FR).

As can be seen, there are several supra-segmental properties that are specifically dependent on the language. These properties can be exploited for the purposes of language identification, especially for natural spoken utterances. Intonation and stress play important roles in discriminating these languages, for example, Telugu is mora-timed language while Hindi is syllable-timed language [6].

4 Literature Survey

LID is a fairly well-explored research area. Several techniques have been developed for accomplishing this task, following the phonetic approach, the phonotactic approach and the prosodic approach. In this section, we mention some of the previous works in the field of LID that we have referred to.

Mohanty [17] takes a parallel phone recognition and language modelling (PPRLM) approach towards LID. It classifies 4 Indian languages using 1 'manner' feature and 10 'place' features, and SVM classifier. A maximum average accuracy of 89.75% is attained at the word level.

Singer et al. [28] build a model using PPRLM, Gaussian Mixture Model (GMM) and discriminative SVM. Then they distinguish 12 languages, achieving equal error rates of 2.7% and 2.8% on the NIST 1996 and 2003 datasets respectively.

In [2], language identification is done on eight languages, using MFCC and SDC (Shifted Delta Coefficients) based features along with GMM and i-vector-based framework. Then, an enhanced self-adjusting extreme learning machine is used for classification, which gives a resultant accuracy of 96.25%. Manchala et al. [15] uses a combination of two kinds of features (MFCCs and formant frequencies extracted using LPCC). Finally, GMM is used to classify 10 languages, which gives a maximum accuracy of 98.8%. Gupta et al. [7] propose an approach based on LPCC and MFCC, and apply SVM and Random Forest as classifiers. It is found that a combination of both the features gives the best results (attaining an accuracy of 92.6%) when used with the Random Forest classifier for 6 Indian languages.

Kumar et al. [11] present a comparative study of different feature extraction techniques used for LID. It uses MFCC, Perceptual Linear Prediction (PLP) coefficients and two-hybrid features namely, Bark Frequency Cepstral Coefficients (BFCC) and Revised PLP, to identify a set of 10 languages. The experiment shows that the hybrid features give better performance than conventional features, and RPLP gives the best results (attaining an average accuracy of 88.75%) with GMM as classifier.

In Mukherjee, Obaidullah, et al. [20] novel features called MFCC-2 are proposed. These are obtained by approximating frame-level MFCC features to the clip level, using mean, standard deviation, etc. Thereafter, a feature reduction technique called Principal Component Analysis (PCA) is used. Finally, the classification is done using a lazy learning approach, which gives a highest accuracy of 98.09% on the languages English, Hindi and Bangla.

Mukherjee et al. [19] distinguish 7 Indian languages using the spectrogram, which is a 3D plot depicting the time, frequency and energy information of a speech signal. Convolutional Neural Network (CNN) based architecture is used for classification and an accuracy of 99.96% is obtained. Also, the introduction of a significant amount of noise leads to a very small decrease in accuracy, proving the model to be fairly robust.

Mukherjee et al. [18] use new features called "MFCC—Statistics Grade", which are, again, clip level approximations of frame-level MFCC features, and use plotting techniques to visualize them. With these image-based features, classification is done on 4 south-Indian languages using CNN which gives a maximum accuracy of 94.51%.

Krishna et al. [21] perform identification of 7 Indian languages using GhostVLAD pooling for aggregation of frame-level features into utterance level representations. Their model achieves an F1-score of 98.43%, which outperforms the previous state of the art technique. They also compare their approach with other pooling techniques like Average pooling, Statistics pooling and NetVLAD pooling, and show that GhostVLAD pooling produces the best results.

In Padi et al. [23], a short-sequence neural network framework is introduced, which takes into account the relevance of parts of a speech sequence by assigning more weightage to parts that are more relevant to the task of language identification. This is achieved by using bidirectional long short term memory (BLSTM) with attention modelling. The model is tested using the RATS dataset, which gives a maximum accuracy of 92.69%.

Beckmann et al. [5] investigate the effectiveness of transfer learning in speech processing. They introduce SpeechVGG, a flexible and transferable deep feature extractor, which is trained on the task of spoken word classification, and tested on four speech processing tasks, one of them being language identification. They distinguish three languages and achieve an accuracy of 97.6%, thus showing that knowledge is transferrable across varying speech processing tasks.

Titus et al. [33] take an interesting approach to tackle the problem of language identification for speakers having varying accents. The proposed system consists of two stages—one acoustic model, which classifies the audio input based on its acoustic characteristics, and one context model, which takes into account factors like the locale (language-location pair) settings of the speech recognizer system. They achieve an average accuracy of 97% across all language combinations.

4.1 Motivation

The above discussion reveals that a fair amount of work has been done on LID. It is to be noted that when it comes to Indian languages, the task becomes more complicated because of the large number of spoken languages in the country. Also, there are quite a few instances of languages being similar, because of them being derived from the same source language. As a result, there is a growing need for research in identification of Indian languages. Our primary motivation behind this work is to build a simple system using MFCC features and SVM classifier, so that we can test their effectiveness for recognition of the targeted set of languages. We also present a comparative study of three different sets of MFCC features in our work. The intention was to present a set of MFCC features that is most pertinent for LID of Indian Languages.

5 Algorithms and Methodology

The overview of the entire work performed here has been shown in Fig. 1. The flowchart shows various blocks outlining the processes used to create the automatic spoken language identifier. Some of the arrows show the various sets used to train,

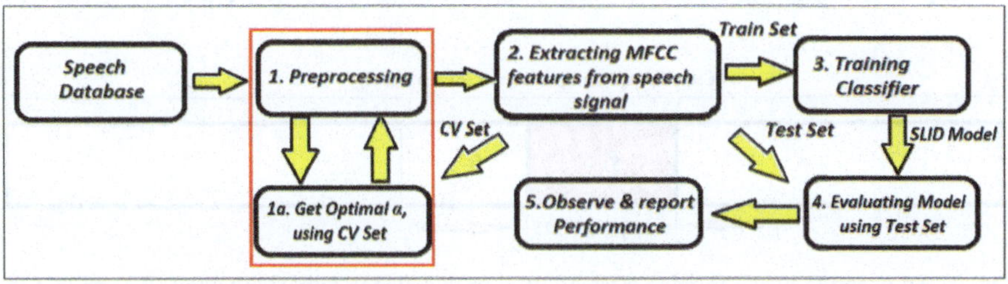

Fig. 1 Flowchart showing all the blocks of the proposed SLID system. The CV set is a set of samples used to optimize α

test and optimize the performance of our work. The CV set (cross-validation set) is used to optimize our pre-processing step only. CV set is a collection of samples used to optimize hyper-parameters (α in our case). The next subsections discuss the proposed methodology in detail.

5.1 Pre-processing

The need for pre-processing arises from our desire to remove irrelevant information from the raw data. This irrelevant information, when present in large amount in the training data, may mislead the classifier. In our case, recorded speech files (.wav) are the raw data. We extract features from these speech files by using windowing approach. The frames of 25–30 ms are generally used to capture phonemes. A human speech sample contains 20–30% of unnecessary silent frames. This is true for all languages. If this factor is not removed; similar frames would be stamped as of different languages, which would ultimately cause inconsistency. So, before features are extracted from the speech signal, the silence frames are first removed (both during training and testing phases).

5.1.1 Silence Removal Algorithm

Let the speech sample be S_k and the length of the sample be $t_k s$. Sampling rate is fs Hz. A_k is the discrete speech signal array with its indices defined in $[1, fs * t_k]$. With window size of 25 ms, and no overlapping between consecutive frames (as shown in Fig. 2) we get N_k frames. For sample S_k we thus have frames f_i where $i \in [1, \frac{t_k}{0.025}]$. So, $N_k = \frac{t_k}{0.025}$. The objective of this method is to remove frames with energy E_f below a certain threshold τ_k (frames with no speech). This, in turn, would produce a continuous speech.

Step 1: Finding τ_k: For the discrete speech signal A_k, total energy E_k is defined as $E_k = M \cdot \sum_j (A_k[j])^2$, where M is a constant. Average frame energy, $< E_k, N_k > = \frac{E_k}{N_k}$. We define, $\tau_k = \alpha < E_k, N_k >$ where α is a scaling parameter.

Step 2: Eliminating silence frame: The following steps are defined $\forall i \in [1, N_k]$.

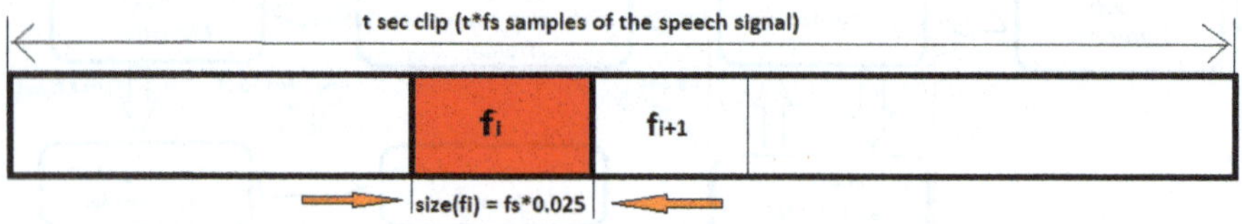

Fig. 2 An example of the framing strategy used in the silence removal method. We see that the window size is 25 ms, with no overlap between consecutive frames

Let $W = \text{size}(f_i) = 0.025 * fs$. We calculate energy of frame f_i as $E_k[i] = M \cdot \sum_j (A_k[j])^2$ where $j \in [(i-1) * W + 1, i * W]$. We remove frame f_i if $E_k[i] < \tau_k$.

Thus, we get the speech signal with the silence frames removed. Figure 3 shows the sample speech signal before and after silence removal. It can be clearly observed that about 30% of the entire speech signal contains frames that do not contain any speech.

For optimizing the value of α, we see the effect of change in α on accuracy with 13 MFCC features (discussed in the next section) and 3 languages, and training with around 50,000 training example for each language (Table 1 and Fig. 4 show these effects). This behaviour is primarily observed because initially the silence frames misguide the classifier, while later as the threshold increases, loss of meaningful information increases. Therefore, a maxima in accuracy is reached as α approaches to 0.01.

So, $\tau_k = 0.01. < E_k, N_k >$ is the optimal τ_k. This optimal τ_k will be used while training and testing the classification model.

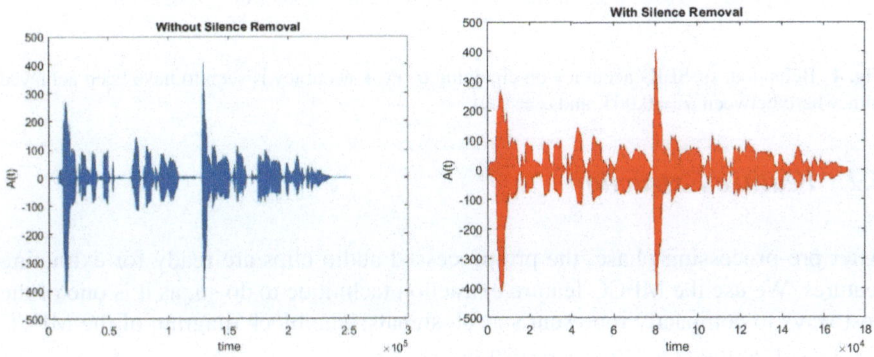

(a) (b)

Fig. 3 Speech signal: **a** before and **b** after removing silence frames. We see that majority of frames having amplitude close to zero in (a) have been removed in (b)

Table 1 Variations of SLID accuracy with change in α

Scaling factor (α)	Cross-validation accuracy (in %)
0.000	66.67
0.001	75.00
0.005	76.67
0.010	76.67
0.030	75.00
0.050	71.67

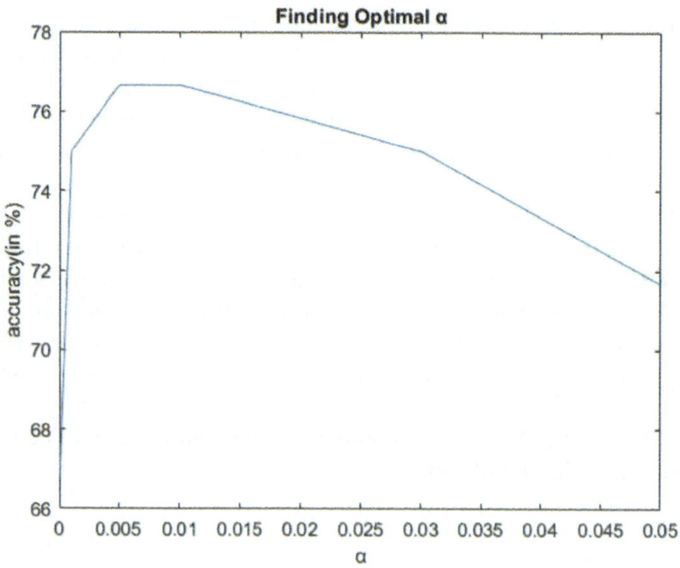

Fig. 4 Behaviour of SLID accuracy on changing α. Peak accuracy is seen to have been achieved somewhere between $\alpha = 0.005$, and $\alpha = 0.01$

5.2 *Feature Extraction*

After pre-processing phase, the pre-processed audio clips are ready for extracting features. We use the MFCC feature extraction technique to do so, as it is one of the best ways to compactly represent speech signals. The block diagram of the MFCC feature extraction algorithm is shown in Fig. 5.

The steps involved in this process are as follows:

(i) **Pre-emphasis** [4]: In this step, a high pass filter is used to increase the amplitude of the high frequency components of the speech signal. This is needed because in the voiced segments such as vowels, the lower frequencies have higher energy

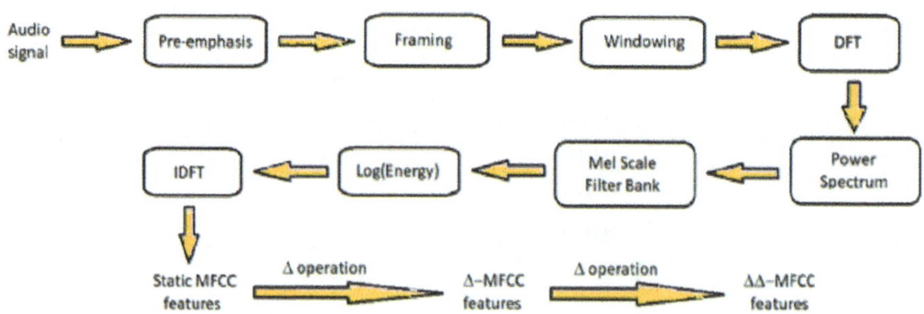

Fig. 5 Flowchart showing all the steps of the MFCC feature extraction process

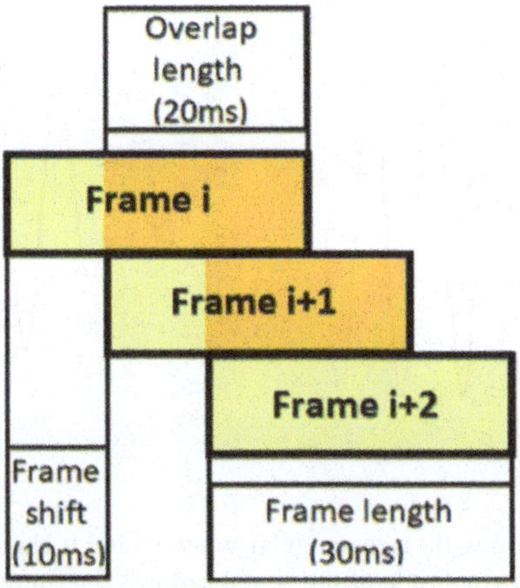

Fig. 6 Example showing the framing of an input speech signal. Each frame is 30 ms long, and overlap between consecutive frames is 20 ms

than the higher ones. The operation of filtering on the digital signal having k samples is given by:

$$y(n) = x(n) - \alpha x(n-1), \tag{1}$$

where $n \in [1, k]$, $x(n)$ is the input signal, $y(n)$ is the resultant signal, and α is a constant satisfying $\mathbf{0.9 < \alpha < 1.0}$ [8].

(ii) **Framing and Windowing**: The properties of an audio signal change over time. Hence, in order to analyze it, we need to slice up the signal, the length of each slice chosen in such a way that it would be reasonable to assume that the properties of the signal remain constant over that region. We slice the speech signal using sliding frames. The length of each frame is kept at **20–30 ms** and frame shift is **10 ms**, so overlap between two consecutive frames would be **10–20 ms**. This is shown in Fig. 6.

Now, we cannot just chop off the signal at the edges. This would result in a sudden change in amplitude at the edges which in turn would create a lot of noise that shows up in the high frequencies. To counter this, the amplitude is made to drop off gradually around the window boundaries by using a window function [8]. The output signal is represented by:

$$y(n) = w(n)x(n) \tag{2}$$

where $w(n)$ is the window function applied to the signal $x(n)$. Most common types of windows are Hamming window and Hanning window. Typically, a window function is of the form:

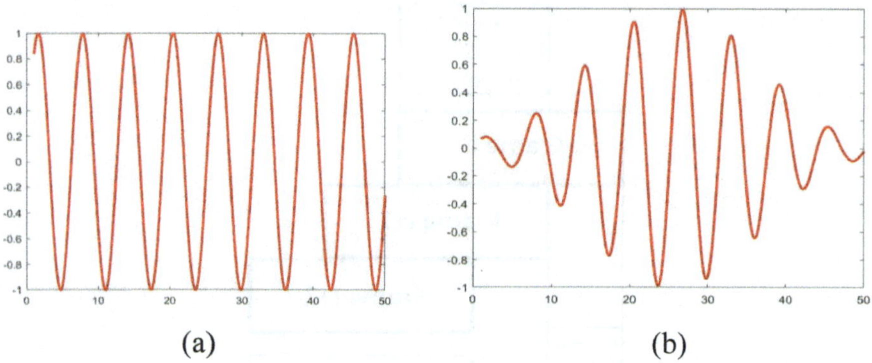

Fig. 7 Example of a sinusoid with: **a** rectangular window and **b** Hamming window applied to it. Note the gradual decrease in amplitude of the signal towards the frame boundaries in (b)

$$w(n) = \beta - (1 - \beta) * \cos\left(\frac{2\pi n}{L}\right) \quad (3)$$

The value of β is 0.54 for Hamming window, 0.5 for Hanning window, and 1.0 for rectangular window. Using a rectangular window does not change the signal. The effect of applying Hamming window on a sinusoid is shown in Fig. 7.

(iii) **Discrete Fourier Transform (DFT)** [31]: Next we extract spectral (frequency domain) information from the windowed signal. For this, DFT is applied.

(iv) **Power spectrum**: Now, the output of the DFT is squared. This gives us the power of the signal at each frequency bandwidth.

(v) **Mel Filter bank** [4]: Human hearing is not equally sensitive at all frequencies. It is less sensitive at higher frequencies, roughly above 1000 Hz. Besides, it has been seen that imitating this property while extracting features increases recognition performance of the model. To do this, a different frequency scale called the mel scale is used. The mel frequency mel(f) is obtained from the frequency f in Hz, by the equation:

$$\text{mel}(f) = 1127 * \ln\left(1 + \frac{f}{700}\right) \quad (4)$$

The linear power spectrum is converted to the mel-scale power spectrum by use of overlapping triangular bandpass filters, as shown in Fig. 8.

(vi) **Log Energy**: Next, we take the logarithm of the mel power spectrum values, since the human response to signal level is logarithmic. Doing so gives us the log energy values.

(vii) **Inverse DFT (IDFT)** [8]: Lastly, IDFT is applied to convert the signal to the time domain. The resulting coefficients are the MFCC. Usually, the first 13 coefficients are required as features.

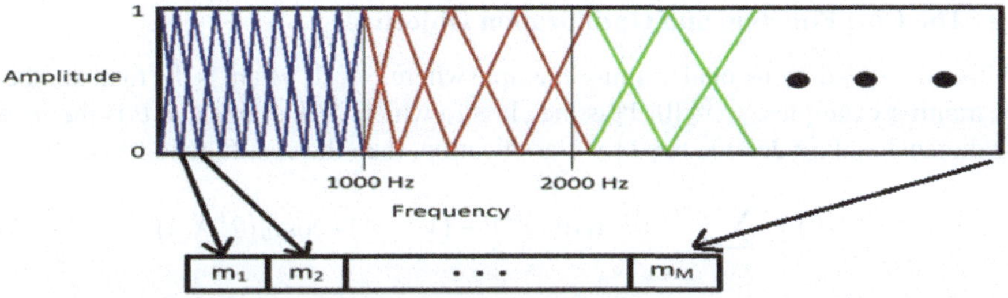

Fig. 8 Conversion of linear power spectrum to mel-scale power spectrum by use of overlapping triangular bandpass filters

(viii) **Velocity and Acceleration coefficients**: The above-mentioned coefficients are static MFCC features. They give us information about the properties of the waveform in a single frame. But these properties do not remain constant from frame to frame, and the changes in the signal over time could give us a useful cue about phone identity. In order to capture the dynamic features of the signal, delta (or velocity) and double delta (or acceleration) MFCC features are added. Delta and Delta-delta features are given by:

$$\Delta(i) = c(i) - c(i-1) \tag{5}$$

$$\Delta^2(i) = \Delta(i) - \Delta(i-1) \tag{6}$$

where, c is the MFCC feature vector, Δ is the delta feature vector, $\Delta^2 = \Delta\Delta$ is the delta-delta feature vector.

Finally, we have 39 features consisting of 13 MFCC, 13 delta MFCC and 13 double delta MFCC. These features are then normalized as $\frac{X_f - \mu_f}{\sigma_f}$ $\forall f \in [1, 39]$, where X_f is the distribution of a feature, with mean μ_f and variance σ_f^2. These features are then used for the purpose of SLID.

6 Description of SVM Classifier

The SVM classifier [16] is an enhancement of the logistic regression. SVM is highly robust and unlike basic logistic regression it is capable of adapting to complex non-linear dataset. Traditionally SVM is used to classify binary classes, but here we use its extended functionality to work for a multi-class system of 6 languages. First X, the train set is augmented with a column of 1 s, i.e. $X = [1, X]$, increasing X' dimension by 1.

(i) **The Cost Function and Optimization Objective**

X^i is a $(n+1)$-dimensional training example where $i \in [1, m]$; m is the total number of training examples, $y^i \in \{0, 1\}$ is the class to which X^i belongs and θ is the $(n+1)$ dimensional model, we need for classification, then the cost function J is:

$$J(\theta) = \sum_i \left(y^i \cdot \cos t_1\left(\theta^T X^i\right) + \left(1 - y^i\right) \cdot \cos t_0\left(\theta^T X^i\right)\right) \quad (7)$$

where $\cos t_0$ and $\cos t_1$ are functions needed that provide weights to positive and negative examples respectively. We need to minimize the cost function to get the optimal θ. The minimization objective [32] is:

$$\min_\theta \left(C \cdot J(\theta) + \frac{1}{2 \cdot m} \sum_{k=1}^{n} \theta_k^2\right) \quad (8)$$

where the second part is used for regularization, i.e. to prevent over-fitting.

(ii) **Robustness of SVM**

SVM classifies data by finding the best hyperplane that separates the data samples of the two classes. The *best* hyperplane for an SVM means the one with the largest *margin* between the two classes. Margin means the maximal width of the slab parallel to the hyperplane that has no interior data points. The *support vectors* are the data points that are closest to the separating hyperplane; these points are on the boundary of the slab. Figure 9a shows the robustness of SVM.

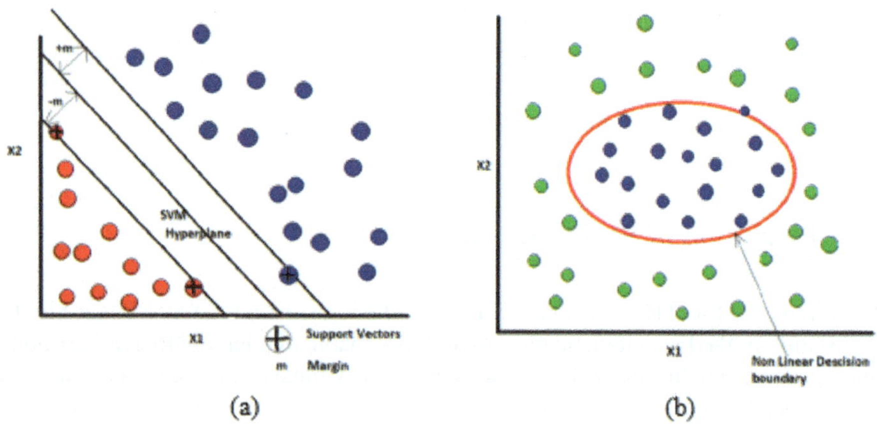

Fig. 9 Illustration of: **a** SVM as a Large Margin Classifier for linear data and **b** Kernels help SVM to separate non-linear data

(iii) **Separating Complex, Non-Linear Data: Kernels** [25]

SVM allows us to adapt to non-linear distribution with the help of kernel functions. Instead of using $\theta^T X^i$ directly, we use $\theta^T V^i$ where $V^i = f(X^i)$, and f is a function $f : X^i V^i$. Therefore, if we select f judiciously we can separate non-linear classes as shown in Fig. 9b. We have defined f to be radial basis function which uses the property of the Normal distribution, i.e. it computes the closeness/similarity of the sample to the landmarks chosen. For $(n + 1)$ dimensional features, landmark L is defined as a $n \times (n + 1)$ matrix (has 1 column of 1 s), and f_j is defined as:

$$f_j(X^i) = V_j^i = \begin{cases} e^{\frac{-(X^i - L_j)^T \cdot (X^i - L_j)}{2 \cdot \sigma^2}} & 1 \le j \le n \\ 1 & j = 0 \end{cases} \tag{9}$$

Here σ^2 is the variance/width of the kernel. So from $n + 1$ dimensional X^i we obtain $n + 1$ dimensional V^i, which is in turn used in the cost function. Clearly as $\boldsymbol{X^i L_j}$, $\boldsymbol{V_j^i} \to 1$ and $\|\boldsymbol{X^i} - \boldsymbol{L_j}\| \gg 0$, $\boldsymbol{V_j^i} \to 0$, which virtually enforces a boundary of a class, eventually separating the two.

(iv) **Converting SVM to a Multiclass Classifier** [13]

We have a classification system of six classes representing six different languages. Thus, SVM which is traditionally binary class classifier must be redesigned to train for a multi-class dataset. We label the languages as $\text{lang}_p \forall p \in [1, 6]$. Therefore, $y^i \in [1, 6]$. Then, we define a 'one-vs-all' approach. We define $\theta^p \forall p \in [1, 6]$, θ^p is the model for the pth language. The multi-class system is trained as follows:

Repeat for k in 1:6: replace $\forall i$, the value of y^i with 1's, such that, $y^k = k$. The rest is set to 0.

We obtain y^k. SVM is trained with X and y^k to obtain θ^p.

6.1 Testing a Classifier and Classifying an Audio Clip

Once $\theta^p \forall p$ is trained, we need to test the model θ on a test set. Our test set contains 100 clips for each language. Unlike training, testing a clip frame-wise would be meaningless as the objective is to classify a clip and not a frame. Therefore, a standard must be defined to classify a clip $C_k \ \forall k \in [1, 600]$ which contains with feature vector $X_j^i \ \forall i \in [1, m_k]$ where m_k is the number of MFCC feature vectors derived from C_k, frame-wise.

Classifying Binary System frame wise: For X_k^i, if $(\boldsymbol{\theta^p})^T \boldsymbol{X_k^i} \ge 0$ then X_k^i is classified to belong to the positive class (i.e. 1), otherwise the negative class (i.e. 0).

Classifying a Multiclass System Clip wise: Here the objective is to find the most probable language lang_k, given an audio clip C_k, i.e. $\boldsymbol{lang_k} = \max_{i \in p} P\left(\frac{i}{C_k}\right)$. This is done by the following algorithm:

Repeat for all X_k {
Repeat for all j ϵ p {
$count_k$ ← number of frames of X_k reported to be positive by the model θ^p
}
$lang_k \leftarrow \max_j(Count_j)$
}

7 Performance Metrices Used

We define Table 2 for all the languages. True/False indicates whether the prediction is correct. On the other hand, Positive/Negative indicates whether the classifier predicts a sample to belong to a class or not. If a clip C_k belongs to class p and is predicted to belong to class p', then the following updates are made to the table of p and p':

$$If\ p = p' \text{ then } \mathrm{TP}_p \leftarrow \mathrm{TP}_p + 1,$$

$$\text{Otherwise } \mathrm{FP}_p{}' \leftarrow \mathrm{FP}_p{}' + 1, \text{ and } \mathrm{TN}_p \leftarrow \mathrm{TN}_p + 1$$

(i) **Accuracy**:

$$A(\text{in }\%) = \frac{\sum_p(\mathrm{TP}_p)}{\text{Total number of clips}} \times 100\% \tag{10}$$

(ii) **Precision**:

$$P = \frac{\sum_p(\mathrm{TP}_p)}{\sum_p(\mathrm{TP}_p) + \sum_p(\mathrm{FP}_p)} \tag{11}$$

Table 2 Confusion matrix obtained by the proposed model while identifying three languages with 13 MFCC features

Predicted	Actual		
	Bangla	English	Hindi
Bangla	96	4	0
English	7	77	16
Hindi	2	3	95

(iii) **Recall**:

$$R = \frac{\sum_p (\mathrm{TP}_p)}{\sum_p (\mathrm{TP}_p) + \sum_p (\mathrm{FN}_p)} \tag{12}$$

(iv) **$\mathbf{F_1}$ Score**:

$$F_1 = \frac{2PR}{P + R} \tag{13}$$

In a multi-class classification problem, Precision, Recall, and F1 score have to be defined for every class.

(v) **Confusion Matrix:** It is a $p \times p$ matrix, with CM_{ij} indicating how many times a test example of class i is predicted to be a clip of class j. Therefore, $\sum_{i=j} CM_{ij}$ (sum of diagonal) gives us the total number of correct predictions.

8 Observations

All the experiments are implemented in MATLAB 2018a under a Windows 10 environment on an Intel Core 2 Duo 2.4 GHz processor with 8 GB of RAM. Feature extraction is performed using the Signal Processing Toolbox [27], and the training and testing of the SVM model are done using functions in the Statistics and Machine Learning Toolbox [30]. From the database of audio clips mentioned in (Section II), clips of both male and female speakers for each language is chosen in equal amounts. The selected clips are separated into 2 sets (train set and test set). The SVM is trained and the train accuracy is recorded to find out instances of underfitting/overfitting. The train set and the test set are kept the same for 13, 26, and 39 dimensional feature vectors. Our train set contains enough data (around 1000 clips for each language). From this huge set, we take a randomly selected subset, which acts as the data to train the model. The metric for the size of the train set is total number of MFCC feature vectors generated per language, which in turn is the measure of the duration of train data (in seconds).

Each of the clip is sampled as follows:

1. Sampling rate: 48,000 Hz, the largest frequency is (48,000 * 0.5 = 24,000 Hz), i.e. the Nyquist frequency [22].

So, each second of clip produces 48,000 samples.

2. Window Size: A 30 ms window is chosen and the window is shifted by 10 ms. (as shown in Fig. 2)

Therefore the 48,000 samples (1 s clip) produces around 100 i.e. (48,000/(48,000 * 0.01)) MFCC feature vectors. Thus the number of feature vectors $M = 100$ * (total duration of training set in secs.).

Table 3 Performance evaluation results of the proposed SLID model while identifying 3 languages with 13 MFCC features

S. No.	Language	Number of test clips	Accuracy (in %)	Precision (in %)	Recall (in %)	F_1 Score (in %)
1	Bangla	100	96.00	91.42	96.00	93.65
2	English	100	77.00	91.67	77.00	83.69
3	Hindi	100	95.00	85.58	95.00	90.08
Total		300	89.33	89.56	89.33	89.14

The above scheme makes the train set generation independent of the number of clips (only dependent on total duration). This prevents the training set from being skewed (as the clips are of varying length).

Once the training is done, models of all our experiments are tested on the same test set (containing 100 clips per language) and their performance is noted.

8.1 Experiments with Three Languages

For identification of three languages, training is done with over 75,000 training examples per language.

(I) **13 MFCC Features (static)**: Train accuracy is 94.47%. The test results achieved are given in Tables 2 and 3.

It is evident from the results shown in Tables 2 and 3, that the classifier does not perform well in identifying English. This might be because of the different accents of the speakers of English, which occurs due to the usage of slightly different phones for the same phoneme that has made the job of the classifier difficult.

(II) **26 MFCC Features (static and velocity (delta))**: Train accuracy is 83.16%. On the test set, we have obtained an accuracy of 61.00%.

(III) **39 MFCC Features (static, velocity and acceleration (delta-delta))**: Train accuracy is 59.95%. Test accuracy obtained is 43.67%.

8.2 Experiments with Six Languages

(I) **Learning Curves**: A classifier fails to improve accuracy beyond a certain limit when the features lack discriminatory power, even if more data is fed to the model. Any further addition of training examples is deemed to be a unnecessary wastage of resources. Thus, it is essential to have knowledge about the saturation point of the classifier for a particular feature set. This is exactly what a learning curve does. It plots the effect of increasing the number of training examples on train and test errors. Train and test errors are computed for 25,000, 50,000, 75,000 and 100,000 training samples (for each language) with 13 features. Table 4 shows the results obtained from the above experiment and Fig. 10 illustrates it. It is clear from Fig. 10 that the performance of the classifier converges when the number of training samples is increased beyond 75,000 per language. Therefore, we train our classifier with 75,000-100,000 training samples (per language) for rest of the experiments.

(II) **13 MFCC Features (static)**: The model is trained with about 75,000 samples per language. Train accuracy obtained is 77.83%. The results obtained are tabulated in Tables 5 and 6.

It is clear from Table 5 that the classifier performs poorly in classifying English and Telugu audio clips. A reason for poor performance in classifying English clips, as speculated earlier, could be due to the wide variety in the accents of the speakers (influenced by their mother tongue). The accent of a Bengali man has very little resemblance with the accent of a South Indian man. This wide variety makes the

Table 4 Observations of learning curves experiment

Training samples (per language)	Train error (in %)	Test error (in %)
25,000	18.75	42.17
50,000	19.42	41.34
75,000	22.16	39.67
100,000	22.22	38.50

Train error and test error has been recorded as a function of number of training examples

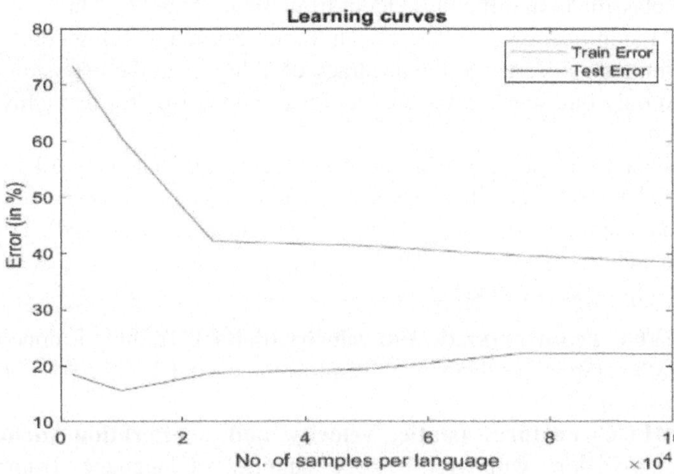

Fig. 10 Illustration of the learning curves of six languages using 13 MFCC features. It is evident that both train error and test error stay almost the same after number of training samples crosses 75,000

Table 5 Performance evaluation results of the proposed SLID model while identifying 6 languages with 13 MFCC features

S. No.	Language	Number of test Clips	Accuracy (in %)	Precision (in %)	Recall (in %)	F_1 Score (in %)
1	Bangla	100	46.00	51.11	46.00	48.42
2	English	100	19.00	86.36	19.00	31.15
3	Hindi	100	76.00	58.02	76.00	65.80
4	Tamil	100	99.00	52.66	99.00	68.75
5	Marathi	100	100.00	70.42	100.00	82.64
6	Telugu	100	22.00	81.48	22.00	34.65
Total		600	60.33	66.68	60.33	55.24

Table 6 Confusion matrix obtained by the proposed model while identifying 6 languages with 13 MFCC features

Predicted	Actual					
	Bangla	English	Hindi	Tamil	Marathi	Telugu
Bangla	46	1	17	32	4	0
No. of languages	9	19	38	23	7	4
No. of languages	1	1	76	12	9	1
No. of languages	1	0	0	99	0	0
No. of languages	0	0	0	0	100	0
No. of languages	33	1	0	22	22	22

problem of classification difficult (evident from Tables 5 and 6). On the other hand, Telugu has a significant resemblance with Tamil. So, many Telugu clips are classified as Tamil. This decreased the accuracy of Telugu and the precision of Tamil. Marathi had only one speaker, which could the main reason for its highly accurate identification.

It is evident from the poor test and train accuracies that these 13 features do not have good enough capability to recognize unseen data and also accommodating seen data into the model (when the number of languages is increased). This may be attributed to underfitting of the model. Thus, we need to look for more useful features that would help make the model better.

(III) **26 MFCC Features (static and velocity (delta))**: Training is done with over 100,000 samples per language. Train accuracy is 62.09%. We obtain a test accuracy of 42.83%.

(IV) **39 MFCC Features (static, velocity and acceleration (delta-delta))**: Training is done with over 100,000 samples per language. Train accuracy is 37.46%. Test accuracy obtained is 30.16%.

The observations of the above experiments are summarized in Table 7 and plotted in Fig. 11. Results shown in Table 7 clearly shows that accuracy decreases drastically

Table 7 List of accuracies achieved for different no. of languages and different no. of features

No. of features	No. of languages		
	13	26	39
3	**89.33**	61.00	43.67
6	**60.33**	42.83	30.16

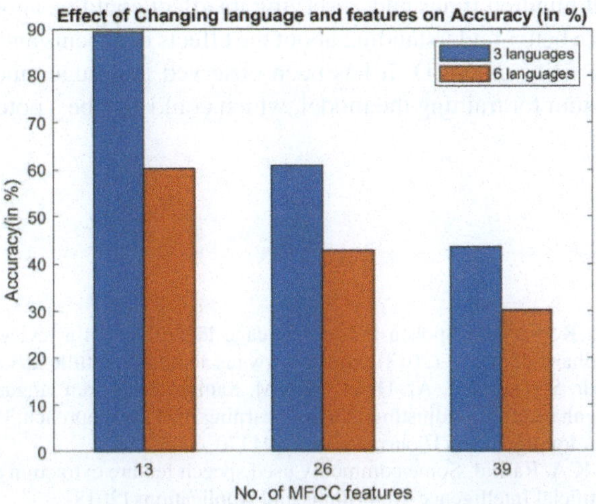

Fig. 11 Illustration of the effect of changing number of languages and number of features on test accuracy. We see that the accuracies obtained using 13 features are greater than those obtained using 26 features, which in turn are better than those obtained using 39 features

when the velocity and acceleration features are added. This leads us to conclude that these features are not good for this particular experiment, and they mislead the classifier. Also, accuracy decreases when number of languages are increased, which is expected, since more languages means more overlap of 'phones' amongst different languages, which confuse the classifier. Moreover, the *one-vs-all* approach skews the training data (much lesser positive samples compared to negative ones) for each language. This also decreases accuracy of the classifier.

9 Conclusion

In this study, we have made an attempt to investigate into the effectiveness of static, delta and delta-delta MFCC features, and SVM classifier for identification of Indian languages. In the process, we found (rather unexpectedly), that including delta and delta-delta features worsened the performance of the classifier. Hence, we can claim that these features, calculated as given in Eqs. (5) and (6), are not useful for identification of the targeted set of languages. There are other more complex methods of calculating these features, which might perform a better job.

Although maximum accuracy is achieved using 13 MFCC features, there is much room for improvement. Further investigation about other feature extraction techniques, use of other classifiers like the GMM, or deep learning models such as CNN, RNN could be considered as a future scope of this work. Increasing the number of speakers several hundred times and observing its effect could be another challenge. This can give us a better understanding about the effects of accents and other speaker-related characteristics on SLID. It has been observed that huge amount of time is taken by the system for training the model, which could also be a potential scope for further research.

References

1. B. Aarti, S.K. Kopparapu, Spoken Indian language identification: a review of features and databases. Sādhanā **43**(4), 53 (2018). https://www.ias.ac.in/article/fulltext/sadh/043/04/0053
2. M.A.A. Albadr, S. Tiun, F.T. AL-Dhief, M.A.M. Sammour, Spoken language identification based on the enhanced self-adjusting extreme learning machine approach. PLOS ONE **13**(4) (2018). https://doi.org/10.1371/journal.pone.0194770
3. S.A. Alim, N.K.A. Rashid, Some commonly used speech feature extraction algorithms. From Natural to Artificial Intelligence-Algorithms and Applications (2018)
4. B. Barai, D. Das, N. Das, S. Basu, M. Nasipuri, VQ/GMM-based speaker identification with emphasis on language dependency. In *Advanced Computing and Systems for Security* (Springer, 2019), pp. 125–141. https://www.researchgate.net/profile/Bidhan_Barai/publication/330414980_VQGMM-Based_Speaker_Identification_with_Emphasis_on_Language_Dependency_Volume_Eight/links/5c5043ad299bf12be3eb7d6a/VQ-GMM-Based-Speaker-Identification-with-Emphasis-on-Language-Depen
5. P. Beckmann,M. Kegler, H. Saltini, M. Cerňak, Speech-VGG: A deep feature extractor for speech processing (2019). https://arxiv.org/pdf/1910.09909.pdf
6. P. Bhaskararao, Salient phonetic features of Indian languages in speech technology. Sadhana **36**(5), 587–599 (2011). https://www.ias.ac.in/article/fulltext/sadh/036/05/0587-0599
7. M. Gupta, S.S. Bharti, S. Agarwal, Implicit language identification system based on random forest and support vector machine for speech, in *2017 4th International Conference on Power, Control & Embedded Systems (ICPCES)*, pp. 1–6 (2017). https://ieeexplore.ieee.org/document/8117624
8. J. Hui, Speech Recognition—Feature Extraction MFCC & PLP (2019). https://medium.com/@jonathan_hui/speech-recognition-feature-extraction-mfcc-plp-5455f5a69dd9
9. *Importance of Language in Society*, pp. 1–35 (n.d.). https://shodhganga.inflibnet.ac.in/bitstream/10603/29223/17/9_chapter 1.pdf
10. B. Jiang, Y. Song, S. Wei, J.-H. Liu, I.V. McLoughlin, L.-R. Dai, Deep Bottleneck Features for Spoken Language Identification. PLoS ONE **9**(7), e100795 (2014). https://doi.org/10.1371/journal.pone.0100795
11. P. Kumar, A. Biswas, A. Mishra, M. Chandra, Spoken language identification using hybrid feature extraction methods. J. Telecommun. **1**(2), 11–15 (2010). https://www.researchgate.net/publication/45909010_Spoken_Language_Identification_Using_Hybrid_Feature_Extraction_Methods/citation/download
12. *Learn Natural Language Processing: From Beginner to Expert* (2020). https://www.commonlounge.com/discussion/3ecabc3d82684d57a62ad8fbc200f43b

13. G. Madzarov, D. Gjorgjevikj, Multi-class classification using support vector machines in decision tree architecture, in *IEEE EUROCON 2009*, pp. 288–295 (2009). https://www.researchgate.net/publication/224564327_Multi-Class_Classification_Using_Support_Vector_Machines_In_Decision_Tree_Architecture
14. S. Maity, A.K. Vuppala, K.S. Rao, D. Nandi, IITKGP-MLILSC speech database for language identification, in *2012 National Conference on Communications (NCC)*, pp. 1–5 (2012). http://cdn.iiit.ac.in/cdn/speech.iiit.ac.in/svlpubs/conference/sudhamay-anil.pdf
15. S. Manchala, V.K. Prasad, V. Janaki, GMM based language identification system using robust features. Int. J. Speech Technol. **17**(2), 99–105 (2014). https://link.springer.com/article/10.1007/s10772-013-9209-1
16. J.M. Moguerza, A. Muñoz, Support vector machines with applications. Stati. Sci. **21**(3), 322–336 (2006)
17. S. Mohanty, Phonotactic model for spoken language identification in Indian language perspective. Int. J. Comput. Appl. **19**, 18–24 (2011). https://doi.org/10.5120/2389-3164
18. H. Mukherjee, A. Dhar, S.M. Obaidullah, S. Phadikar, K. Roy, Image-based features for speech signal classification. Multimedia Tools Appl., 1–17 (2020). https://doi.org/10.1007/s11042-019-08553-6
19. H. Mukherjee, S. Ghosh, S. Sen, O. Sk Md, K.C. Santosh, S. Phadikar, K. Roy, Deep learning for spoken language identification: can we visualize speech signal patterns? Neural Comput. Appl. **31**(12), 8483–8501 (2019). https://doi.org/10.1007/s00521-019-04468-3
20. H. Mukherjee, S.M. Obaidullah, K.C. Santosh, S. Phadikar, K. Roy, A lazy learning-based language identification from speech using MFCC-2 features. Int. J. Mach. Learn. Cybernet. **11**(1), 1–14 (2020). https://doi.org/10.1007/s13042-019-00928-3
21. N. Krishna, A. Patil, M.S. Prince, S. Sai, P. Garapati, *Identification of Indian Languages using Ghost-VLAD pooling* (2020). https://www.researchgate.net/publication/339065645_Identification_of_Indian_Languages_using_Ghost-VLAD_pooling
22. *Nyquist-Shannon Sampling Theorem*, Wikipedia (n.d.). https://en.wikipedia.org/wiki/Nyquist–Shannon_sampling_theorem
23. B. Padi, A. Mohan, S. Ganapathy, Towards relevance and sequence modeling in language recognition. IEEE/ACM Trans. Audio Speech Lang. Process. **28**, 1223–1232 (2020). https://doi.org/10.1109/TASLP.2020.2983580
24. A.D. Patil, Spoken language identification using machine learning, in *Project report submitted to M S Ramaiah Institute of Technology* (Issue May 2012). http://timewarp.adarshpatil.in/misc/LID_final_report.pdf
25. A. Patle, D.S. Chouhan, SVM kernel functions for classification. Int. Conf. Adv. Technol. Eng. (ICATE) **2013**, 1–9 (2013). https://doi.org/10.1109/ICAdTE.2013.6524743
26. U. Shrawankar, V.M. Thakare, Techniques for feature extraction in speech recognition system: a comparative study (2013). *ArXiv Preprint* ArXiv:1305.1145. https://arxiv.org/ftp/arxiv/papers/1305/1305.1145.pdf
27. *Signal Processing Toolbox*, Mathworks (n.d.). https://www.mathworks.com/help/signal/
28. E. Singer, P. Torres-Carrasquillo, T. Gleason, W. Campbell, D. Reynolds, Acoustic, phonetic, and discriminative approaches to automatic language identification, in *Eighth European Conference on Speech Communication and Technology*, vol. 1 (2003). https://www.researchgate.net/publication/221489129_Acoustic_phonetic_and_discriminative_approaches_to_automatic_language_identification
29. *Speech and Music Technology Lab IIT Madras*, IIT Madras Speech Corpus (n.d.). https://www.iitm.ac.in/donlab/tts/database.php
30. *Statistics and Machine Learning Toolbox*. Mathworks (n.d.). https://www.mathworks.com/help/stats/index.html
31. G. Strang, Linear algebra and its application, in *Linear Algebra 4th Edition*, pp. 211–221, Chap. 3.5 (n.d.-a). http://facultymember.iaukhsh.ac.ir/images/Uploaded_files/[Strang_G.]_Linear_algebra_and_its_applications(4)[5881001].PDF

32. G. Strang, Linear algebra and its application. In *Linear Algebra 4th Edition*, pp. 180–195, Chap. 3.3 (n.d.-b). http://facultymember.iaukhsh.ac.ir/images/Uploaded_files/[Strang_G.]_Linear_algebra_and_its_applications(4)[5881001].PDF
33. A. Titus, J. Silovsky, N. Chen, R. Hsiao, M. Young, A. Ghoshal, *Improving Language Identification for Multilingual Speakers* (2020). https://arxiv.org/pdf/2001.11019.pdf

Performance Evaluation of One-Class Classifiers (OCC) for Damage Detection in Structural Health Monitoring

Akshit Agarwal, Varun Gupta, and Dhiraj

Abstract Structural Health Monitoring (SHM) has become an area of continuous research with the ever-increasing demand for the safety of civil structures. The damage in civil structures can be detected using multimodal data from sensors, which presents instances of both damaged and undamaged data. The availability of damaged data in real life is difficult to obtain from a healthy structure and hence the problem of damage detection needs to be attempted using normal healthy data, and it becomes synonymous with the anomaly or novelty detection. One-Class classifiers work on the principle that the abundance of healthy data can be used to model an envelope of conditions which, if violated by any data instance, can be termed as damage or outlier detection. We have attempted an array of classifiers on a benchmark structure dataset (IASC-ASCE) both from supervised and unsupervised machine learning domain and propose a comparison between their success rates in determining damage in civil structures. We used classical techniques such as One-Class Support Vector Machines (OC-SVM), One-Class Isolation Forest (OC-IF), One-Class K-means clustering (OC-KMC), One-Class K-nearest neighbors (OC-KNN), Density-Based Spatial Clustering of Applications with Noise (DBSCAN), One-Class Principal Component Analysis (OC-PCA), Local Outlier Factor (LOF) and One-Class Gaussian Distribution (OC-GD). These techniques were tested on the IASC (International Association for Structural Control)–ASCE (American Society of Civil Engineers) SHM benchmark problem for a range of noise levels and range of force intensities to cover wide variations in the generated dataset using MATLAB based simulation. Our study helps us conclude that OC-SVM, Isolation Forest, and OC-PCA are the most robust algorithms for the anomaly detection task.

Keywords Anomaly detection · Machine learning · Structural health monitoring · Support vector machines · Isolation forest · K-means clustering · K-nearest

---

A. Agarwal · V. Gupta
BITS, Pilani, India

Dhiraj (✉)
CSIR-CEERI, Pilani, India
e-mail: dhiraj@ceeri.res.in

P. Kumar et al. (eds.), *Machine Learning for Intelligent Multimedia Analytics*, Studies in Big Data 82,
https://doi.org/10.1007/978-981-15-9492-2_13

neighbors · Density-based spatial clustering of applications with noise · Principal component analysis · Local outlier factor · Gaussian distribution

1 Introduction

Mankind has been living and building civil structures for themselves for more than 11,000 years. Every year, the earth suffers from a great deal of recurring natural disasters like floods, earthquakes, and landslides, etc. It causes loss of numerous lives and a significant impact on the economy. Such incidents have reminded mankind the importance of structural safety and have led to the development of various damage detection and identification methods. Structural Health Monitoring (SHM) is one such scientific group of techniques that aims to analyze the health of the structure over a period of time and can be used to prevent further deterioration of the structure and loss of lives.

Structural Health Monitoring (SHM) [1] is commonly defined as the process of implementing a damage detection and characterization strategy for engineering structures. It is aimed at providing precise diagnosis about the constituent materials and their "state" at any given point of time. Due to continuous monitoring of the structure and better computation techniques available at our disposal, it is also possible to do a prognosis on the structure. The advancement in Electronics and Internet-of-Things (IoT) has enabled us to integrate the various sensors in the structure design for SHM. The performance evaluation study of different one-class algorithms has not been performed on the IASC-ASCE benchmark structure [2]. So, we aim to perform anomaly detection [3] on the stated benchmark structure and present a comparative study on the same.

Anomaly detection is of great interest to diverse fields, including data mining and machine learning, and plays a critical role in a wide range of applications, such as medical health, credit card fraud, and intrusion detection. It aims to identify those regions from data whose behaviours or patterns do not conform to expected values. The unexpected behaviours, which are significantly different from those of the remainder of the given data, are commonly called anomalies [4].

There are numerous one-class classification techniques available in the Machine Learning domain for anomaly detection. This paper aims to evaluate the performance of the various One-Class Classifiers (OCC) [5] available for anomaly detection on a reference benchmark structure in the Earthquake Engineering Research Laboratory at the University of British Columbia (UBC) [2]. The data was collected using a Matlab based simulation, which modeled the response of the real structure under different noise levels and force excitation scenarios and generated the data for different damaged and undamaged cases. The damaged states were produced by various mechanisms, as described in Sect. 3.2. The classifiers were trained on the undamaged data and then evaluated on the damaged data. This is because one class classifiers work on the property that only one class, i.e., the undamaged class, is available for training, and the classifier should then be able to predict if there has been some damage done to the structure.

2 Related Works

Anomaly detection is a very popular and heavily-researched field these days [6]. There have been numerous publications [7] trying to solve the anomaly detection problem in different ways. The problem of anomaly detection in time series data [8] has been solved using window sliding [9] as well as the use of deep neural networks [10]. There is also a use of probabilistic methods like Hidden Markov Models [11] or Bayesian Networks [12], where the probability of test data is compared with the threshold obtained using training data and points from the test dataset are accordingly classified as anomalous. But these methods do not scale well and are computationally expensive as well.

Besides studies evaluating a single algorithm only, outlier ensembles [13, 14] is a technique of combining multiple unsupervised anomaly detection algorithms in order to boost their joint anomaly detection performance. Since unsupervised anomaly detection [15] does not rely on labeled data, this task is very challenging and often restricted to simple combinations [16]. There has also been a lot of work done using neural networks on SHM [17].

Anomaly detection is critical given the increase in cyber attacks in wireless communication these days. Thus, it becomes important to detect network anomalies more accurately. Techniques dealing with network anomaly detection use autoencoders because of their ability to capture the non-linear correlations between features so as to increase the detection accuracy [18].

There are a lot of anomaly detection techniques which deal with low-dimensional data. However, developing effective anomaly detection methods for complex and high-dimensional data [19] like images remains a challenge. Since Generative Adversarial Networks (GANs) [20] are able to model the complex high-dimensional distributions of real-world data, they offer a promising approach to address this challenge [21, 22].

This paper deals with a comparative study of the performance of various one-class classifiers on the benchmarking dataset and concludes that some of these algorithms adjust really well to the anomaly detection on a one-class problem because of certain reasons which will be explained through this paper.

3 Dataset and Techniques

The multimedia dataset which includes the magnitude versus time response of the structure was collected using a Matlab based simulation, which modeled the response of the real structure under different noise levels and force excitation scenarios. It was similar to the other multimedia signals obtained from the accelerometers in the structure. The data was obtained in its raw form as Amplitude-Time variations.

3.1 Model Structure

Benchmarking is a way of discovering the best performance attained by a different set of algorithms. A "benchmark" dataset or structure is used for the purpose of benchmarking. Naturally, conducting a performance evaluation study incorporates a benchmark. For our purpose of developing and testing Structural Health Monitoring (SHM) [1], we use a civil reference structure [2]. The experimentation phase of the study includes the Phase-I IASC-ASCE structure.

The structure shown in Fig. 1 is a 4-story, 2-bay by 2-bay steel frame scale-model structure in the Earthquake Engineering Research Laboratory at the University of British Columbia (UBC) [23]. The structure stands tall at 3.6 m and has a plan of 2.5 m × 2.5 m. It is equipped with hot rolled grade 300 W steel that has a nominal yield stress of 300 MPa (42.6 KPSI). Its columns are oriented to have stronger bending towards the x-direction and weaker bending towards the y-direction. The floor beams, on the other hand, are oriented to have stronger bending vertically, i.e., about the y-axis. The absence of bending stiffness in the braces makes their orientation irrelevant. The round dots in Fig. 1 represent the placement of the accelerometers. There are four accelerometers on each floor, and there are four floors, giving a total of 16 accelerometers.

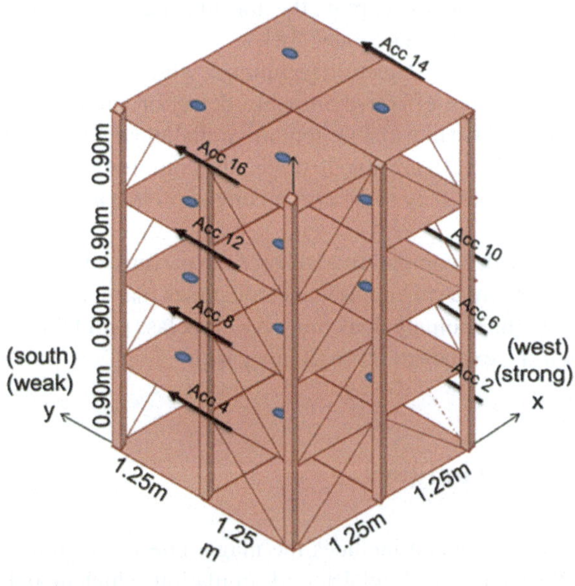

Fig. 1 Undamaged pattern for our study (UD)

Undamaged Pattern (UD) has all the braces in position (Fig. 1). There are no loose beam-connections in the structure. It is the training pattern for our study.

3.2 Damage Patterns Description

The benchmarking structure, in addition to the Undamaged dataset (UD), incorporates six different damage patterns. These damage patterns advance from the easiest to detect cases to those that are more demanding in terms of detection. The damage patterns were designed to test the classifiers' ability to detect, localize, and quantify the damage caused due to earthquakes. These damaged patterns are defined as follows:

1. **Damage Pattern 1 (D1)**: Absence of any kind of stiffness in the braces of the first story of the structure (i.e., the braces will still continue contributing to the mass, but they will not provide any kind of resistance within the structure) (Fig. 2i).
2. **Damage Pattern 2 (D2)**: Absence of stiffness in the braces of the first and third stories of the structure (Fig. 2ii).

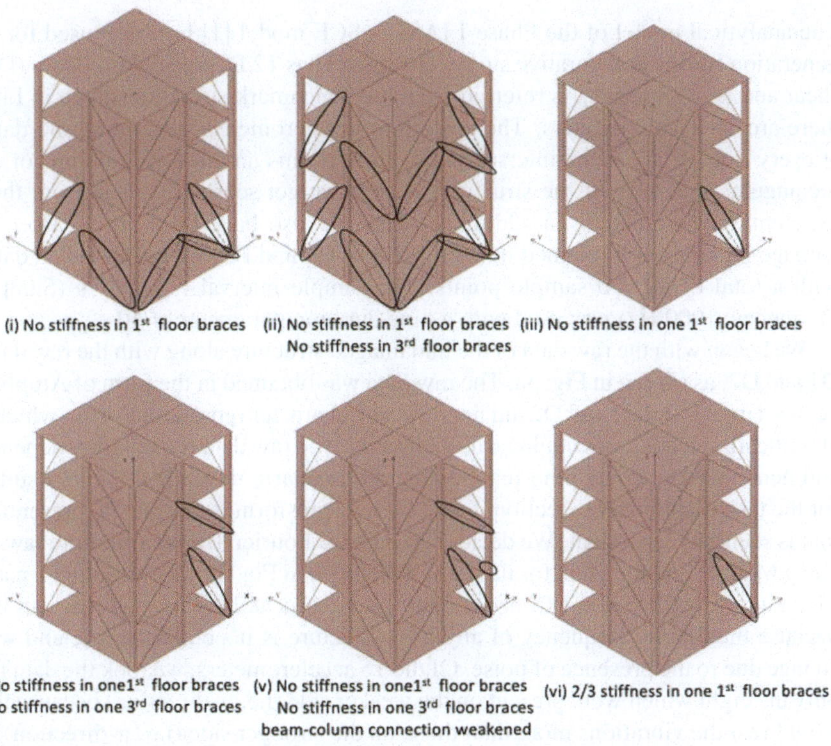

Fig. 2 Six damaged patterns from the benchmark problem

3. **Damage Pattern 3 (D3)**: Absence of stiffness in a particular brace (north brace on the west face of the structure; noted with an ellipse) of the first story (Fig. 2iii).
4. **Damage Pattern 4 (D4)**: Absence of stiffness in the first story (north brace on the west face) and also in the third story (west brace on the north brace) (Fig. 2iv).
5. **Damage Pattern 5 (D5)**: This damage pattern is same as D4 but with the north floor beam at the first level on the west face of the structure (i.e., the beam from [(2.5 m, 0, 0.9 m) to (2.5 m, 1.25 m, 0.9 m)] partially unscrewed from the northwest column at (2.5 m, 0, 0.9 m).Therefore, the beam-column connection there can only transmit forces and cannot sustain any bending moments (Fig. 2v).
6. **Damage Pattern 6 (D6)**: Two-third stiffness (i.e., a one-third stiffness loss) in one brace in the first story (the same brace damaged in D3, i.e., the north brace on the west face) (Fig. 2vi).

The black elliptical shapes in Fig. 2 indicate the absence of stiffness in certain braces in each damage case and has been done to make the reader easily visualize the damage patterns.

3.3 Dataset

The analytical model of the Phase-I IASC-ASCE model [1] has been used for data generation for the comparative study. The model has 12 Degrees of Freedom (DOF) shear and is symmetric. As referred from the benchmark structure shown in Fig. 1, there are 16 accelerometers. There are four accelerometers generating the dataset at every floor level, at the intersection of floor beams and middle columns of each rectangular wall face of the structure. Hence, we get sensor readings from the 16 different accelerometers. In addition to UD, we also have D1, D2, …, D6 as the damage patterns, as explained in Sect. 3.2. For D1 and D2, we have made a dataset with a total of 40,000 sample points. The sample interval of 0.001 s (Sampling Frequency, 1000 Hz) was used with a time duration parameter of 40 s.

We began with the raw data of the undamaged structure along with the raw data of D1 and D2, as shown in Fig. 3a. The raw data was obtained in the form of Amplitude versus Time. UD, D1, and D2 did not have any character representation on which the classification techniques could be used directly. The raw data is excitation dependent and hence dataset containing features that are scenario dependent is more suitable for the task of anomaly detection. Fast Fourier Transform is one such representation that is scenario dependent. We decided to take the Fourier Transform of the raw data using Matlab, and the plots for the same are shown in Fig. 3b. The structure's natural frequency was filtered out from the data using the peak heights, and this was done because the natural frequency of any civil structure is it's characteristic and won't change due to the presence of noise. Of the 16 accelerometers, we took the data from only the eight which were present on the weaker side (i.e., *y*-direction) because these would feel the vibrations more than those on the stronger side (i.e., *x*-direction). We decided to take into consideration only the first four natural frequencies because, after

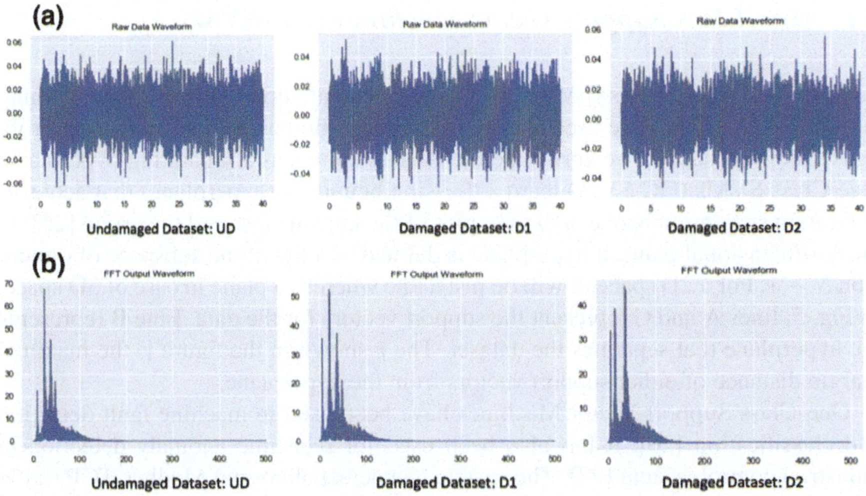

Fig. 3 (**a**) Raw data form of UD, D1, D2, (**b**) Fourier Transform plots of UD, D1, D2

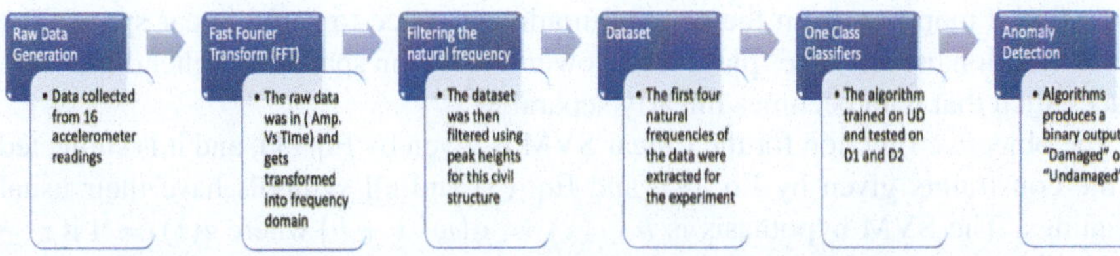

Fig. 4 Flowchart for the data processing, training and output for one-class classifiers

that, the natural frequency becomes weak enough to capture the structure-property, and hence we were not left with a huge amount of data. A pipeline for the process is shown in Fig. 4.

4 One-Class Classification Techniques

The anomaly detection is one of the classical problems that the scientific community has researched in detail [6]. The dataset often used for the training in the anomaly detection consists of a class that represents normal behavior. The data for the other class is often unavailable due to rare occurrences. A class of classifiers known as One-Class classifiers is used for the task of anomaly detection. The One-Class Classification (OCC) is a technique that uses samples mainly from the normal data class. It may not or may have few samples from the "abnormal" or anomalous class. In this section, we aim to discuss some of the commonly used algorithms for the task of anomaly detection.

4.1 One-Class Support Vector Machines (OC-SVM)

Support Vector Machines (SVM) are traditional semi-supervised machine learning algorithms [24]. They are used widely for classification as well as regression problems. They use support vectors to define what is known as a separating hyperplane. One-Class SVMs (OC-SVM) learn a decision boundary (hyperplane) that achieves maximum separation between the samples of the known class and the origin [25]. In the N-dimensional plane, a hyperplane is defined as a flat affine subspace of dimension $N - 1$. For a 2D space, it will be just a line whereas a plane in case of 3D space. In Fig. 5, lines A and C represent the support vectors for the data. Line **B** represents the hyperplane that separates the dataset. The r shown in the figure is the maximal margin distance of either support vectors from the hyperplane.

One-Class Support Vector Machines have been used in machine fault detection and classification [26]. It has also been used for real-time anomaly detection in industrial control system [27]. The system can detect abnormal Modbus TCP traffic and helps in ensuring the critical infrastructures' security.

Primarily, it requires the dataset to be linearly separable. This condition can be overcome using what is known as the Kernel Trick [28, 29]. In the Kernel trick, we try to find a mapping from the original nonlinear space to some linear space. The transformation usually takes place from lower dimension spaces to higher dimension spaces such that data becomes linearly separable.

The objective function for the Linear SVM is given by Eq. (1), and it is subjected to the constraints given by Eq. (2), and Eq. (3) and all symbols have their usual meanings. The SVM hypothesis is $h_{w,b}(x) = g\left(w^T x + b\right)$ where $g(z) = 1$ if $z \geq 0$, -1 otherwise. In Eq. (1) w represents the weight vector, b represents the bias, ξ_i represents the slack variable for the ith training example, and C is a regularization parameter for the minimization. The functional margin of a hyperplane with respect

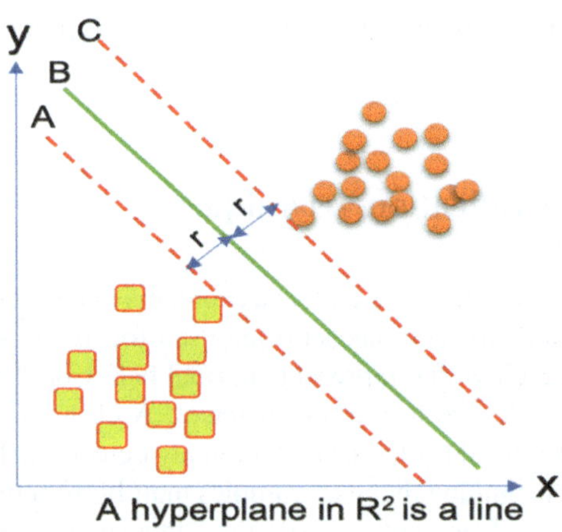

Fig. 5 A decision boundary in 2-D space

to the ith training example is represented by Eq. (2) where y_i is a sign of the hypothesis function $h_{w,b}(x)$ [30].

Mathematically we can write the quadratic optimization problem for the SVM as follows:

$$\min_{w,b,\xi_i} \frac{\|w^2\|}{2} + C\sum_{i=1}^{n} \xi_i \tag{1}$$

subject to:

$$y_i\left(w^T(x_i) + b\right) \geq 1-\xi_i \quad \text{for all } i = 1, \ldots, n \tag{2}$$

$$\xi_i \geq 0 \quad \text{for all } i = 1, \ldots, n \tag{3}$$

Support Vectors are the points in the dataset that are either on the hyperplane or are the closest to the hyperplane. These points decide the position and orientation of the hyperplane. These Support Vectors are used to maximize the margin of the classifier.

One-Class Support Vector Machines (OC-SVM) are used in anomaly detection. We try to separate the data from one particular class, the target class. The algorithm tries to build a hyperplane around the "normal" data. If there are points in the dataset that are not "normal," they are classified as anomaly or outlier (the orange colored points in Fig. 5). Thus the function returns $+1$ in a "small" region (capturing the data points) and -1 everywhere.

The quadratic programming minimization function [31] is slightly different from the original, but it is very similar to it.

The *parameter* υ [31] decides the smoothness. It performs two tasks:

1. Setting an upper bound on the fraction of the outliers (training examples regarded out-of-class).
2. It acts as a lower bound on the number of training examples used as Support Vectors.

The modified objective function for the OCSVM is given by Eq. (4) constrained to Eqs. (5) and (6).

Formally,

$$\min_{w,\xi_i,\rho} \frac{\|w^2\|}{2} + \frac{1}{\upsilon n}\sum_{i=1}^{n} \xi_i - \rho \tag{4}$$

subject to:

$$(w.(x_i)) \geq \rho - \xi_i \quad \text{for all } i = 1, \ldots, n \tag{5}$$

$$\xi_i \geq 0 \quad \text{for all } i = 1, \ldots, n \tag{6}$$

where υ represents the upper bound on the fraction of training error and ρ being the offset.

4.2 One-Class Isolation Forest (OC-IF)

The Isolation Forest or iForest, in principle, is similar to the Random Forest algorithm for classification. The method builds an ensemble of iTrees for a given dataset, and then it tries to mark anomalies in the test data. The algorithm exploits a very basic property of the Random Forest, path length.

Definition: Path Length $h(x)$ of an observation x is measured by the number of edges x traversed in an isolation Tree from the root node until the traversal is terminated at an external node [32].

The average path length for a normal point on Isolation Forest is usually very high because of the similarity in the data points. Separating two normal points from each other requires a large number of random splits to be made for the classification. On the other hand, an anomalous point has very different properties from the normal points, and they can be easily separated from them using a small number of random splits. Therefore, the average length of the anomalous point is usually low.

In these trees, partitions are created by first randomly selecting a feature and then selecting a random split value. The split value also decides the maximum and minimum value of the selected features [32]. The main idea here is to use the fact that the outliers are a rare occurrence in the data and are inherently different from the normal dataset points. So, naturally, they lie far away from the normal points and are thus very easy to classify. Performing random partitioning enables the outliers to be identified closer to the root of the decision tree, thereby making use of fewer splits necessary. The algorithms have two variables for adjustment: number of trees $\boldsymbol{t}$ to build and Sub-sampling size $\boldsymbol{\psi}$. The number of trees $\boldsymbol{t}$ controls the ensemble size. Studies have shown that path length converges well before $\boldsymbol{t} = 100$ [32]. The training data size is controlled by $\boldsymbol{\psi}$. Fei et al. [32] recommends using $\boldsymbol{\psi} = 256$ and says that the sub-sampling size is enough to perform anomaly detection. They also mention that increasing $\boldsymbol{\psi}$ leads to an increase in memory size and processing time without any significant gain in the performance of the detection mechanism.

The algorithm uses an anomaly score for anomaly detection. It is calculated using the path length and number of data points in a dataset $\boldsymbol{n}$.

Isolation Forest anomaly score is given by Eq. (7) as below:

$$(s(x, n)) = 2^{-\frac{E(h(x))}{c(n)}} \tag{7}$$

where:

$h(x)$ is the path length of point x,

n is the number of external nodes,

$E(h(x))$ is the average of $h(x)$ from a collection of isolation trees and

$c(n)$ is the average of $h(x)$ given n.

Now, every observation's anomaly score is calculated using the above method. There are three possibilities for every observation, and they are given as [32]:

1. The score calculated is near to **1** (Flagged as Anomaly)
2. The score is very less than **0.5** (Normal data observation)
3. Most of the observations are **close to 0.5**. In this case, we can conclude that the entire sample does not have any clear cut distinct anomaly detection.

Isolation forest is used by the Anti-Abuse AI team at LinkedIn to create, deploy and maintain model detects and prevents different types of abuse. These include and are not limited to the creation of fake accounts, member profile scraping, automated spam and account takeovers.

4.3 One-Class Local Outlier Factor (LoF)

Local Outlier Factor (LoF) [14] is a density-based unsupervised algorithm for anomaly detection. The algorithm looks at a certain number of neighbors as discussed in [33] to find out density and then compares this with other points' density. This algorithm is different from the rest of the algorithms because it considers the local outliers instead of global outliers.

A parameter k is used for the calculation and is defined as the number of neighbors the calculation is considering. The algorithm uses something known as k-distance, which is simply the distance to the nearest kth neighbor [33]. Local Reachability Density (LRD) is used to calculate the LoF. LRD, in turn, uses reachability distance. Reachability Distance is the maximum of the real distance of any two points a, b in the training dataset, and the k-distance of the second point b in the consideration. It acts as a smoothing factor. Equation (8) gives the reachability distance.

$$\text{Reachability} - \text{distance}(a, b) = \max(k - \text{distance}(b), \text{dist}(a, b)) \tag{8}$$

LRD is given by Eq. (9). To calculate LRD, we take the average of the reachability-distance of the k neighbors of the point. LRD is the inverse of the calculated quantity.

LRD physically gives us an intuition of how much we need to travel to reach the next point or cluster of points as it is based on the density calculations [33]. Finally, we calculate the LoF using the LRD. LoF is the ratio of average LRD of k neighbors to LRD of point in consideration.

$$\mathrm{LRD} = \frac{1}{\left(\frac{\sum_{i=1}^{k} \mathrm{reachability{-}distance}(n,i)}{k}\right)} \tag{9}$$

If the densities of neighbors are almost equivalent to the density of the point in consideration, we will almost get ~1. On the other hand, if the ratio is $\gg 1$, then that means the given point is an outlier as the density does not match up with the neighbors.

It can be represented as:

$$x = \begin{cases} \text{Normal Point, } \mathrm{LoF}(x) \approx 1 \\ \text{Outlier Point, } \mathrm{LoF}(x) \gg 1 \end{cases}$$

Local Outlier Factor has been used to do fault diagnosis in steel plates [34]. The algorithm is used to find the anomalous samples. Local outlier Factor has also been leveraged in Intrusion Detection System [35] in Computer Networks.

4.4 Density-Based Spatial Clustering of Applications with Noise (DBSCAN)

DBSCAN [36–38] is based on the principle of clustering similar data points together and thus falls under the category of unsupervised machine learning algorithms. It is essentially used to separate clusters of high density from those of relatively low density. DBSCAN is mainly used for model building and clustering. It has been used for Crystallography of X-Ray, Pattern Recognition in Satellite Imagery [39].

DBSCAN works on two parameters: **eps** and **minPoints**, which represent the maximum distance between any two points within the same cluster and the minimum number of points to form a dense region, respectively. DBSCAN groups together points which are at maximum **eps** distance from each other, and the group contains at least **minPoints**. It also marks points in low-density regions as outliers. We have made use of this property to detect outliers using just one class.

The algorithm usually defines three types of data points (Fig. 6):

1. **Core Point**: A point is defined as a core point if and only if it has more than minPts points with eps.
2. **Border Point**: A point is defined as a border point if it has fewer than minPts within eps, but the point is in the neighborhood of a core point.

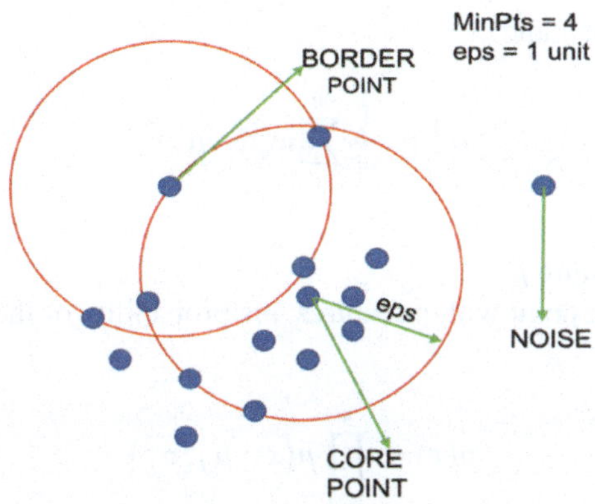

Fig. 6 Illustration of DBSCAN

3. **Noise or outlier**: A point acts as noise or an outlier if it is not a core point or a border point.

The undamaged data of the benchmarking model was used to train a DBSCAN classifier. To detect anomalous behavior (damage in the structure), some damaged samples were taken, and their distance from each of the core points of the clusters was calculated using Euclidean distance. It was grouped with the most similar cluster if the distance to any of the core points was less than eps. Otherwise, it was classified as an anomalous point.

4.5 One-Class Gaussian Anomaly Detection (OC-GD)

Gaussian anomaly detection is based on the principle of a normally distributed Gaussian function. It is usually dependent on two parameters: μ (which represents the mean of the distribution) and σ (which represents the standard deviation). It is an unsupervised machine learning algorithm.

The algorithm tries to fit the sample training data points in a bell-shaped curve by calculating the probability $p(x)$ as follows [31]:

$$(p(x)) = p\left(x; \mu, \sigma^2\right) = \frac{1}{\sqrt{2\pi}\sigma} e^{-\left(\frac{(x-\mu)^2}{2\sigma^2}\right)} \tag{10}$$

where $p(x)$ is the probability of x in Gaussian Distribution

where

$$\mu_j = \frac{1}{m} \sum_{i=1}^{m} x_i^j \tag{11}$$

is the mean of feature j and

$$\sigma_j^2 = \frac{1}{m}\sum_{i=1}^{m}(x_i^j - \mu_j)^2 \tag{12}$$

is the variance of feature j.

Thus for a sample point with n features, the probability of the point is calculated as:

$$p(x) = \prod_{j=1}^{n} p\left(x_j; \mu_j, \sigma_j^2\right) \tag{13}$$

The undamaged data of the benchmarking model was used to generate a Gaussian distribution curve. To detect anomalous behavior (fault in the structure), some damaged samples were taken, and their probabilities were calculated using the above-mentioned formulae. A **threshold** parameter was decided based on the performance of the technique on different values, and the best value was chosen as the threshold. If the probabilities of test points were **less than the threshold** value, they were classified as anomalous points.

Gaussian Detection has been used for instance detection in problems such as Object Recognition, Event Detection and Defect localization. It has been used in Wire-Rope Defect Detection. The algorithm has also been used in recognition of novel bacteria [40].

4.6 One-Class KMeans (OC-KMC)

K-means [41] is one of the well-known clustering algorithms and comes under the category of unsupervised machine learning algorithms. K-Means is often used for document clustering, identifying crime prone areas, insurance fraud detection and in fault detection. It aims to divide n sample points into k clusters such that each cluster has a representative called a centroid of that cluster.

K-means essentially works on one major parameter, **k**, which is the number of clusters that should be formed by the algorithm. It marks points that do not lie in any of the clusters as anomalous points.

Anomaly detection was performed by dividing the training data points into k groups (the value of k was adjusted, and the classifier was trained for different values), and for every test data point, its distance from each of the centroids was computed. If this data point did not belong to any of these clusters, it was considered an anomaly. The accuracy was calculated as the ratio of a number of damaged sample points classified as anomalous to the total number of damaged sample points taken.

4.7 One-Class KNN (OC-KNN)

KNN is one of the well-known classification algorithms and comes under the category of supervised machine learning algorithms. One-class KNN has been used for single-class classification to classify vehicle model recognition from images [42]. It aims to classify a point using the majority vote of the classes of its ***k* nearest neighbors**.

KNN works on one major parameter, ***k*** (the value of k was adjusted, and the classifier was trained for different values), which is the number of nearest neighbors to be considered while predicting the class for a point.

Anomaly detection was performed by training a classifier using the training data points, and for all the test sample points, an anomaly score was calculated based on the Euclidean distance from its k nearest neighbors [43]. A threshold for anomaly score was decided using certain values, and the threshold selected as the value for which the model performed the best. If the anomaly score was greater than the threshold, the test data point was classified as anomalous. The accuracy was calculated as the ratio of a number of damaged sample points classified as anomalous to the total number of damaged sample points taken.

4.8 One-Class PCA (OC-PCA)

Principal Component Analysis, abbreviated as PCA [44], is a well-known technique in machine learning and is used to study the inner structure of the data. It also explains the variance in data. It is used majorly for the two tasks **dimension reduction** and **classification**.

PCA works by analyzing data that contains multiple features and finds correlations between them. It tries to remove features that do not have much effect on the variance, thereby reducing the dimensionality of the data. Mathematically, it is an **orthogonal linear transformation** that transforms the data from the original coordinate system to a new coordinate system. It is done in such a way that the parts of the dataset where there was large variance, those particular parts become the axis in the transformation. The **number of components** is a parameter that decides the cardinality of the transformed plane.

Anomaly detection was performed by training a PCA model with **n_components = 2**, less than the original number of features (4). The training data was fit on this model, and dimensionality reduction was applied. **Mahalanobis** distance $D_M(x)$ given by Eq. (14), was computed using the mean and inverse covariance matrix of the training data. The Mahalanobis distance was then used to compute a **threshold** value. For the test data points, if their Mahalanobis distance was greater than the threshold, they were classified as anomalous points. Mahalanobis distance is defined as:

$$D_M(\vec{x}) = \sqrt{(\vec{x} - \mu)^T S^{-1} (\vec{x} - \mu)} \tag{14}$$

where, $\mu = (\mu_1, \mu_2, \mu_3, \ldots, \mu_N)^T$ and represents mean and $\vec{x} = (x_1, x_2, x_3, \ldots, x_N,)^T$ and Covariance matrix S.

5 Results and Analysis

We have used the various techniques discussed in Sect. 4 to do the performance evaluation on the dataset extracted from the IASC-ASCE structure (Sect. 3.2). Two different sets of experiments were performed, namely **Experiment-I** and **Experiment-II**. The Experiment-I consisted of two parts: (i) Experiment-I (a) and (ii) Experiment-I (b). Experiment-II also consisted of two parts: (i) Experiment-II (a) and (ii) Experiment-II (b).

A ***data point*** in the dataset is defined as a tuple containing the first four natural frequencies for the structure at a given time. The 40,000 points were sampled in the raw dataset for UD, D1, and D2. We filtered out the points using FFT on all the datasets and extracted the first four natural frequencies. Only the first four frequencies were extractable from the dataset. We collected only the unique datapoints from the complete dataset to avoid repetition. We were able to collect 1893 points in UD, 2326 points in D1, and 3194 points in D2. We split UD data samples in the ratio of **80:20 i.e. 1514:379** samples for the train and test purposes. The 1514 points of UD were available for training and 379 points of UD for the test in Experiment-I (a). For Experiment-I (b), we use a training dataset that has noise data points in it. We wanted to simulate the behavior of algorithms in the presence of noise in the training dataset. Hence, we replaced 10% data with the noise in the Experiment-I (b) and Experiment-II (b). In Experiment-I (b), the training data consists of 90% of the 1514 data samples of UD i.e. 1361 points from the UD, and 10% of 151 data points data samples randomly sampled from the D1 thus totaling to same 1514 number of training samples as used for Experiment-I (a). The test set for both the Experiment-I (a) and Experiment-I (b) is 380 points from UD and 380 points from D1. In Experiment-II (a), the training data consists of 1514 data samples points only from UD. Similar to Experiment-I (b), in Experiment-II(b) also the training data is a mix of 90% of data samples of UD i.e. 1361 and 10% of data samples from D2 i.e. 151, hence total 1514 data points. For testing a mix of 380 samples from UD and 380 samples from D2 are used.

Visualization of the training data and the test data helps in getting insights that can be missed without them: the following three figures, i.e., Figs. 7, 8 and 9 are the representation of the training dataset consisting of points from the Undamaged Dataset (UD) and have a single test point. Figure 7 consists of the test point from the Undamaged Dataset. The subfigures 7.1–7.7 contain a 4-Dimensional plot. The 4th Dimension is represented in the form of the heat map. The subfigures 7.1–7.7 contain the test point in the form of a black square, whereas the training data is represented as colored dots. The subfigure 7.8 contains the result for OC-PCA, and the brown lines represent the hyperplane on the data. The blue line (test point) is below the hyperplane, and it represents that the point is classified as Undamaged

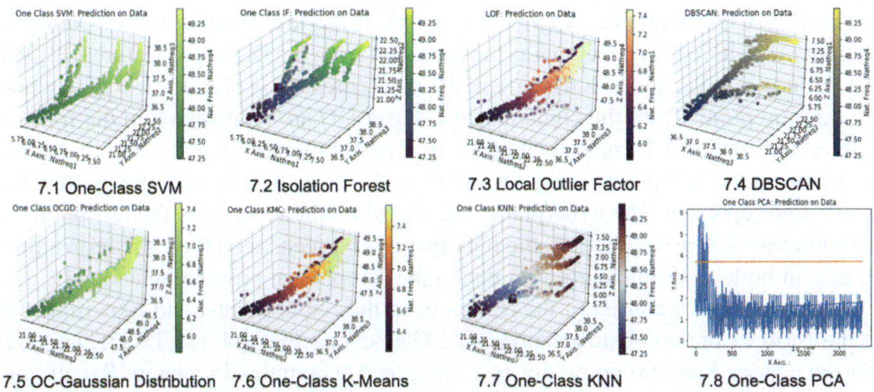

Fig. 7 Prediction of a sample test point from UD

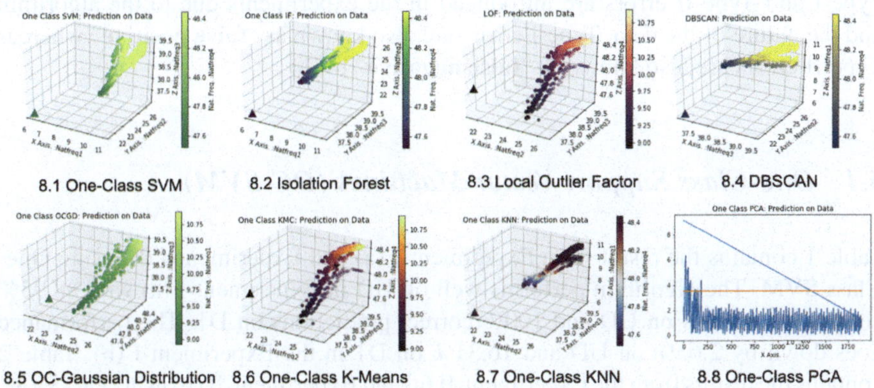

Fig. 8 Prediction of a sample test point from D1

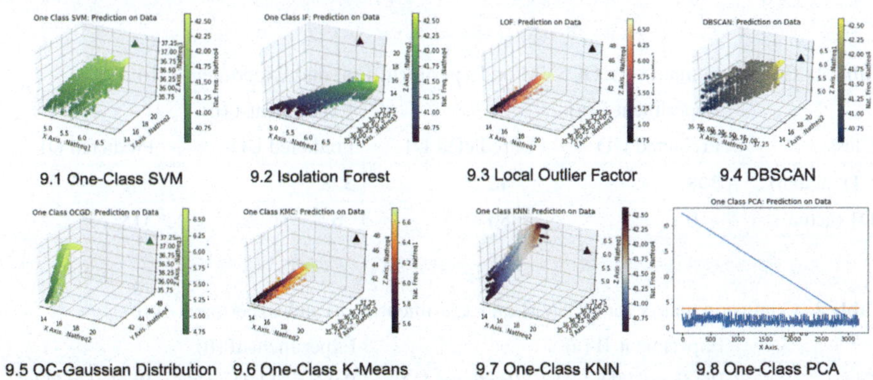

Fig. 9 Prediction of a sample test point from D2

by the One-Class Classifier. Figure 8 consists of the test point from the Damaged Dataset D1, whereas Fig. 9 contains the test point from the Damaged Dataset D2. The test point is represented as a triangle in the subfigures 8.1–8.7 and 9.1–9.7, whereas the training points from the Undamaged Dataset UD are represented in the form of colored dots on the 4-Dimensional heatmap. The subfigures 8.8 and 9.8 both contain the graph of the test points using OC-PCA as the classifier. In both the figures, the brown line represents the hyperplane, and the blue line represents the test point. In the subfigure 8.8, as well as 9.8, the blue line can be seen well above the brown line. Hence, in both cases, the algorithm marks the sample test point as an anomaly.

Sections 5.1–5.8 consist of the results from the Experiment-I and Experiment-II in the form of the confusion matrix. The Classes are marked as UD, D1, and D2. All the classes had 380 points for test. Section 5.9 contains Precision, Recall, and F1-Score for the Experiment-I and Experiment-II. The analysis in Sect. 5.9 helps in deciding the robust and optimal algorithm for the purpose of anomaly detection. Type-I and Type-II errors are introduced in the experiments due to the algorithm and the nature of the data. Type-I errors are also known as "**false positive**" whereas Type-II errors are also known as "**false negative**" [45].

5.1 *One-Class Support Vector Machines (OC-SVM)*

Table 1 contains the result from Experiment-I (a) and Experiment-I (b) using One-Class SVM. The algorithm performs well on UD in Experiment-I (a) with 88.95% correct predictions on UD and 100% correct predictions on D1. The performance goes down by 2.63% on UD and 16.31% on D1 in the Experiment-I (b). Table 2 contains the results from the Experiment-II (a) and Experiment-II (b) using OC-SVM. The performance of the algorithm is almost similar for UD in both Experiment-II (a) and Experiment-II (b) as Experiment-I (a) and Experiment-I (b), respectively.

Table 1 Confusion matrix for OCSVM on Experiment-I (a) and Experiment-I (b)

	Experiment-I (a)		Experiment-I (b)	
Class	Predicted UD	Predicted D1	Predicted UD	Predicted D1
UD (actual)	338	42	328	52
D1 (actual)	0	380	62	318

Table 2 Confusion matrix for OCSVM on Experiment-II (a) and Experiment-II (b)

	Experiment-II (a)		Experiment-II (b)	
Class	Predicted UD	Predicted D2	Predicted UD	Predicted D2
UD (actual)	338	42	325	55
D2 (actual)	0	380	53	327

The algorithm identifies D2 correctly with an accuracy of 100% in Experiment-II (a). There is a 13.94% drop in the prediction accuracy on D2 in Experiment-II (b). The algorithm overall performs optimally on D1 in Experiment-I (a) and D2 in Experiment-II (a). Almost 90% accuracy in all the experiments on UD in Tables 1 and 2 suggests that the algorithm performs well. There is an almost 16% drop in Experiment-I (b) and a 14% drop in Experiment-II (b) on D1 and D2, respectively. The algorithm thus can be concluded as an optimal algorithm for anomaly detection.

5.2 *Isolation Forest (iForest)*

Table 3 contains the result from Experiment-I (a) and Experiment-I (b) using Isolation Forest. The algorithm performs very well on UD with an accuracy of 98.15% and 98.94%, respectively, in both Experiment-I (a) and Experiment-I (b). The Type-II errors go down from 25.26% in Experiment-I (a) to 16.84% in Experiment-I (b). Table 4 contains the result from Experiment-II (a) and Experiment-II (b) using Isolation Forest algorithm. The performance in Experiment-II (a) and Experiment-II (b) is also optimal and similar to results from Table 3. There is a 100% accuracy on D2 in Experiment-II (a), but there is a 14.73% decrease in the D2 classification in Experiment-II (b). From the data in Tables 3 and 4, we can conclude that this algorithm performs robustly on UD. The algorithm performs better in the presence of noise in the case of Experiment-I on the data points from D1. There is a slight decrease in the performance in Experiment-II on the points from the dataset D2. We can conclude that this algorithm performs optimally in both the presence and absence of noise in the training dataset.

Table 3 Confusion matrix for isolation forest on Experiment-I (a) and Experiment-I (b)

	Experiment-I (a)		Experiment-I (b)	
Class	Predicted UD	Predicted D1	Predicted UD	Predicted D1
UD (actual)	373	7	376	4
D1 (actual)	96	284	64	316

Table 4 Confusion matrix for isolation forest on Experiment-II (a) and Experiment-II (b)

	Experiment-II (a)		Experiment-II (b)	
Class	Predicted UD	Predicted D2	Predicted UD	Predicted D2
UD (actual)	376	4	371	9
D2 (actual)	0	380	56	324

Table 5 Confusion matrix for LoF on Experiment-I (a) and Experiment-I (b)

	Experiment-I (a)		Experiment-I (b)	
Class	Predicted UD	Predicted D1	Predicted UD	Predicted D1
UD (actual)	330	50	340	40
D1 (actual)	0	380	359	21

Table 6 Confusion matrix for LoF on Experiment-II (a) and Experiment-II (b)

	Experiment-II (a)		Experiment-II (b)	
Class	Predicted UD	Predicted D2	Predicted UD	Predicted D2
UD (actual)	330	50	341	39
D2 (actual)	0	380	315	65

Table 7 Confusion matrix for DBSCAN on Experiment-I (a) and Experiment-I (b)

	Experiment-I (a)		Experiment-I (b)	
Class	Predicted UD	Predicted D1	Predicted UD	Predicted D1
UD (actual)	370	10	376	4
D1 (actual)	0	380	363	17

5.3 Local Outlier Factor (LoF)

Table 5 contains the result from Experiment-I (a) and Experiment-I (b) using Local Outlier Factor. The algorithm performs well on UD with an accuracy of 86.84% in Experiment-I (a) and 89.47% in Experiment-I (b). There is a 100% accuracy on D1 in Experiment-I (a). The algorithm performs very poorly in Experiment-I (b) on D1 with only 5.52% accuracy. Almost all the test dataset points from the D1 are misclassified as UD in Experiment-I (b) making **High Type-II error**. Table 6 contains the result from Experiment-II (a) and Experiment-II (b) using Local Outlier Factor. The algorithm performs in a similar manner for UD on the Experiment-II (a) and Experiment-II (b) as above. The behavior for D2 is also similar to above for both Experiment-II (a) and Experiment-II (b) with 100% and 17.11% accuracy, respectively. We can conclude from Tables 5 and 6 that the algorithm is **feasible** in the **absence of noise points** in the training dataset. The algorithm rather performs poorly in the presence of noise for the Damaged Datasets D1 and D2.

Table 8 Confusion matrix for DBSCAN on Experiment-II (a) and Experiment-II (b)

	Experiment-II (a)		Experiment-II (b)	
Class	Predicted UD	Predicted D2	Predicted UD	Predicted D2
UD (actual)	370	10	376	4
D2 (actual)	0	380	342	38

5.4 *Density-Based Spatial Clustering of Applications with Noise (DBSCAN)*

Table 7 contains the result from Experiment-I (a) and Experiment-I (b) using DBSCAN. The algorithm performs well on UD with an accuracy of 97.36% in Experiment-I (a) and 98.94% in Experiment-I (b). There is a 100% accuracy on D1 in Experiment-I (a). The accuracy is reduced drastically to 4.47% for the Experiment-I (b). The algorithm is wrongly classifying the D1 in Experiment-I (b) for almost all the data points. Table 8 contains results for Experiment-II (a) and Experiment-II (b) using DBSCAN. The algorithm performs very well on UD, similar to the above results for both Experiment-II (a) and Experiment-II (b). There is 100% accuracy on D2 for Experiment-II (a), whereas there is only 10% accuracy for Experiment-II (b). This really low accuracy on D1 and D2 in Experiment-I (b) and Experiment-II (b) respectively suggests that the algorithm is not suitable as the training dataset where there is a possibility of the presence of noise. The algorithm is, however, suitable for the case where there is an absence of noise in the training dataset.

5.5 *One-Class Gaussian Anomaly Detection (OC-GD)*

Table 9 contains the result from Experiment-I (a) and Experiment-I (b) using One-Class Gaussian Distribution. The algorithm performs well on UD in Experiment-I (a) and performs poorly on D1 in the Experiment-I (a). The performance on UD **improves by 5%** as well as on D1 in Experiment-I (b). Table 10 contains the result from Experiment-II (a) and Experiment-II (b) for the same algorithm. The results for UD in Experiment-II (a) remain the same because there is no change in the train and test conditions w.r.t. the class UD. However, the algorithm performs perfectly on the damaged datapoints from D2 in the Experiment-II (a). The Experiment-II (b)

Table 9 Confusion matrix for OC-GD on Experiment-I (a) and Experiment-I (b)

	Experiment-I (a)		Experiment-I (b)	
Class	Predicted UD	Predicted D1	Predicted UD	Predicted D1
UD (actual)	345	35	364	16
D1 (actual)	133	247	143	237

conditions make the detection of damaged datapoints from D2 tougher. About 38% of points from D2 in Experiment-II (b) are misclassified as UD and cause Type-II error. The results from Table 9 for Experiment-I (a) helps us conclude that the algorithm performs poorly on the damaged dataset D1 and thus is not suitable for the anomaly detection in the case of training done on undamaged dataset alone. The **introduction of Type-II error** in Table 10 in Experiment-II (b) helps us conclude that this algorithm is not suitable if the absence of noise in the training dataset cannot be ensured. Hence, the algorithm can be concluded as **unsuitable** for the anomaly detection task in **both** the presence and absence of noise in the training dataset.

5.6 *One-Class KMeans (OC-KMC)*

Table 11 contains the result from Experiment-I (a) and Experiment-I (b) using One-Class KMeans. The algorithm performs poorly on UD in Experiment-I (a) but is able to classify every point from D1 as an anomaly. The rate of Type-I error increases in Experiment-I (b). More UD points are classified wrongly as D1. The accuracy remains 100% on D1 in Experiment-I(b) as well. Table 12 contains the result from Experiment-II (a) and Experiment-II (b) for One-Class KMeans. The algorithm performs accurately for class D2 in both the experiments in Table 12. The performance for UD in Experiment-II (a) is poor, as almost 27% of points from UD are classified as D2. The misclassification rate rises to 63% in Experiment-II (b) for the points from UD. The data from Tables 11 and 12 helps us conclude that the algorithm performs well on the points from the damaged datasets D1 and D2. The algorithm is not optimal for the anomaly detection inspite of 100% accuracy on D1 and D2 due to Type-I errors. The algorithm misclassifies the undamaged datapoints as anomalous. The **high false-positive rate** makes it unsuitable for anomaly detection.

Table 10 Confusion matrix for OC-GD on Experiment-II (a) and Experiment-II (b)

	Experiment-II (a)		Experiment-II (b)	
Class	Predicted UD	Predicted D2	Predicted UD	Predicted D2
UD (Actual)	345	35	380	0
D2 (Actual)	0	380	140	240

Table 11 Confusion matrix for OC-KMC on Experiment-I (a) and Experiment-I (b)

	Experiment-I (a)		Experiment-I (b)	
Class	Predicted UD	Predicted D1	Predicted UD	Predicted D1
UD (actual)	278	102	245	135
D1 (actual)	0	380	0	380

Table 12 Confusion matrix for OC-KMC on Experiment-II (a) and Experiment-II (b)

	Experiment-II (a)		Experiment-II (b)	
Class	Predicted UD	Predicted D2	Predicted UD	Predicted D2
UD (actual)	278	102	138	242
D2 (actual)	0	380	0	380

5.7 *One-Class KNN (OC-KNN)*

Table 13 contains the result from Experiment-I (a) and Experiment-I (b) using One-Class K-Nearest Neighbours. The algorithm performs well on UD in Experiment-I (a) and is 100% accurate on D1 in the same experiment. The performance improves marginally by 1.58% on UD in Experiment-I (b), but the performance becomes poor on D1. 67% points from D1 are misclassified as UD in the Experiment-I (b). Table 14 contains the result from Experiment-II (a) and Experiment-II (b) on the One-Class K-Nearest Neighbours. The performance on UD in Experiment-II (a) is the same as Experiment-I (a). All the D2 test points are correctly classified as anomalous by the algorithm. The results in Experiment-II (b) show that there is a 3.68% increase in the performance on UD. However, 50% points are classified as UD in Experiment-II (b) thereby introducing the Type-II errors. The results from Experiment-I(a) and Experiment-II (a) suggest that the algorithm can be used for the anomaly detection task if there is an absence of noise points (outliers) in the training dataset. However, the **high false-negative rate (Type-II errors)** in both Experiment-I (b) and Experiment-II (b) suggests that the algorithm should not be used in the case where there is a possibility of the presence of noise in the training dataset.

Table 13 Confusion matrix for OC-KNN on Experiment-I (a) and Experiment-I (b)

	Experiment-I (a)		Experiment-I (b)	
Class	Predicted UD	Predicted D1	Predicted UD	Predicted D1
UD (actual)	346	34	352	28
D1 (actual)	0	380	257	123

Table 14 Confusion matrix for OC-KNN on Experiment-II (a) and Experiment-II (b)

	Experiment-II (a)		Experiment-II (b)	
Class	Predicted UD	Predicted D2	Predicted UD	Predicted D2
UD (actual)	346	34	360	20
D2 (actual)	0	380	190	190

Table 15 Confusion matrix for OC-PCA on Experiment-I (a) and Experiment-I (b)

	Experiment-I (a)		Experiment-I (b)	
Class	Predicted UD	Predicted D1	Predicted UD	Predicted D1
UD (actual)	375	5	380	0
D1 (actual)	0	380	63	317

Table 16 Confusion matrix for OC-PCA on Experiment-II (a) and Experiment-II (b)

	Experiment-II (a)		Experiment-II (b)	
Class	Predicted UD	Predicted D2	Predicted UD	Predicted D2
UD (actual)	375	5	380	0
D2 (actual)	0	380	56	324

5.8 *One-Class PCA (OC-PCA)*

Table 15 contains the result from Experiment-I (a) and Experiment-I (b) using One-Class PCA. The algorithm performs optimally in Experiment-I (a) due to 98.68% correct predictions on UD and 100% correct predictions on D1. With Experiment-I (b), OC-PCA is able to predict 100% accurately on UD. There is a 16% percent decrease in the performance on D1 in Experiment-I (b). Table 16 contains the result for the Experiment-II (a) and Experiment-II (b) using OC-PCA. The results in Table 16 are comparable to the results in Table 15. The algorithm behaves in a similar method for UD in the Experiment-II (a) and Experiment-II (b) as Experiment-I (a) and Experiment-I (b) respectively. The algorithm performs well on D2 in Experiment-II (a) but has a 14.73% percent decrease in the performance in Experiment-II (b). The algorithm performs robustly on UD in both the experiments. With a slight decrease in performance from the 100% accuracy on D1 in Experiment-I (a) and on D2 in Experiment-II (a), the algorithm is still able to identify D1 in Experiment-I (b) and D2 in Experiment-II (b) robustly. The algorithm can be concluded as robust for the anomaly detection task in both the presence and absence of noise in the training dataset.

5.9 *Performance Analysis Using Precision-Recall Metrics*

To assess the performance of the One-Class Classifiers on the IASC-ASCE dataset for anomaly detection, we use Recall, Precision, and F1 score for the above experiments. Precision is defined as the fraction of the relevant instances that were retrieved in the total instances. Recall is defined as the fraction of the total instances retrieved that are actually relevant to the class. These concepts led to the formation of the F1 score given in Eq. 15 as below:

$$F1 = \frac{2 * P * R}{P + R} \tag{15}$$

where P is Precision, and R is Recall.

Table 17 contains the Precision, Recall, and F1 score for Experiment-I (a). Table 18 contains the Precision, Recall, and F1 score for Experiment-I (b). The F1 score for UD from Tables 17 and 18 has been used to plot the Fig. 10. The plot in Fig. 10 can be used to compare the performance of the algorithm on UD for two different training datasets. We can see that for iForest and Gaussian Distribution, the algorithms work better in Experiment-I (b) by 3.4% and 3.3%, respectively. For the rest of the cases, the F1-score is better in Experiment-I (a). OC-PCA has the highest F1 score of 0.995 in the Experiment-I (a) in Table 17. Figure 11 shows the F1 score for Experiment-I on the dataset D1. It can be seen from the plot that the performance of the LoF, DBSCAN, and OC-KNN on D1 drops. OC-KNN is not a good method for anomaly detection due to 2 major reasons. First, the anomaly scores would differ greatly on the basis of the parameter k. Second, if there are anomalies in the training data, they

Table 17 Precision, recall and F1 score for Experiment-I (a)

Dataset →	UD			D1		
Algorithm	Precision	Recall	F1-Score	Precision	Recall	F1-Score
OC-SVM	1	0.89	0.942	0.9	1	0.947
Isolation forest	0.8	0.98	0.881	0.98	0.75	0.850
LoF	1	0.87	0.930	0.88	1	0.936
DBSCAN	1	0.97	0.985	0.97	1	0.985
Gaussian distribution	0.8	0.78	0.790	0.8	0.78	0.790
K-means	1	0.73	0.844	0.79	1	0.883
KNN	1	0.91	0.953	0.92	1	0.958
OC-PCA	1	0.99	0.995	0.99	1	0.995

Table 18 Precision, recall and F1 score for Experiment-I (b)

Dataset →	UD			D1		
Algorithm	Precision	Recall	F1-Score	Precision	Recall	F1-Score
OC-SVM	0.84	0.86	0.850	0.86	0.84	0.850
Isolation forest	0.85	0.99	0.915	0.99	0.83	0.903
LoF	0.49	0.89	0.632	0.34	0.06	0.102
DBSCAN	0.51	0.99	0.673	0.81	0.04	0.076
Gaussian distribution	0.72	0.96	0.823	0.94	0.62	0.747
K-means	1	0.64	0.780	0.74	1	0.851
KNN	0.58	0.93	0.714	0.81	0.32	0.459
OC-PCA	0.86	1	0.925	1	0.83	0.907

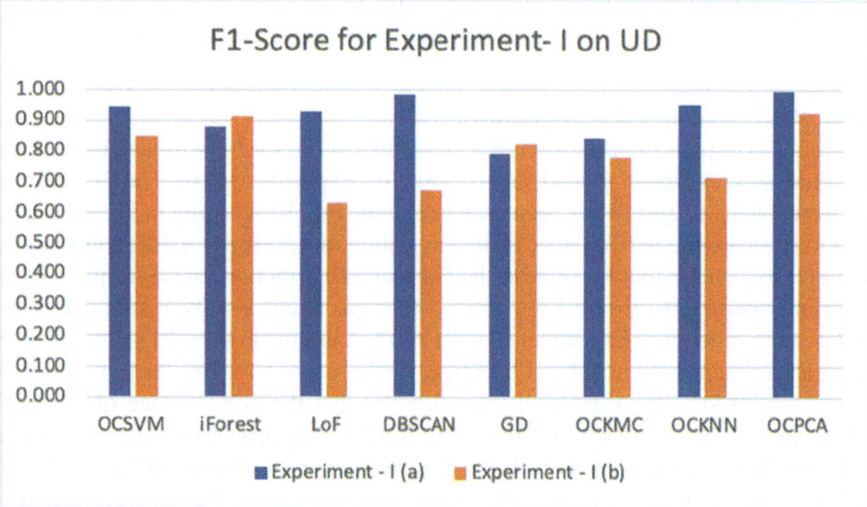

Fig. 10 F1-Score for Experiment-I on UD

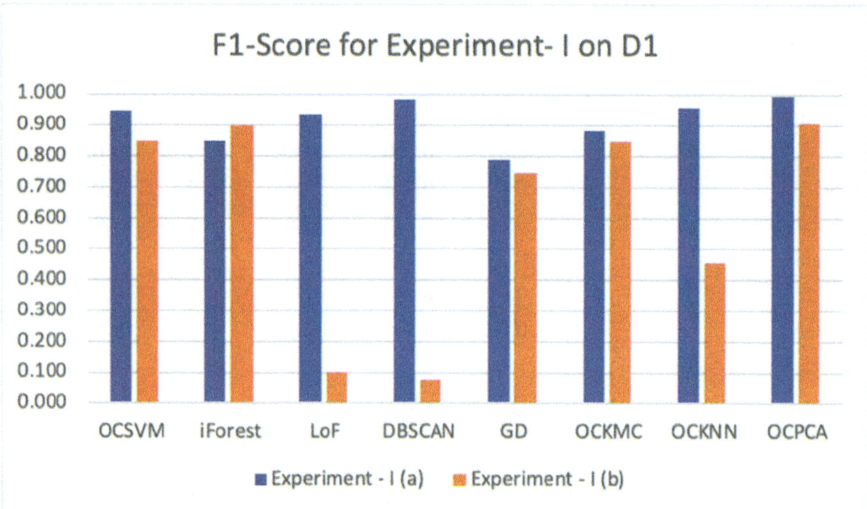

Fig. 11 F1-Score for Experiment-I on D1

would also affect the anomaly scores of the test data points to a great extent, and some points which should have been classified as anomalous might not be through this algorithm. Notably, Isolation Forest is the only algorithm in Fig. 11 that has an improved F1-Score, marginally by 5.3% in the case of Experiment-I (b). OC-SVM, iForest, and OC-PCA perform robustly on the Experiment-I (a) and Experiment-I (b).

Table 19 Precision, recall and F1 score for Experiment-II (a)

Dataset →	UD			D1		
Algorithm	Precision	Recall	F1-Score	Precision	Recall	F1-Score
OC-SVM	1	0.89	0.942	0.9	1	0.947
Isolation forest	1	0.99	0.995	0.99	1	0.995
LoF	1	0.87	0.930	0.88	1	0.936
DBSCAN	1	0.97	0.985	0.97	1	0.985
Gaussian distribution	1	0.91	0.953	0.92	1	0.958
K-means	1	0.73	0.844	0.79	1	0.883
KNN	1	0.91	0.953	0.92	1	0.958
OC-PCA	1	0.99	0.995	0.99	1	0.995

Table 20 Precision, recall and F1 score for Experiment-II (b)

Dataset →	UD			D1		
Algorithm	Precision	Recall	F1-Score	Precision	Recall	F1-Score
OC-SVM	0.86	0.86	0.860	0.86	0.86	0.860
Isolation forest	0.87	0.98	0.922	0.97	0.85	0.906
LoF	0.52	0.9	0.659	0.62	0.17	0.267
DBSCAN	0.52	0.99	0.682	0.9	0.1	0.180
Gaussian distribution	0.73	1	0.844	1	0.63	0.773
K-means	1	0.36	0.529	0.61	1	0.758
KNN	0.65	0.95	0.772	0.9	0.5	0.643
OC-PCA	0.87	1	0.930	1	0.85	0.919

Table 19 contains the Precision, Recall, and F1 score for Experiment- II (a). Table 20 contains the Precision, Recall, and F1 score for Experiment-II (b). The F1 score for UD from Tables 19 and 20 has been used to plot Fig. 12. The F1-score for D2 has been compiled in the form of Fig. 13 with the help of Tables 19 and 20. The plot in Fig. 12 can be used to compare the performance of the algorithm on UD for two different training datasets. 7 out of 8 algorithms have an F1-Score more than 0.9 in both the Experiment-II (a) and Experiment-II (b). The lowest OC-KMC also has 0.844 and 0.883 in Experiment-II (a) and Experiment-II (b), respectively. Similar to Experiment-I, LoF and DBSCAN perform very poorly on D2 in Experiment-II (a) and Experiment-II (b). We can see OCKMC performs rather poorly in Experiment-II (a) in Fig. 13. OC-PCA and iForest perform the most robustly in Experiment-II (a) and Experiment-II (b).

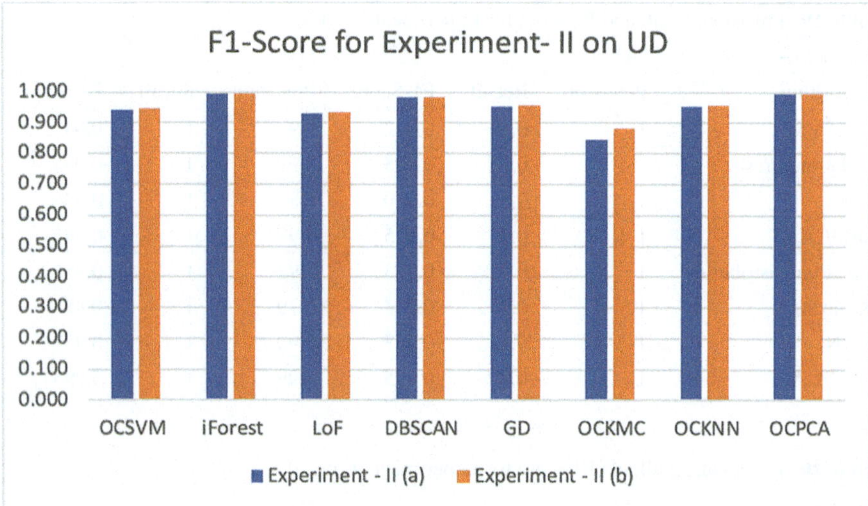

Fig. 12 F1-Score for Experiment-II on UD

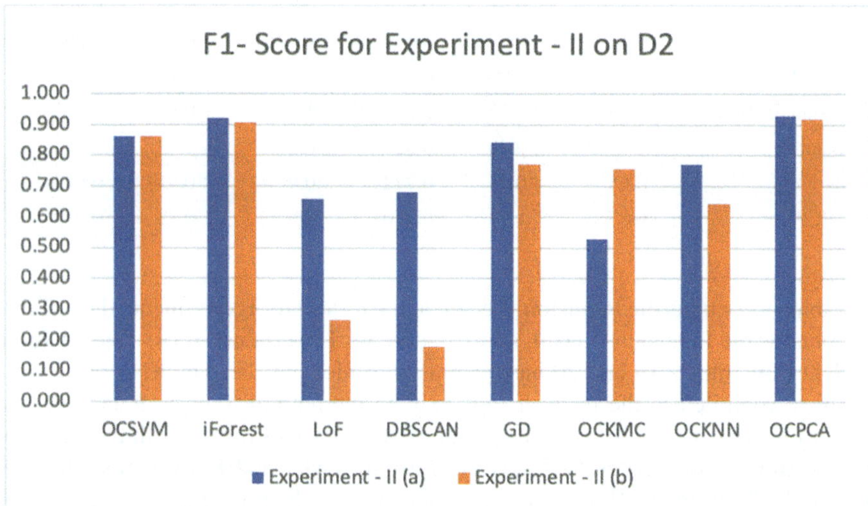

Fig. 13 F1-Score for Experiment-II on D2

6 Conclusions

The performance evaluation of One-Class classifiers has been done in this paper for the purpose of damage detection in Structural Health Monitoring. Structural health monitoring is an active area of research. It primarily aims to provide the diagnosis of the current health conditions of the building and its constituent materials using advanced sensors and computational methods. It can also be extended to provide the prognosis of the structure. The benchmark structure for the evaluation used is the IASC-ASCE structure. The benchmark structure is present in the Earthquake Engineering Research Laboratory at the University of British Columbia (UBC). The structure has four floors and 12 Degrees of Freedom. There are four accelerometers on every floor, two along X-axis and two along Y-axis. The data collected from the 16 accelerometers forms the Undamaged Dataset (UD) as the structure is not damaged in this scenario. As a part of the benchmark problem, there are six different methods of data collection in addition to the undamaged structure. In these methods, the damage introduced to the structure increases gradually. These datasets are known as D1, D2, …, D6, respectively.

For the experiments, we collected a total of 40,000 points from UD, D1, and D2. Due to the overlapping nature of UD, D1, and D2, we could not use the raw data for evaluation purposes. For a damage level, the natural frequencies of the structure remain constant. Fast Fourier Transformation (FFT) was applied to the raw data to obtain several frequencies. The structure's natural frequency was filtered out using different height peaks. The first four natural frequencies were taken to form the dataset for the experimentation. Undamaged Dataset (UD) was primarily taken for training purposes. It was split into 8:2 for training and testing purposes. There were two sets of experiments that were done using the transformed data. The first experiment was performed on D1 and done in two parts. In the first part, the training was done on UD only (1513 points). In the second part, the training set comprised of 90% undamaged points (1361 points) and 10% damaged points (152 points). In both the cases, the testing set comprised of 380 points from UD and 380 points from D1. Likewise, the second experiment was also done in two parts, but on D2, wherein the first part had a training dataset comprising of UD only (1513 points). In the second part, the training set comprised of 90% undamaged points (1361 points) and 10% damaged points from D2 (152 points). In both the cases, the testing set comprised of 380 points from UD and 380 points from D2. The One-Class Classifiers are usually trained on the normal data points only and contain no data from the positive class. But in reality, often, there are anomalous points in the dataset, and to compensate for that, we mixed some data points from D1 and D2 for the training purposes and to observe the robustness of the algorithms.

For the evaluation purposes, we considered eight classical One-Class Classifiers (OCC). We have used One-Class Support Vector Machines (OC-SVM), Isolation Forest, Local Outlier Factor (LoF), Density-Based Spatial Clustering of Applications with Noise (DBSCAN), One-Class Gaussian Distribution (OC-GD), One-Class K-means Clustering (OC-KMC), One-Class KNN (OC-KNN) and One-Class Principal Component Analysis (OC-PCA) in both the experiments. Preliminary analysis of the dataset shows that the data is easily separable into damaged data points and undamaged data points. This observation explains the fact that the recall and precision were very high for many algorithms. The results were collected for both the experiments in the form of Confusion Matrices along with Precision, Recall, and F1 Score. From the first part of both the experiments, we can infer that all algorithms performed well to correctly identify the test data points as damaged or undamaged. From the second part in both experiments, we can observe the deteriorated performance of the LoF, DBSCAN, OC-KNN because of a lower recall value, i.e., their inability to correctly predict damaged data points in case of the mixed training set. The three algorithms that consistently perform well are OC-SVM, isolation forest, and OC-PCA for the task of anomaly detection on the current benchmark structure and the related dataset.

7 Contribution and Novelty of Paper

Structural Health Monitoring is one of the highly researched areas these days because of its critical role in examining and detecting the safety of civil structures. There has been a lot of research and development in this field using a wide range of techniques from different domains. The publications as a result of these studies bring out new approaches for anomaly detection but there is also a need to classify which technique works best for a particular scenario.

This paper presents a comparative study of some of the recent One-Class classifiers used for the anomaly detection problem. The classifiers are trained and tested on a benchmark dataset (IASC-ASCE). The evaluation methodology involves two different experiments. In the first experiment, we measure the performance of the algorithms using an ideal training dataset that is free from any noise or outliers. It helps us to compare the theoretical accuracy of One-Class classifiers. We add a few noisy samples in the second experiment to evaluate the performance in a non-ideal scenario.

A performance comparison is drawn based on their F1 scores to conclude which classifiers are more suited to anomaly detection in ideal and non-ideal cases for One-Class classification.

8 Scope of Improvement

These series of experiments were the Phase-I in the localization of the anomaly detection and localization task on the IASC-ASCE benchmark model. In the current series of experiments, we utilized only the Fast Fourier Transform (FFT) on the raw dataset from the accelerometers. We further want to exploit techniques like Continuous Wavelet Transform (CWT) and the Discrete Wavelet Transform (DWT) on the raw dataset to get a different insight into the data [46]. The Discrete Wavelet Analysis has been used earlier to detect and localize the damage after an earthquake. We further aim to exploit the power of Deep Learning in the form of One-Class Neural Networks (OC-NN) on the undamaged dataset to tackle anomaly detection. The raw data is sequential in nature, suggesting the use of Sequential Model training on the dataset as well.

References

1. S. Alampalli, M. Ettouney, Structural identification, damage identification, and structural health monitoring. Proc. SPIE Int. Soc. Opt. Eng. (2007). https://doi.org/10.1117/12.715000
2. E. Johnson, H. Lam et al., Phase I IASC-ASCE structural health monitoring benchmark problem using simulated data. J. Eng. Mech. **130**(1), 3–15 (2004)
3. V. Chandola, A. Banerjee, V. Kumar, *Anomaly Detection: A Survey* (University of Minnesota, Minneapolis, 2007)
4. X. Xu, H. Liu, M. Yao, Recent progress of anomaly detection. Complexity, 2019
5. M. Sabokrou, M. Khalooei, M. Fathy, E. Adeli, Adversarially learned one-class classifier for novelty detection, in *Proceedings of the IEEE Conference on Computer Vision and Pattern Recognition* (2018), pp. 3379–3388
6. The Science of Anomaly Detection, Numenta, Redwood City, CA, 2015
7. S. Agrawal, J. Agrawal, Survey on anomaly detection using data mining techniques (KES, 2015)
8. M. Landauer, M. Wurzenberger, F. Skopik, G. Settanni, P. Filzmoser, Time series analysis: unsupervised anomaly detection beyond outlier detection, in *International Conference on Information Security Practice and Experience* (Springer, Cham, 2018), pp. 19–36
9. Y. Yu, Y. Zhu, S. Li, D. Wan, Time series outlier detection based on sliding window prediction. Math. Probl. Eng. (2014). https://doi.org/10.1155/2014/879736
10. E. Busseti, I. Osband, S. Wong, *Deep Learning for Time Series Modeling* (Stanford University, Stanford, 2012)
11. N. Gornitz, M. Braun, M. Kloft, *Hidden Markov Anomaly Detection* (Berlin Institute of Technology, 2014)
12. D. Hill, B. Minsker, E. Amir, *Real-time Bayesian Anomaly Detection for Environmental Sensor Data* (University of Illinois, 2009)
13. E. Schubert, R. Wojdanowski, A. Zimek, H.P. Kriegel, On evaluation of outlier rankings and outlier scores, in *The 12th SIAM International Conference on Data Mining (SDM)* (SIAM/Omnipress, 2012), pp. 1047–1058
14. A. Zimek, R.J.G.B. Campello, J. Sander, Ensembles for unsupervised outlier detection: challenges and research questions a position paper. SIGKDD Explor. Newsl. **15**(1), 11–22 (2014). https://doi.org/10.1145/2594473.2594476
15. B. Zong, Q. Song, M.R. Min, W. Cheng, C. Lumezanu, D. Cho, H. Chen, Deep autoencoding Gaussian mixture model for unsupervised anomaly detection, in *International Conference on Learning Representations* (2018)
16. M. Goldstein, S. Uchida, A comparative evaluation of unsupervised anomaly detection algorithms for multivariate data. PLoS ONE **11**(4), e0152173 (2016). https://doi.org/10.1371/journal.pone.0152173
17. S. Kanarachos, J. Mathew, A. Chroneos, M. Fitzpatrick, Anomaly detection in time series data using a combination of wavelets, neural networks and Hilbert transform, in *2015 6th International Conference on Information, Intelligence, Systems and Applications (IISA)*, Corfu, 2015, pp. 1–6
18. Z. Chen, C.K. Yeo, B.S. Lee, C.T. Lau, Autoencoder-based network anomaly detection, in *2018 Wireless Telecommunications Symposium (WTS)*, Phoenix, AZ, 2018, pp. 1–5. https://doi.org/10.1109/WTS.2018.8363930
19. A. Zimek, E. Schubert, H.-P. Kriegel, A survey on unsupervised outlier detection in high-dimensional numerical data. Stat. Anal. Data Min. **5**, 363–387 (2012)
20. I. Goodfellow, J. Pouget-Abadie, M. Mirza, B. Xu, D. Warde-Farley, S. Ozair et al., Generative adversarial nets. Adv. Neural. Inf. Process. Syst. **27**, 2672–2680 (2014)
21. H. Zenati, M. Romain, C. Foo, B. Lecouat, V. Chandrasekhar, Adversarially learned anomaly detection, in *2018 IEEE International Conference on Data Mining (ICDM)*, Singapore, 2018, pp. 727–736. https://doi.org/10.1109/ICDM.2018.00088

22. T. Schlegl, P. Seeböck, S.M. Waldstein, U. Schmidt-Erfurth, G. Langs, Unsupervised anomaly detection with generative adversarial networks to guide marker discovery, in *International Conference on Information Processing in Medical Imaging*, 2017, pp. 146–157
23. A. Shi, X. Yu, Structural damage detection using artificial neural networks and wavelet transform, in *2012 IEEE International Conference on Computational Intelligence for Measurement Systems and Applications (CIMSA) Proceedings*, Tianjin, 2012, pp. 7–11
24. V. Chandola, A. Banerjee, V. Kumar, Anomaly detection: a survey. ACM Comput. Surv. **41**(3), 1–58 (2009). https://doi.org/10.1145/1541880.1541882
25. B. Schölkopf, J.C. Platt, J.C. Shawe-Taylor, A.J. Smola, R.C. Williamson, Estimating the support of a high-dimensional distribution. Neural Comput. **13**(7), 1443–1471 (2001)
26. H. Shin, D.-H. Eom, S.-S. Kim, One-class support vector machines—an application in machine fault detection and classification. Comput. Ind. Eng. **48**, 395–408 (2005)
27. W. Shang, P. Zeng, M. Wan, L. Li, P. An, Intrusion detection algorithm based on OCSVM in industrial control system. Secur. Commun. Netw. **9**, 10(July 2016), 1040–1049 (2016)
28. A. Anaissi, N.L.D. Khoa, S. Mustapha, M.M. Alamdari, A. Braytee, Y. Wang, F. Chen, Adaptive one-class support vector machine for damage detection in structural health monitoring, in *Proceedings of the Pacific-Asia Conference on Knowledge Discovery and Data Mining* (Springer, 2017), pp. 42–57
29. P.F. Evangelista, M.J. Embrechts, B.K. Szymanski, Some properties of the Gaussian kernel for one class learning, in *Artificial Neural Networks–ICANN 2007* (Springer, 2007), pp. 269–278
30. M.A. Hearst, S.T. Dumais, E. Osuna, J. Platt, B. Scholkopf, Support vector machines, in *IEEE Intelligent Systems and their Applications*, vol. 13, no. 4, pp. 18–28, July–Aug 1998
31. V.P. Singh, Normal distribution, in *Entropy-Based Parameter Estimation in Hydrology*. Water Science and Technology Library, vol 30 (Springer, Dordrecht, 1998)
32. F.T. Liu, K.M. Ting, Z.H. Zhou, Isolation forest, in *Eighth IEEE International Conference on Data Mining, 2008. ICDM'08* (IEEE, 2008), pp. 413–422
33. M.M. Breunig, H.P. Kriegel, R.T. Ng, J. Sander, LOF: identifying density-based local outliers, in *ACM Sigmod Record*, vol. 29, №2 (2000, May), pp. 93–104
34. Z. Zhao, J. Yang, W. Lu, X. Wang, Application of local outlier factor method and back-propagation neural network for steel plates fault diagnosis, in *The 27th Chinese Control and Decision Conference (2015 CCDC)*, Qingdao, 2015, pp. 2416–2421
35. N. Paulauskas, Ą.F. Bagdonas, Local outlier factor use for the network flow anomaly detection. Secur. Comm. Netw. **8**, 4203–4212 (2015)
36. T.M. Thang, J. Kim, The anomaly detection by using DBSCAN clustering with multiple parameters, in *2011 International Conference on Information Science and Applications*, Jeju Island, 2011, pp. 1–5
37. H. Saeedi Emadi, S.M. Mazinani, A novel anomaly detection algorithm using DBSCAN and SVM in wireless sensor networks. Wireless Pers. Commun. **98**, 2025–2035 (2018). https://doi.org/10.1007/s11277-017-4961-1
38. M. Ester, H.-P. Kriegel, J. Sander, X. Xu, A density-based algorithm for discovering clusters in large spatial databases with noise, in *Proceedings of 2nd International Conference on Knowledge Discovery and Data Mining* (AAAI Press, Portland, 1996), pp. 226–231
39. S. Dang, Performance evaluation of clustering algorithm using different datasets. IJARCSMS **3**, 167–173 (2015)
40. M. Kemmler, E. Rodner, J. Denzler, One-class classification with Gaussian processes (2010), pp. 489–500. https://doi.org/10.1007/978-3-642-19309-5_38
41. M.F. Lima, B.B. Zarpelão, L.D.H. Sampaio, J.J.P.C. Rodrigues, T. Abrão, M.L. Proença, Anomaly detection using baseline and K-means clustering, in *SoftCOM 2010, 18th International Conference on Software, Telecommunications and Computer Networks, Split*, Dubrovnik, 2010, pp. 305–309
42. D. Munroe, M. Madden, Multi-class and single-class classification approaches to vehicle model recognition from images (2020)
43. S. Ramaswamy, R. Rastogi, K. Shim, Efficient algorithms for mining outliers from large data sets, in *Proceedings of the 2000 ACM SIGMOD International Conference on Management of Data (SIGMOD'00)* (ACM Press, New York, 2000), pp. 427–438

44. L. Huang, X. Nguyen, M. Garofalakis, M. Jordan, A.D. Joseph, N. Taft, In-network PCA and anomaly detection, in *Proceeding of Advances in Neural Information Processing Systems*, vol. 19 (2007)
45. A. Banerjee, U.B. Chitnis, S.L. Jadhav, J.S. Bhawalkar, S. Chaudhury, Hypothesis testing, type I and type II errors. Ind Psychiatry J. **18**(2), 127–131 (2009)
46. N.G. Pnevmatikos, G.D. Hatzigeorgiou, Damage detection of framed structures subjected to earthquake excitation using discrete wavelet analysis. Bull. Earthquake Eng. **15**, 227–248 (2017). https://doi.org/10.1007/s10518-016-9962-z

Brain Tumor Classification in MRI Images Using Transfer Learning

Aaditya Pundir and Er. Rajeev Kumar

Abstract Bio-medical image processing is a challenging and most emerging field in the computer based medical diagnosis. Brain tumors are most common and harmful disease these days, over a thousands people suffering annually with brain tumors. Among different types of brain tumors Gliomas are most common (around 30% of all brain tumor). Magnetic Resonance imaging (MRI) is a well received imaging approach to identify brain tumors. The classification of brain MRI is a difficult task. Due to large number of MRI images manual analysis is tedious and pron to errors. So there is a need of automatic classification method. The proposed method focus on a binary classification problem that seeks to differentiate between normal and tumorous MRI images by using the idea of deep transfer learning with the help of a pre-trained VGG16 model. The proposed method achieve an accuracy of 91.8% and consume less time as compared to previous traditional methods. Image augmentation techniques are also used to increase the size of training data set. The proposed method suggest that transfer learning based approach achieve good results despite limited data availability in medical field.

Keywords Deep learning · Convolutional neural network · Transfer learning
MRI · Brain tumor · Image classification

1 Introduction

Brain is the most important part of human body. It works as a control center because it control all the functions of human body. Brain tumor indeed the most dangerous disease that affect normal functioning of brain and even affect the life expectancy of patient. A tumor is a result of uncontrolled cell growth. Diagnosis of brain tumor is a challenging task. However several methods are available in order to confirm the presence of brain tumor inside brain. Some of these techniques are CT scan,

A. Pundir (✉) · Er. Rajeev Kumar
Department of Computer Science and Engineering, National Institute of Technology Hamirpur, Hamirpur, India
e-mail: adityapundir1995@gmail.com

P. Kumar et al. (eds.), *Machine Learning for Intelligent Multimedia Analytics*, Studies in Big Data 82,
https://doi.org/10.1007/978-981-15-9492-2_14

MRI, Pathology etc. Among these techniques Magnetic Resonance imaging (MRI) is widely used due to its high spatial resolutions [1]. Also the process of MRI does not produced any harmful radiation unlike CT scan and X-rays. The MRI images are divided into several different kind based on the pulses sequences such as T1 weighted, T2 weighted and Flair. These category of MRI images contains different contrast for different tissue types. The analysis of these MRI images is a difficult and time consuming task due to the complexity of these images. Manual analysis is pron to error and can result in a very critical situation if any mistake is happen because these analysis related to human life. So there is a need of some automatic method to determine tumor preciously. Various automatic MRI image classification techniques are proposed in recent years. In proposed method a transfer learning based approach is suggested. CNN is most popular deep learning based approach used in image analysis. A convolutional neural network uses multiple layers of convolution which result in better feature extraction of images. The name of CNN is derived from the convolution layer. First CNN is trained with available training data after that testing is done in order to check the expected output is predicted by the model or not. During training image feature such as pixels intensity, color, shape etc. are extracted. CNN is capable in automatically discovering features from input images. In proposed method a pre-trained CNN model VGG16 is used. VGG16 model is developed by A. Zisserman and K. Simonyan in 2014. In transfer learning based approach knowledge gained from one task is utilize to solve another related task. The main advantages of transfer learning based approach is that there is no need to train a CNN model from scratch. So with a little amount of data transfer learning perform better as compared to other traditional methods in term of time and resources used. In proposed method a binary classification model is proposed to classify brain MRI into two category-one is Normal (MRI image without any tumor) and other one is Tumor (MRI image with tumor). This paper is organised in five sections, in section two some previous work related to this area are reviewed. In section three the proposed approach is described. In section four results achieved using proposed method are given and in section five a conclusion is given.

2 Related Work

Various methods are proposed by researchers for classification of tumor tissues in brain MRI images. Some of these techniques are based on traditional image processing methods while other are inspired by machine learning models. One basic technique used by researchers was based on edge detection algorithm for tumor segmentation and classification [2]. Edge detection algorithms finds points in the image where image brightness vary sharply as compared to other points, Based on these point image is segmented into regions. An approach based on region growing algorithm is also proposed by researchers [3]. In region growing approach a initial pixels is chosen and added to a region. Now all other pixels adjacent to this pixel are also added into region until pixels of different intensity is not found, After this a new seed pixel is selected and entire process is repeated once again. Machine learning based approaches are also there, where [4] a Support Vector Machine (SVM) is used in order to detect tumor in MRI images. SVM is an example of supervised machine learning algorithm and can be used for binary and multi class classification. In this paper K-means algorithm is used along with SVM for classifying tumor. An another SVM based method is proposed by the researchers [5], where gray scale, texture and symmetrical feature are extracted from MRI images and used by SVM for classification. This SVM based classifier achieves an accuracy of 74% with linear kernel function. In [6] author uses a K-nearest neighbors algorithm for segmentation of tumor. In this method the distance of a chosen pixel is calculated from K different classes and based on the result for which the distance between gray point and class is minimum, the pixel is assign to that particular class. The method proposed in [7] uses an optimized Random Forest algorithm for segmentation of brain tumor. Random forest method is based on the uses of multiple decision tree. Here each decision tree predict the class of input data based on the decision rule define at each node of tree. Result shows this method perform better than SVM based classification. In [8] K-Means clustering is used for segmentation of tumor. In clustering image pixels are grouped into cluster based on the similarity. In K-means pixels are divided into K cluster based on the nearest mean value. In [9] K-Means method with Object Labeling Algorithms for detection of brain tumor. Some CNN based method are also proposed by various researcher. In one method patch based feature extraction is used for brain tumor segmentation by CNN [10]. CNN is one of the most popular method used in the analysis of multimedia data. In this method several hidden layers are used in order at achieve final result. In patch based image feature extraction method pattern from each image is extracted in order to train the model. In [11] a faster R-CNN model is used for the analysis and segmentation of brain images. Result shows that R-CNN perform better than SVM and K-NN algorithms. A sophisticated CNN based approach is proposed for brain tumor classification in MRI images [12], where brain MRI images are classified into five different categories. This method requires a lot of training data which is not generally available in medical field, and also this require a lot of time and hardware resources in order to train a CNN from scratch. Transfer learning model was also used for the analysis of brain tumor by using Alexnet [13]. In transfer learning method an already trained model for one problem is used to solve another different but related problem. This method can be useful when training data is not enough to train the model from scratch. So in that case an already train model is used. Alexnet is a popular CNN model developed by Alex Krizhevey. Five different CNN architectures are proposed by author for brain tumor classification with an accuracy of 84.19% [14]. Chato and Latifi [15] proposed methods to predict overall survival of Patients suffering with tumor using machine learning and deep learning based methods. A LLRBFNN (Local Linear Radial Basis Function Neural Network) algorithm based on wavelet transformation and PSO (Particle Swarm Optimization) is used by the researchers for the classification of brain tumor in MRI [16]. A CNN based method [17] with two each convolutional and max pooling layer is used by the researchers with two fully connected layers for brain tumor classification. In [18] a detail review is given about various techniques used for tumor detection in MRI Images.

3 Proposed Method

In this segment proposed transfer learning technique is discussed for the classification of brain tumor in MRI images.

3.1 The Dataset

The dataset consist of both normal and tumor MRI images obtained from following sources-2500 MRI images with HGG (High Grade Gliomas) and LGG (Low Grade Gliomas) are obtained from the collection REMBRANDT of The Cancer Imaging Archive (TCIA) by University of Arkansas for Medical Sciences (UAMS) [19]. 2500 healthy brain MRI of normal subject obtained from ImageBank repository of University of Edinburgh, Scotland [20]. All MRI are in DICOM format. All these MRI are transformed into the standard format like PNG for further processing. So a total 5000 images are used for the proposed research work. Among 5000 images, 3000 images are used for training (1500 normal image and 1500 with tumor) and 1000 each for validation and testing (500 normal images and 500 with tumor). Images in training, validation and testing data set are selected randomly out of total 5000 images. In proposed work only T2 weighted and Flair MRI images are used for preparing the dataset. The generated data set is equally balanced. All images are resized to 224×224 pixel resolutions, Because the input size of VGG16 model is $224 \times 224 \times 3$. Figure 1 shows some MRI images from prepared dataset.

3.2 Overall Architecture

The proposed approach is based on the concept of transfer learning. In transfer leaning the knowledge gained by solving one task is utilize to solve another related task. In this research work VGG16 pre trained CNN model is used [21]. VGG16 is a popular CNN model developed by K. Simonyan and A. Zisserman in 2014. Figure 2 present the VGG16 architecture which was originally trained using ImageNet dataset for 1000 different classes. There are 16 layers in VGG16, out of which 13 are convolution layer and 3 fully connected (FC) neural network layer. Broadly there are five convolution block in VGG16 model. First and second block consist of two convolution layer each and last three block contains three convolution layer. All convolution layer uses a filter of size 3×3. Small filter size is used because it results in better feature extraction. These filters are slide over entire image with sliding window technique

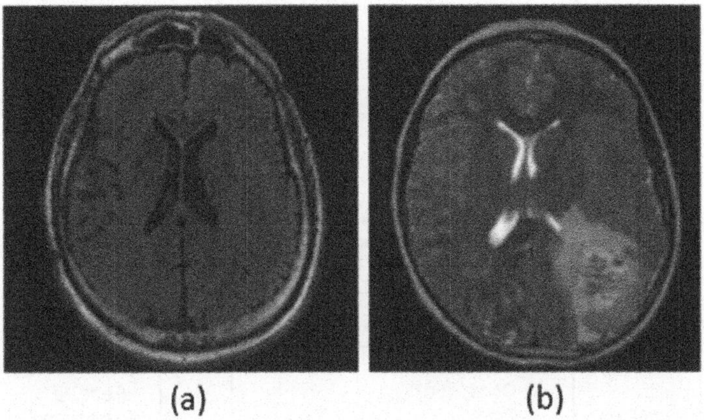

Fig. 1 Example MRI images from the dataset: **a** normal (healthy image), **b** MRI image with tumor

with stride value one. Each block is followed by a max pooling operation. The sliding window size for pooling layer is 2×2. In the end after the last block (block5) three fully connected layers are used. The last fully connected layer uses a softmax function for classification task while other layers uses ReLU activation function. The input size for VGG16 model is $224 \times 224 \times 3$, where 3 represents RGB channel in color image. As MRI images are originally grayscale images, so in proposed work the same greyscale values is replicated 3 times for creating 3 channels (RGB). In first convolution block there are two convolution layer each with 64 filters. The number of filters for block1, block2, block3, block4 and block5 are 128, 256, 512 and 512 respectively. The convolution layer is the main part of a CNN architecture. Convolution layers are used to extract the feature from the input image that later used by fully connected layers for classification. Convolution is a mathematical operation (dot product) between image pixel value and filter value. Filter is slide over entire image and extract the feature value for each sliding window. Second convolution block also contain two convolution layer. The last three convolution block contains three convolution layers each. Every convolution block is followed by a max pooling layer.

Scherer [22] in his research work shows that max pooling perform better on image dataset as compared to any other pooling. The main reason behind using pooling operation in CNN architecture is that pooling operation reduce the size of input data. Thus decrease the computation time, number of parameters. Max pooling takes the maximum value for each sliding window over entire image. After convolution block5 there are three fully connected layers. All the layers uses activation function ReLU except the final fully connected layer which uses softmax activation function. An activation function defines the output of a particular neuron for given input. ReLU is a popular activation function used to increase non-linearity in data. One advantage of using ReLU as compared to other activation function is that ReLU does not have

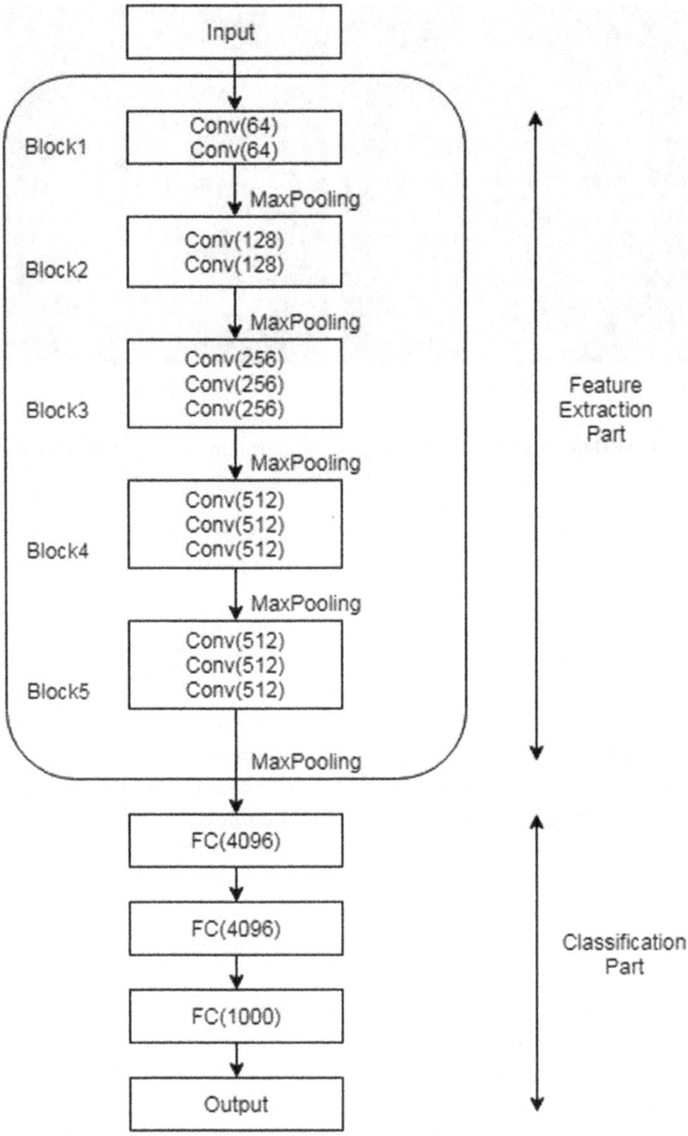

Fig. 2 VGG16 architecture

any bounded output value (except for negative input). Figure 3 shows several types of nonlinear functions.

In proposed research work there are two output class one for normal (healthy) MRI images and other one is for MRI images with tumor. So in modified VGG16 architecture final fully connected layer with 1000 neuron is replaced with a single neuron with sugmoid activation function. For binary classification problem a single

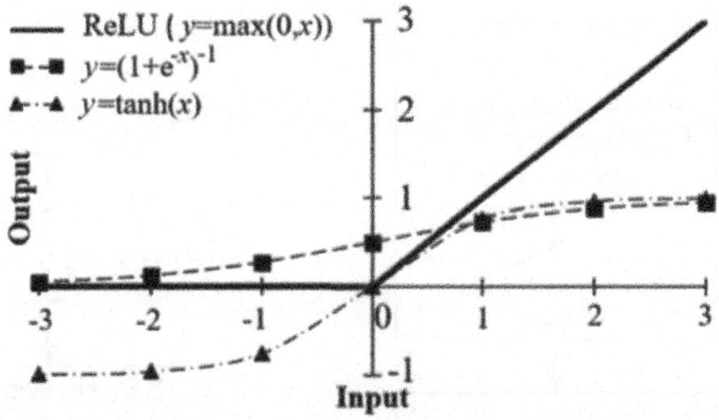

Fig. 3 Nonlinear activation functions

neuron can be used as output. Because if the probability of first class is known then estimated probability of second class is equal to one minus the probability of first class. For binary classification problem sigmoid function works same as a softmax function. In proposed work all the convolution block are remain freeze except the last one (block5) during training process. Because the lower layer (Layers near to input) extract more general feature from the image and these feature are already learned by the model on ImgeNet dataset. So in modified model only convolution block 5 is trained using the training data, because top layer (Layers near to output) extract more problem specific feature. So in proposed approach only few layers of VGG16 are trained (Fine Tuned) instead of training full model, this save a lot of training time and resources. Two hidden layers with 512 neuron each are added after block5. One dropout layer with dropout value 0.5 is used in order to avoid over fitting problem. Figure 4 represent the flow chart of proposed method.

3.3 Hyperparameters

In proposed work a minibatch size of 30 is used during training. The learning rate of 0.001 is used with RMSProp as an optimizer [23]. Small learning rate is recommended because if the learning rate is too high then network may become diverge [24]. The stride sizes for convolution and pooling layer are set to 1 and 2 respectively. Dropout rate is set to 0.5 at dropout layer. Two FC layers with 512 neuron each are added after feature extraction part. At final layer a single neuron is used with sigmoid activation function.

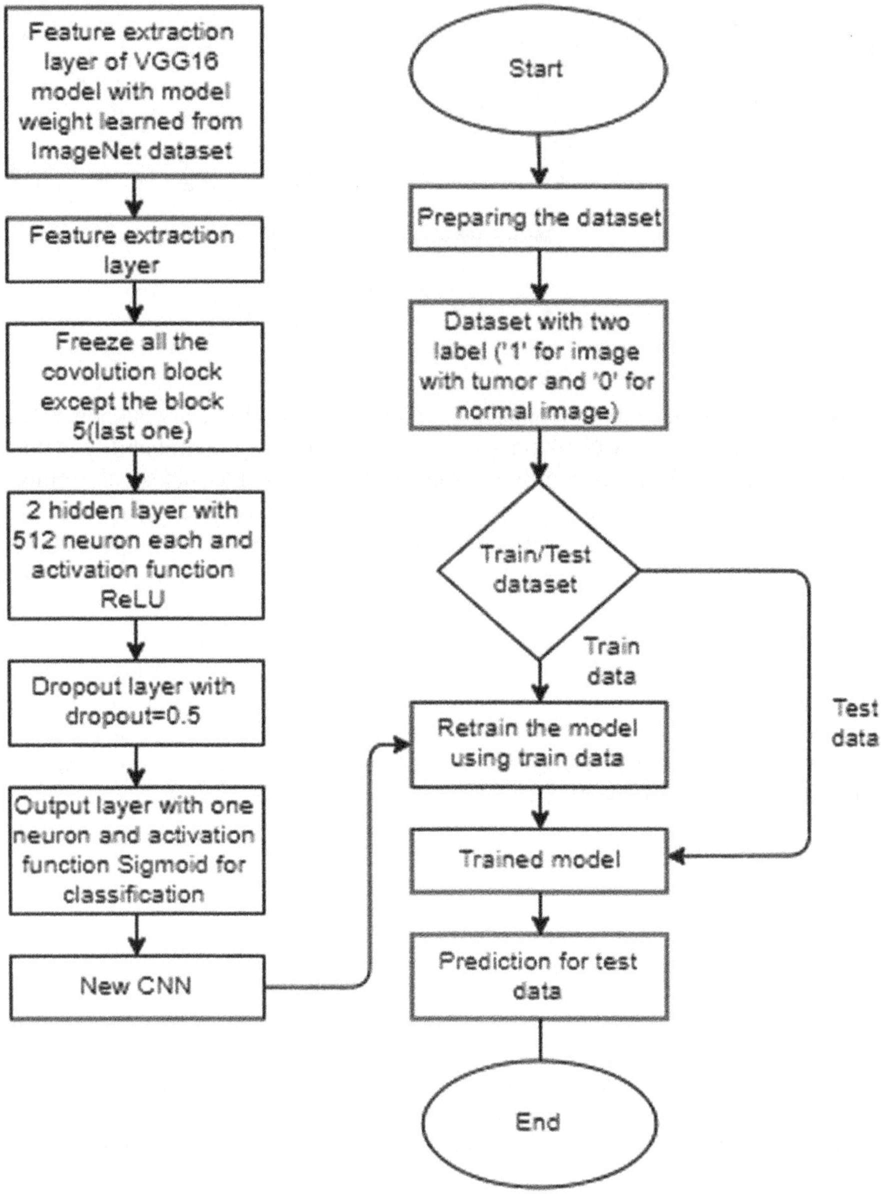

Fig. 4 Flowchart of proposed method

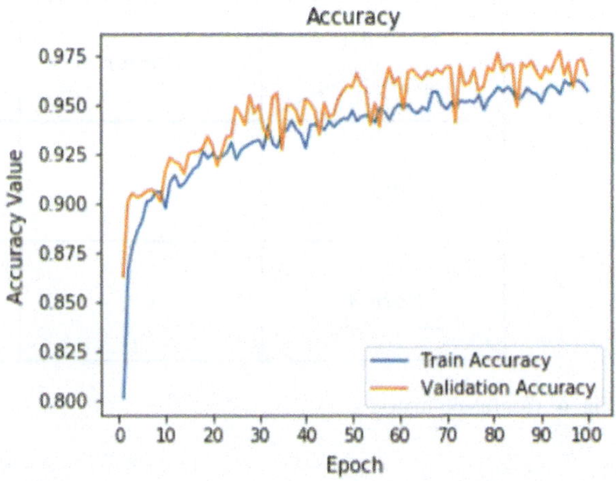

Fig. 5 Accuracies for each epoch

4 Results and Discussions

The proposed model is developed by using different Python libraries. For faster training model is trained using Google Colab online platform with 12 GB NVIDIA Tesla K80 GPU [25].

4.1 Training and Validation Result

The one of the advantages of using a CNN model as compared to other model for image analysis is that in CNN based model no separate feature extraction algorithm is needed, because the feature extraction is done by CNN itself. For training 3000 images (1500 normal MRI and 1500 with tumor) are used. Thus training accuracy is calculated for 3000 images. For validation 1000 images (500 normal and 500 tumor MRI images) are used. The number of epoch and iterations per epoch is 100. Image augmentation techniques are also used in order to increase the size of training data. The accuracy achieved by the proposed model during training and validation are exceptional. The highest accuracy during training is 95.22% at 93th epoch, while the highest accuracy for validation is 97.13 at 91th epoch. Figure 5 shows the training and validation accuracy with increasing epoch value.

4.2 Testing Results

To test the performance of the trained and validated CNN model 1000 MRI images (500 normal and 500 with tumor) that are not used during validation and training

Actual / Predicted	Tumor	Normal
Tumor	431	13
Normal	69	487

Fig. 6 Confusion matrix

are used. To measure the performance various performance metrics are used such as testing accuracy, Precision, sensitivity and F1 score. Figure 6 shows the confusion matrix for all 1000 images used in testing. In the confusion matrix of the proposed framework, in true positive (TP)ca se the actual class of image being evaluated is tumor and the type predicted by the model is also tumor. For the true negative (TN) scenario, the actual image type is normal (healthy image) and the type projected by the model is also normal. In false positive (FP) case the actual type of image being evaluated is normal while the type projected by the model is tumor. In false negative (FN) case the actual class of data was class tumor while the class expected by the model is normal. Table 1 represent the testing results for proposed model.

The results achieved by proposed model are remarkable with a test accuracy of 91.8%. The performance metrics for classification are calculated in following ways:

$$\text{Accuracy} = \frac{\text{TP} + \text{TN}}{\text{TP} + \text{TN} + \text{FP} + \text{FN}}$$

$$\text{Sensitivity} = \frac{\text{TP}}{\text{TP} + \text{FN}}$$

$$\text{Precision} = \frac{\text{TP}}{\text{TP} + \text{FP}}$$

Table 1 Various performance metrics

Classification results		
S. No.	Performance metrics	Value (%)
1	Accuracy	91.8
2	Sensitivity	86.2
3	Precision	97.07
4	F1 Score	91.31

$$F1\ \text{Score} = \frac{2 * \text{Precision} * \text{Sensitivity}}{\text{Precision} + \text{Sensitivity}}$$

A brief summary of testing results is as follows:

- Test dataset contains 500 MRI of Normal brain and 500 MRI of brain infected with tumor.
- 431 images of Tumor MRI have been correctly identified as Tumor.
- 69 images of Tumor MRI have been incorrectly identified as Normal.
- 487 images of Normal MRI have been correctly identified as Normal.
- 13 images of Normal MRI have been incorrectly identified as Tumor.

5 Contribution and Novelty of the Proposed Work

As per International Agency for Research on Cancer's [26] the instances and risk of mortality due to brain tumor in Asian Continent and notably in South Asia are highest in the world. By identifying these tumor in early stage of tumor cell development the risk of mortality can be reduced by providing suitable treatment to the patient. Identification of brain tumor by traditional method is a time consuming task and also require subject expert like radiologist in this case to identify the tumor in brain MRI images. Proposed work utilize a transfer learning based method which limits the need of subject expert and along with various other benefits. The key asset of using a transfer-based learning methodology is that, as a pre-trained model is utilized, the training period is shortened. CNN-based method is indeed very useful in medical image processing because in medical imaging extraction of features requires expert level skills, but here model itself extracted feature automatically in CNN. The other advantage in utilizing a method based on transfer learning in medical image analysis is that the free data availability in the medical sector is very limited, and there is a need for a better performing approach while having fewer data. So the proposed method indicate that machine learning methods focused on transfer learning achieves strong outcomes given limited data availability in medical field and play a critical role in healthcare sector.

6 Conclusion

The proposed work develop a transfer learning based tumor classification system for brain MRI images. A dataset of 5 K MRI images of 224×224 pixel resolution are used for training, validation and testing. The proposed model achieved a remarkable accuracy of 91.8% during testing. Results shows that the proposed method perform better than the methods discussed in Sect. 2. The main advantage of using a transfer learning based approach is that, the training time is reduced because a pre-trained model is used. In medical image analysis CNN based method seems to be very useful because feature extraction of medical image require an expert level knowledge, but in CNN feature are extracted automatically by the model itself. The other advantage of using a transfer learning based method in medical image analysis is that the availability of data is limited in medical field, so there is a need of approach which perform better despite having less data. So the proposed method automatically classify the tumor and normal MRI images and help in early stage diagnosis of brain tumor.

This method can be improved from a problem of binary classification to a multi-class classification. Furthermore some transfer learning based model can be created to predict the overall chance of survival of patient along with tumor. The proposed research use VGG16 model, in future specific pre trained model will be used and there outputs can be correlated with present architecture.

References

1. C. Roth, S. Deshmuk, *Fundamentals of Body MRI* (Elsevier, Amsterdam, 2016)
2. A. Aslam, E. Khan, M.M. Sufyan Beg, Improved edge detection algorithm for brain tumor segmentation, in *Second International Symposium on Computer Vision and the Internet (VisionNet'15)* (Elsevier, Amsterdam, 2015), pp. 430–437
3. W. Deng, W. Xiao, H. Deng, J. Liu, MRI brain tumor segmentation with region growing method based on the gradients and variances along and inside of the boundary curve, in *2010 3rd International Conference on Biomedical Engineering and Informatics*, vol. 7 (IEEE, 2010), pp. 930–938
4. G. Birare, V.A. Chakkarwar, Automated detection of brain tumor cells using support vector machine, in *9th International Conference on Computing, Communication and Networking Technologies* (IEEE, 2018), pp. 1–4
5. H.B. Nandpuru, S.S. Salankar, V.R. Bora, MRI brain cancer classification using support vector machine, in *2014 IEEE Conference on Electrical, Electronics and Computer Science* (IEEE, 2014), pp. 1–6
6. X. Xie, A k-nearest neighbor technique for brain tumor segmentation using minkowski distance. J. Med. Imaging Health Inform. (2018)
7. L. Lefkovits, S. Lefkovits, L. Szilágyi,Brain tumor segmentation with optimized random forest, in *International Workshop on Brainlesion: Glioma, Multiple Sclerosis, Stroke and Traumatic Brain Injuries*, Washington, DC, 2016
8. A. Min, Z.M. Kyu, MRI images enhancement and tumor segmentation for brain, in *International Conference on Parallel and Distributed Computing, Applications and Technologies (PDCAT)*, 2017
9. V. Dhanve, M. Kumar, Detection of brain tumor using k-means segmentation based on object labeling algorithm, in *2017-IEEE International Conference on Power, Control, Signal and Instrumentation Engineering* (IEEE, 2017), pp. 944–951
10. Z. Cui, J. Yang, Y. Qiao. Brain MRI segmentation with patch-based CNN approach, in *Chinese Control Conference, Chengdu, China, 2016* (2016)
11. R. Ezhilarasi, P. Varalakshmi, Tumor detection in the brain using faster r-CNN, in *Second International conference on I-SMAC (IoT in Social, Mobile, Analytics and Cloud) (I-SMAC 2018)* (2018)
12. M. Narmada, Balasooriya, A sophisticated convolutional neural network model for brain tumor classification, in *IEEE International Conference on Industrial Information Systems* (IEEE, 2017)
13. B. Srinivas, G.S. Rao, Transfer learning model for MRI brain tumor classification. Int. J. Innov. Eng. Manage. Res. **8**, 32–37 (2019)

14. N. Abiwinanda, M. Hanif, S. Tafwida Hesaputra, A. Handayani, T.R. Mengko, Brain tumor classification using convolutional neural network. *World Congress on Medical Physics and Biomedical Engineering* (2018)
15. L. Chato, S. Latifi, Machine learning and deep learning techniques to predict overall survival of brain tumor patients using MRI images, in *IEEE 17th International Conference on Bioinformatics and Bioengineering* (2017)
16. T. Gopi Krishna, S. Mishra, Detection and classification of brain tumor from MRI medical image using wavelet transform and PSO based LLRBFNN algorithm. Int. J. Comput. Sci. Eng. **6** (2018)
17. R. Uday, K.D. Yamini, Brain tumor classification in MRI images. Int. J. Adv. Res. Comput. Sci. (2018)
18. H.S. Hamoud, G. Sulong, Tumor detection through MRI images: a review literature. J. Theor. Appl. Inf. Technol. (2014)
19. K.W. Clark, B.A. Vendt, K. Smith, J.B. Freymann, J. Kirby, P. Koppel, S.M. Moore, S.R. Phillips, D.R. Maffitt, M. Pringle, L. Tarbox, F. Prior, The cancer imaging archive (TCIA): Maintaining and operating a public information repository. J. Digit. Imaging **26**(6), 1045–1057 (2013)
20. D.A. Dickie, D.E. Job, D. Rodriguez, A. Robson, C. Pernet, I.J. Deary, J.M. Wardlaw, Brain imaging of normal subjects (brains age-specific MRI atlases from young adults to very elderly (v1.0)). University of Edinburgh, 2016. bRAINS Imagebank
21. K. Simonyan, A. Zisserman, Very deep convolutional networks for large-scale image recognition. *arXiv 1409.1556*, Nov 2014
22. D. Scherer, A. Muller, S. Behnke, Evolution of pooling operation in convolutional architectures for object recognition, in *Artificial Neural Network-ICANN* (Springer, Berlin, 2010), pp. 92–101
23. Y. Dauphin, H. Vries, J. Chung, Y. Bengio, Rmsprop and equilibrated adaptive learning rates for non-convex optimization. *arXiv*, 35, Feb 2015
24. D.R. Wilson, T.R. Martinez, The need of small learning rate for large problem, in *Proceedings of International Joint Conference on Neural Networks*, Washington, DC (IEEE, 2001), pp. 115–119
25. Google. Google colaboratory, in Building machine learning and deep learning models on Google cloud platform. https://colab.research.google.com/. Accessed 2020
26. R. Leece, J. Xu, Q.T. Ostrom, Y. Chen, C. Kruchko, Global incidence of malignant brain and other cental nervous system tumors by histology. Neuro-Oncology **19**, 1553–1564 (2017)

Semantic-Based Vectorization Technique for Hindi Language

Shikha Mundra, Ankit Mundra, Josh Agarwal, and Pankaj Vyas

Abstract Word embeddings capture contextual information about individual words in a vector form. These representations of words help learning algorithms to achieve better performance in natural language processing tasks by grouping similar words. Earlier vectorization techniques have mainly focused onto words only as they do not consider contextual information of word. Recently popular known representation learning known as word2vec word embedding has shown a significant improvement in state of art. As the technique is language dependent, most of the work done has been focused on the English language only. Research in non-Latin languages such as Hindi has been of limited because the unavailability of proper corpus and hence not so efficient word embeddings. We aim to embed words from a Hindi corpus scrapped from various newspapers and other literary sources. Using bidirectional LSTM, we train a neural model that embeds entire sentential contexts and target words in the same low-dimensional space, which is optimized to reflect inter-dependencies between targets and their entire sentential context as a whole. To test the model, we will try to predict the missing word in a sentence by passing the preceding and succeeding parts of sentences to the model.

Keywords Word2Vec · Embedding · Skip gram

1 Introduction

Machine learning algorithm operates on numeric feature which requires input as a 2D matrix where rows behave as instance and column as feature. To apply conventional machine learning or recent deep learning, we need to transform data to a 2D

S. Mundra
Department of Computer Science and Engineering, Manipal University Jaipur, Jaipur, India

A. Mundra (✉) · J. Agarwal · P. Vyas
Department of Information Technology, Manipal University Jaipur, Jaipur, India
e-mail: ankit.mundra@jaipur.manipal.edu

P. Kumar et al. (eds.), *Machine Learning for Intelligent Multimedia Analytics*,
Studies in Big Data 82,
https://doi.org/10.1007/978-981-15-9492-2_15

numerical feature matrix which is also known as text representation. Some of the text representation techniques achieved state of art have been represented in Fig. 1.

Research on textual data representation is focused and progressive from long years back. Several methods have been experimented at word level and sentence level to represent them as a vector. Most common methods are statistical method out of them; the most popular is known as bag of word model in which presence of word in a vocabulary space will be marked by 1 and rest all marked with 0. This method is the most simplest and earliest method and has achieved state of art in this field but major problem with this technique is it does not consider order of words and size of the vector become too large result in a sparse vector. It mainly focuses on most frequent word; simultaneously, it focuses on irrelevant frequent words also like the, a, an, by. To overcome this drawback, a popular technique called as term frequency inverse document frequency(tf-idf) helped in decreasing weightage of irrelevant word and achieved an improvement over bag of word but also suffer from same problem of word order and sparse vector [1]. Unfortunately, these two technique produced quite good result in case of multiple document shares words but reduced its performance when they are semantically same but do not share most words; in other words, it does not perform well when sentence is semantically same but syntactically different.

Analyzing the limitation of not including context while creating feature and for including context *n*-gram techniques is used which take sequence of *n* words, e.g., for bigram sequence of 2 word is considered and for trigram sequence of 3 word is considered but the drawback is it checks for repetition of exact sequence of words which is not true every time in free text as we can write a sentence or a document in multiple ways which infer same meaning [2, 3]. Along with the vector, which is very sparse hence to reduce the size of vector and include contextual meaning as well, distributed text representation is explored further which are based on neural network architecture.

Recent distributed representation as word2vec and glove has achieved state of art in detecting document as semantically similar even if they are syntactically different. Popularly distributed representation algorithm word2vec is completely based on self-supervised shallow neural network and glove is based on co-occurrence matrix.

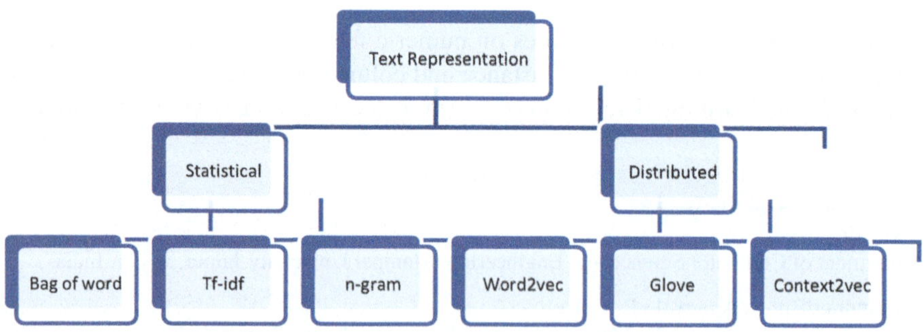

Fig. 1 Popular text representation methods

Further in this chapter, we have focused on two models of word2vec as CBOW and skip gram. We have chosen word2vec as it is easy to train word2vec onto self-scraped Hindi corpus to create Hindi word embedding. Along with this, we have also explored recent context2vec-based bilstm and unsupervised way of text vectorization.

Also, in this chapter, we present a comparison between word2vec [4] and context2vec model. The main aim of this paper is to find word embeddings that carry the right meaning from context. Firstly, word2vec is chosen as the base model to extract the word embeddings, and then to do the comparison, context2vec is applied. Hindi corpus is chosen as less work is done on it. Word2vec was the first model that extracted word embeddings for a very large corpus [5]. Skip gram model of word2vec is used in this paper to represent a word as a vector on N dimensions. In skip gram model, the focus word is taken to large corpora and one by one all the words inside the window are also taken in the corpora. After that, it is taken to feed a neural network to predict the probability of all the surrounding words to appear around the main word inside the window.

Some of the points where skip gram lagged was that it cannot handle polysemy (one word with multiple meanings), e.g., tank used as military tank and water tank; it also fails to figure out combined word phrases, e.g., "New York" and sentential context is normally just a collection of nearby word's word embeddings or average of the word embeddings inside the window [6, 7].

On the other hand, because of all these disadvantages of word2vec, we have used the context2vec model. Context2vec is an unsupervised model that captures generic word embeddings of large sentential context using bidirectional LSTM. Context2vec gives better results for the sentential context than the average word embedding representation of word2vec. The average word embedding representation of the word2vec model is replaced by some improved method in the context2vec model.

2 Literature Review

Proposed by Yuret [8], substitute vectors show contexts as a probabilistic distribution over the possible gapped words for the target slot but it is limited to the fixed-size window as it can only be produced with an *n*-gram language model. Context embeddings as syntactic dependency [9] rely on the accessibility of a parser that is of high quality. Instead of viewing this as a solo representation for the whole sentential context, it can be reviewed as a "bag of dependencies."

Some of the newly researched works replace the internal context representation of word2vec's CBOW. Ling [10] suggested a window that is a simple one-dimensional linear projection of the embeddings and represent it into a low-dimensional vector. Ling et al. (2015) suggested "CBOW with attention," which is used for finding the features that are close or relevant in a context window.

Dyer and Kawakami [11] proposed words in context using LSTMs that are bidirectional and multi-lingual supervision. A good amount of work is done to show a sequence of words by using recurrent neural networks [12]. The ways/methods used

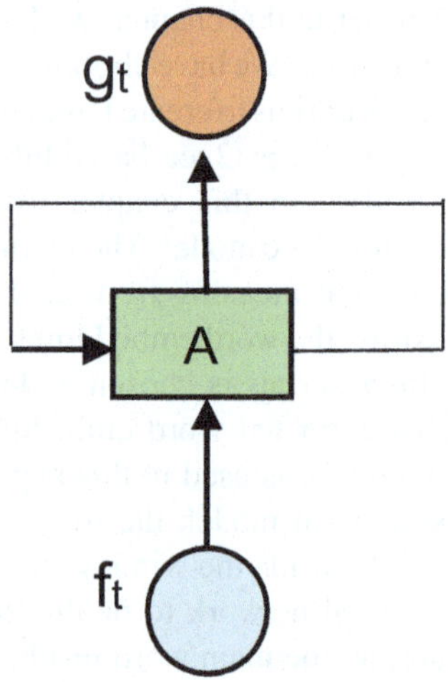

Fig. 2 RNN loop structure

to learn sentence representation have a lot in common with sentential context representation. Distinct target slots in the sentence can produce entirely distinct context representation.

2.1 Recurrent Neural Networks (RNN)

It is observed that we humans do not think from Scratch each and every moment. While going through this chapter, you might have understood each word as per your knowledge of previous words. But it is not possible with the traditional neural networks. For example, if we needed to classify each event that is happening in an act. To use it's reasoning traditional neural network to use is reasoning about the previous events in the act to inform later ones [13, 14].

RNN is responsible for handling this major shortcoming, as these networks are with loops in them, allowing information to prevail.

A is a piece of neural network, takes an input g_t and outputs a value f_t. In a loops, information can be passed from one step of the network to the another (next) (Fig. 2).

2.2 Long-Term Dependencies in RNN

One of the requests of RNNs is that they might be able to connect preceding information to the current task, like using preceding pictures frames might deliver the understanding of the current frame. But it depends that RNNs can do this.

For example, in a model, we are trying to predict the last word in "the sun rises from the east"; it is obvious that next word is going to be east. In this case, the gap between the relevant information and the position that it is needed is small; therefore, this task can be achieved by RNNs. But what if needed more context? For example, try to predict the last word in the text I grew up in India... I speak fluent Hindi." Latest information advices that the next word is probably the name of a language, but what if we want to know which language? We need the context of India, from a further back [15]. Here, the gap between the relevant information and the point where it is needed might become too large. As the gaps increases, RNNs are not able to connect the information for this purpose LSTMs are used.

2.3 *Long Short-Term Memory Networks (LSTMs)*

Here in Fig. 3,

X_t = Input Vector
H_{t-1} = Previous Cell Output
C_{t-1} = Previous Cell Memory
H_t = Current Cell Output
C_t = Current Cell Memory

$$f_t = \sigma(X_t * U_f + H_{t-1} * W_f)$$
$$\bar{C}_t = \tanh h(X_t * U_o + H_{t-1} * W_c)$$
$$I_t = \sigma(X_t * U_i + H_{t-1} * W_i)$$
$$O_t = \sigma(X_t * U_o + H_{t-1} * W_o)$$
$$C_t = f_t * C_{t-1} + I_t * \bar{C}_t$$
$$H_t = O_t * \tanh h(C_t)$$

W, U = Weight vectors for forget gate(f), candidate(c), i/p gate(I), o/p gate (O)

Note: These are different weights for different gates, for simplicity's sake I mentioned W and U.

LSTMs are a special kind of RNN, which are skilled of learning long-term dependencies [16]. Their aim is to avoid long-term dependency issue. It is their default behavior to remember information for long periods of time. A chain of repeating modules of neural networks are there in the recurrent neural networks. In traditional RNNs, this module will have a very simple structure, like single tanh layer. Just like RNNs, LSTMs also have the chain like structure, but this structure is a bit different as it has four neural network layer interrelating in a very extraordinary way [3, 6, 7, 16].

The cell state and its various gates are the core concept of LSTM. The cell state is like a bridge which helps in delivering the significant information down the sequence chain. It can be classified as a "CPU" of the network. Therefore, earlier information can be transferred to later steps and thereby helps in decreasing the paraphernalia of short-term memory. The gates help in adding or removing of information as the cell state goes on its journey. The neural networks which decide that what information is

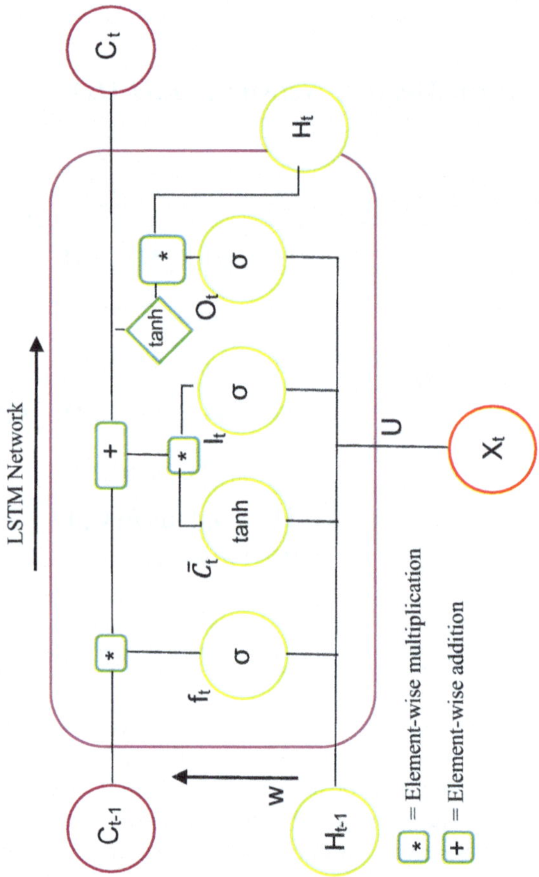

Fig. 3 LSTM network

needed to be passed to the cell state are known as gates. During training, these gates learn which information is appropriate to keep or to disremember.

3 Dataset

In order to implement the proposed technique, a Hindi language dataset has been scrapped from various news websites. It contains almost 10,000 lines and 78,499 words. It has 15,070 unique words out of which 3591 are used for embedding. Special and numeric characters have been removed to keep it consistent with the Hindi language.

4 Methodology

The task of our model is to learn a generic embedding function for variable-length sentences. To accomplish this, we use two effective approaches, first is the skip gram method of the Word2Vec model and the second one is a neural network architecture, which is based on Word2Vec's CBOW architecture [4]. The second approach replaces CBOW's naive method of context encoding of averaging word embeddings [17], with a much more robust neural model, using bidirectional LSTM. Figure 4a, b shows a comparison in Word2Vec CBOW and context2Vec LSTM models.

4.1 Pseudocode for Our Methodology

Step 1: We have Scrape a large about of data from multiple website belongs to Devanagari script such as Hindi news.
Step 2: While preprocessing, we have removed incomplete sentences manually.
Step 3: We have removed all sentences which does not belong to Devanagari script using language identification.
Step 4: We have applied Context2Vec onto clean monolingual data to get vector of each word.
Step 5: We have compared the Context2Vec and word2Vec as mentioned in result section.

4.2 The Skip Gram Model

To build word embeddings, the skip gram model (Fig. 5) is trained with the objective of finding word representations of neighboring word of input word. This model is

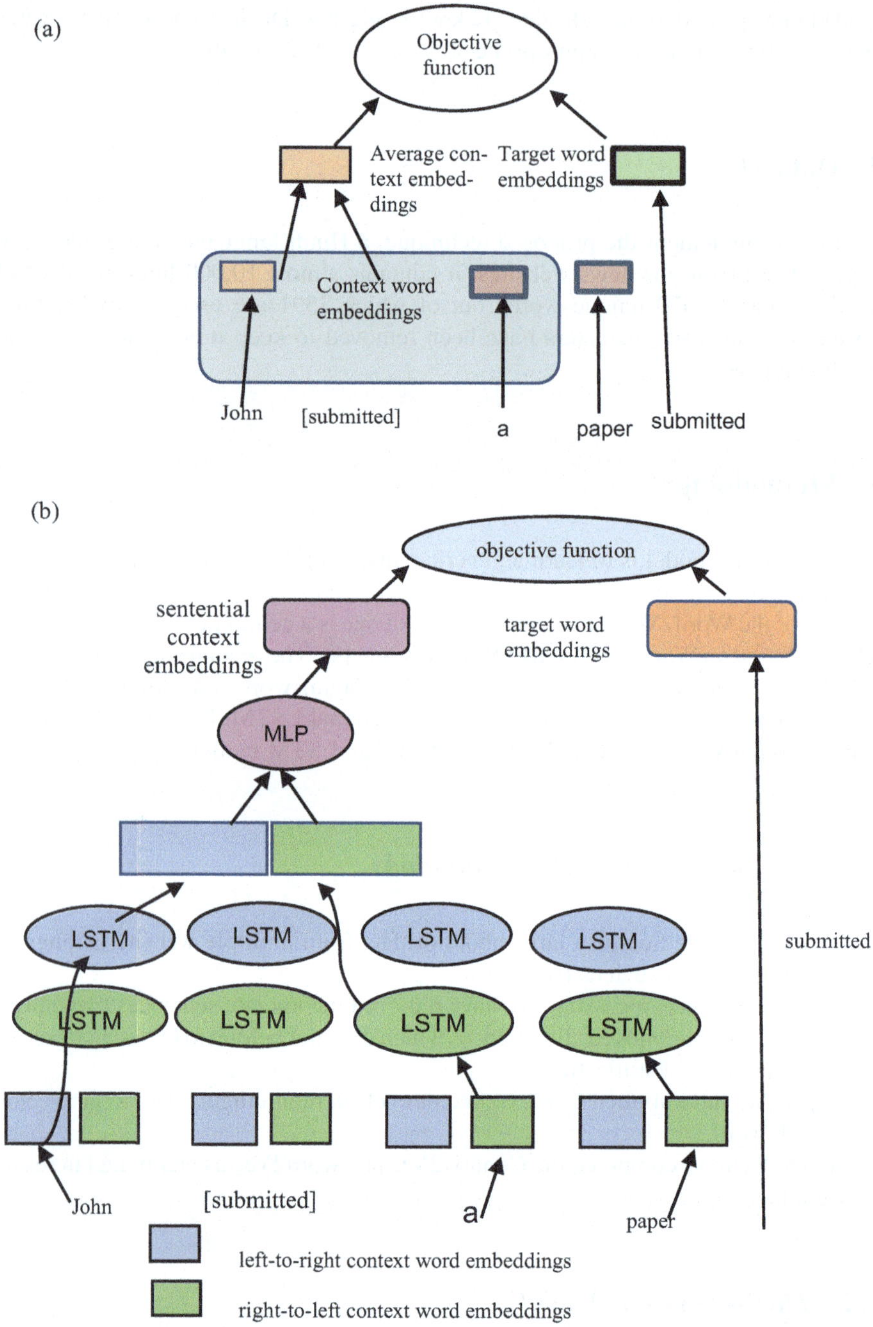

Fig. 4 **a** word2vec CBOW model architecture; **b** context2vec model architecture

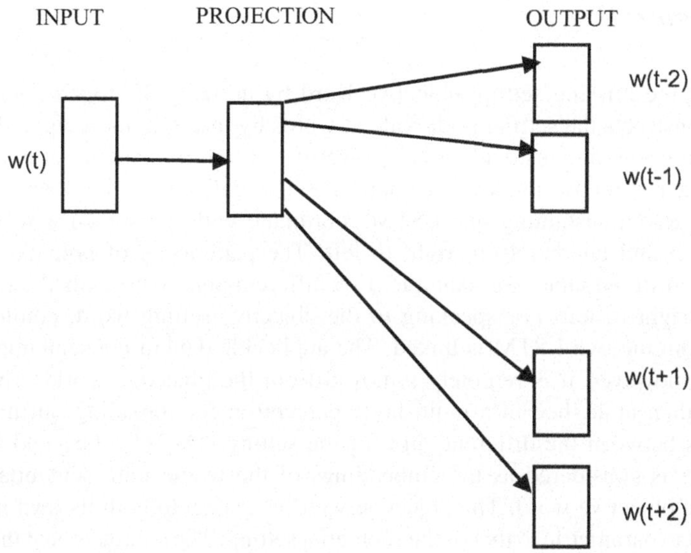

Fig. 5 Skip gram model

useful for predicting the surrounding words in a sentence or a document. A sentence consists of multiple words in a sequence and the task of the skip gram model is to predict whether two word appear in a similar context.

As skip gram is based on shallow network architecture, the training is performed between input and output in such a way that loss decreases continuous using gradient descent method. Suppose a sentence S consists of 6 words in a sequence in this w_2 will be the input and w_1 and w_3 will be the output during training.

$$S = \{w_1,\ w_2,\ w_3,\ w_4,\ w_5,\ w_6\}$$

We train our skip gram model with pair $[(w_2,\ w_1),\ 1]$ here w_2 is the input and w_1 is the context and positive label 1 indicate that this pair is a relevant pair. For faster training purpose, negative sampling is used as $[(w_2,\ \text{random}),\ 0]$ where w_2 is input word and mapped to random output word. In these, 0 labels show w_2 and any random word are not contextually related or they are not semantically similar words. We have trained skip gram model of word2vec onto our Hindi corpus and found better in comparison to statistical text representation technique as mentioned in result section.

4.3 Context2Vec

It predicts the missing setting objective word by utilizing the log-straight model. Also, it constructs the setting portrayals of words by inserting more up to date ones into the base low-dimensional space. However, we will, in general, utilize a much compelling parametric model to gather the sentential setting. We have utilized a Bi-LSTM RNN, sustaining one LSTM coordinate with the sentence words from left to right and another from right to left. The parameters of both the systems are totally differentiable, for example, two differentiable word embeddings left to right and right to left. For speaking to the objective setting word, coming about vectors from the two LSTM is linked. The applicable data in the sentential setting is caught; however, it is remotely removed from the objective world. The linked vector is then nourished in a multi-layer perceptron for spreading out non-paltry conditions between the different sides of the setting [18–21]. The yield from the MLP layer is considered as the embeddings of the whole joint sentential setting around the objective word. The objective word is spoken to with its own inserting, equivalent in parametric to that of the sentential setting. The parameters of the system are found out utilizing word2vec's negative examining target work, with a positive pair being an objective word and its sentential setting. To become familiar with the system parameters, word2vec's negative examining target work, making positive pair being an objective word alongside its whole sentential setting, and separate k negative combines as arbitrary objective words, inspected from a (smoothed) unigram appropriation over the jargon, matched with a similar setting [22]. Utilizing this both setting and target word, embeddings are scholarly. Our essential objective is setting portrayal as opposed to utilizing word2vec and other comparative word vectorization models which consider the objective of word embeddings as their fundamental yield [23, 24]. Our model accomplishes its goal by allocating comparable embeddings to sentential settings and their related objective words. Further, like the case in word2vec models, this in a round about way brings about doling out comparative embeddings to target words that are related with comparable sentential settings, and on the other hand to sentential settings that are related with comparative objective words. We will appear in the accompanying areas of how these properties make our model helpful.

4.3.1 Formal Specification and Analysis

To acquire a sentence-level setting portrayal, a Bi-LSTM RNN is utilized. Let lLS be an LSTM perusing the expressions of a given sentence from left to right, and let rLS be an inverted one perusing the words from right to left. Given a sentence $w_{1:n}$, our "shallow" bidirectional LSTM setting portrayal for the objective w_i is characterized as the accompanying vector link (Eq. 1)

$$\text{biLS}(w_{1:n}, i) = \text{lLS}\ (l_{1:i-1}) \oplus \text{rLS}\ (r_{n:i+1}) \quad (1)$$

where l/r speak to unmistakable left-to-right/right-to-left word embeddings of the sentence words. This definition is somewhat unique in relation to standard bidirectional LSTM, as we do not encourage the LSTMs with the objective word itself (e.g., the word in the position I). Next, we apply the accompanying non-direct work on the link of the left and right setting portrayals (Eq. 2)

$$\mathrm{MLP}(x) = \mathrm{L}_2(\mathrm{ReLU}(\mathrm{L}_1(x))) \tag{2}$$

where MLP represents multi-layer perceptron, ReLU is the rectified linear unit initiation capacity, and $L_i(x) = W_i(x) + \mathrm{bi}$ is a completely associated straight activity. Let $c = (w_1; \ldots; w_{i1}; ; w_{i+1}; \ldots; w_n)$ be the sentential setting of the word in position I. We characterize context2vec's portrayal of c as:

$$\vec{c} = \mathrm{MLP}\ (\mathrm{biLS}(w_{1:n},\ i)) \tag{3}$$

Next, we signify the implanting of an objective word t as ~ t We utilize the equivalent inserting dimensionality for target and sentential setting portrayals. To learn target word and setting portrayals, we utilize the word2vec negative examining target work:

$$S = \sum_{t,c}\left(\log\sigma(\vec{t}\,.\,\vec{c}) + \sum_{i=1}^{k}\log\sigma\left(-\vec{t_i}\,.\,\vec{c}\right)\right) \tag{4}$$

In Eq. 4, where the summation goes over each word token t in the preparation corpus and its relating (single) sentential setting c and is the sigmoid capacity. $t_1; \ldots; t_k$ is the negative examples, autonomously examined from a smoothed variant of the objective words unigram dispersion: $\mathrm{p}(t)/(\#t)$, to such an extent that $0 < 1$ is a smoothing factor, which builds the likelihood of uncommon words. Toll and Goldberg demonstrated that when the target work in Eq. 4 is applied to single-word settings, it is streamlined as shown in Eq. 5.

$$\vec{t}.\vec{c} = \mathrm{PMI}_\alpha(t, c) - \log(k) \tag{5}$$

where $\mathrm{PMI}(t;\ c) = \log \mathrm{p}(t;\ c) \cdot \mathrm{p}(t) \cdot \mathrm{p}(c)$ is the pointwise shared data between the objective word t and we cushion each information sentence with uncommon BOS and EOS words in positions 0 and $n + 1$, separately. Figure 3 shows a 2D outline of context2vec's inserted space and similitude measurements. Triangles and circles signify sentential setting embeddings and target word embeddings, individually. Triangles and circles signify sentential setting embeddings and target word embeddings, individually for the setting word c. The examination exhibited in Levy and Goldberg is substantial for each co-event grid that portrays the joint conveyance of two irregular factors. It very well may be applied to our case, where the setting is not only a solitary word yet a whole sentential setting of an objective word. In like manner, we can see the objective setting installing got by our calculation as a factorization of the PMI grid between all conceivable objective words and all conceivable distinctive sentential settings. Not at all like the instance of single-word settings, it is not attainable to expressly figure here this PMI framework because of the exponential number of conceivable sentential settings. Nonetheless, the target work that we streamline still intends to best estimate it. In view of the above investigation, we can expect the inward result of our objective and setting embeddings to rough PMI $(c; t)$. We note that likewise, with bigger qualities, there will be progressively inclination toward putting uncommon words nearer to their related settings in this space.

Through research and analysis, we have already marked out stop words; we remove them using the regex class in Python. We then maintain two separate dictionaries—one which is used for mapping words to indexes and other for mapping indexes to words. A random matrix is initialized, each index contains a context vector of every unique word. Using the dictionaries and matrix, we can now find out which word represents the vector and vice-versa. The embeddings are learned through the skip gram model. We evaluate the true word matrix using the cosine and Euclidean distances both, to find the nearest words. Again, the context2vec embeddings are learned on the corpus using a multi-layer Bi-LSTM, in which a variable-length sentence can be encoded. Both left and right of the set target word are passed through two different LSTMs for the word prediction task.

5 Results and Conclusion

Hindi trained model has been experimented for lexical substitution task using context2vec as shown in Fig. 6 and word similarity check as shown in Fig. 7 using word2vec and context2vec technique. From experimentation, it is found that context2vec performs better than word2vec as it also considers previous and forward context with variable length which is not the case with word2vec as word2vec is based on fixed window size only. In terms of matrix sparsity, distributed representation has contributed much in reducing the size of vector in comparison to statistical vectorization methods [25].

6 Future Work

Seeing the better performance of the LSTM model, we plan to train it on a larger Hindi corpus and use it in different tasks such as machine translation, text summarization and also extend the functionality of missing word prediction.

Machine translation tasks are very useful; one such small task includes conversion of date and time. People tend to write dates in multiple formats depending upon their needs, understanding and way of writing. Till now, the baseline models used fail to understand the way of writing completely; they fail to gather the sentential

```
>> लीची की फसल लगभग [] हो चुकी है
<EOS>: 0.5
मग: 0.498636394739151
प्रभावित: 0.4890299141407013
फुट: 0.4886992275714874
अनबन: 0.4886934161186218
निलंबित: 0.4856671392917633
बम: 0.47936511039733887
वृद्धि: 0.47902828454971313
दौड़: 0.47640588879585266
मिमी: 0.47510015964508057
```

```
>> लोगों के खिलाफ [] दर्ज कराया है
मुकदमा: 0.5440834164619446
मामला: 0.5439561605453491
केस: 0.5114154815673828
टीए-डीए: 0.5007340908050537
<EOS>: 0.5
तफ्तीश: 0.49163687229156494
रोक: 0.4840041697025299
प्राथमिकी: 0.4833263158798218
तहरीर: 0.47555333375930786
निर्णय: 0.47537845373153687
```

(a)

```
>> सरकार की गलत [] की वजह से देश में महंगाई दिन प्रतिदिन बढ़ रही है
संशोधन: 0.5476641654968262
गर्मी: 0.5193084478378296
साफ: 0.5178591012954712
नीतियों: 0.5141640305519104
बारिश: 0.5040026307106018
बिक्री: 0.5037804245948792
जाम: 0.502662718296051
<EOS>: 0.5
स्कूलों: 0.49947652220726013
दिक्कत: 0.4986676871776581
```

(b)

Fig. 6 a–h Missing word predictions

context when converting date to target format. Hence, we expect our model with some configuration can help to solve the issue.

In regard to text summarization, the current problems faced relating to abstractive summarization, in which the sentential context of the document is gathered and re-written in other words keeping the sentential context same. Since our model proves to glean sentential context quite accurately, we expect that by creating using a dictionary of words, with their meanings written and analyzing the sentential context in which each word is written, this problem can be solved.

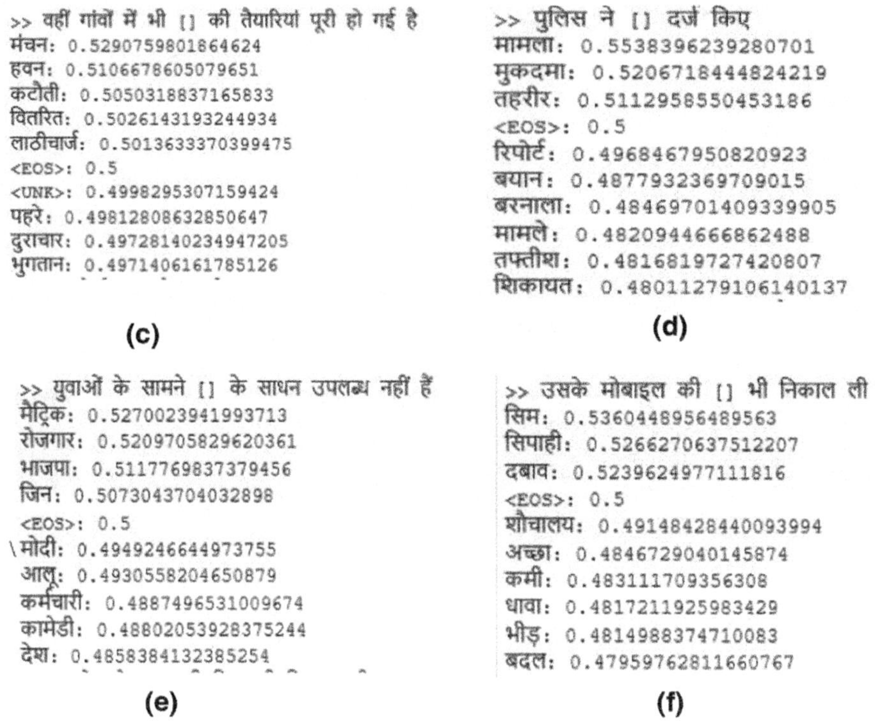

Fig. 6 (continued)

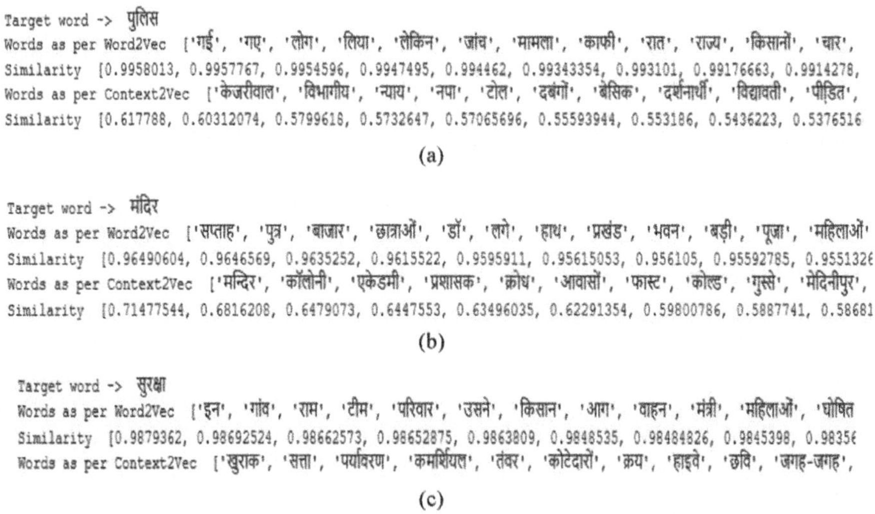

Fig. 7 **a–c** Comparisons between word2vec and context2vec

References

1. Using tf-idf to determine word relevance in document queries. Proc. First Instr. Conf. Mach. Learn. **242**, 133–142 (2003)
2. P. Majumder, M. Mitra, B.B. Chaudhuri, N-gram: A language independent approach to IR and NLP. Int. Conf. Univ. Knowl. Lang. Nov (2002)
3. A.K. Somani, R.S. Shekhawat, A. Mundra, S. Srivastava, V.K. Verma, eds. *Smart Systems and IoT: Innovations in Computing: Proceeding of SSIC 2019*, vol. 141 (Springer Nature, 2019)
4. Q. Le, T. Mikolov, Distributed representations of sentences and documents, In Inter. Conf. Mach. Learn. 1188–1196, Jan (2014)
5. J.L. Netto, A.A. Freitas, Automatic text summarization using a machine learning approach (2002)
6. N. Majumder, S. Poria, A. Gelbukh, E. Cambria, Deep learning-based document modeling for personality detection from text. IEEE Intell. Syst. **32**(2), 74–79 (2017)
7. K. Greff, R.K. Srivastava, J. Koutník, B.R. Steunebrink, J. Schmidhuber, LSTM: A search space odyssey. IEEE Trans. Neural Netw. Learn. Syst. **28**(10), 2222–2232 (2016)
8. D. Yuret, FASTSUBS: An efficient and exact procedure for finding the most likely lexical substitutes based on an n-gram language model. IEEE Signal Process. Lett. **19**(11), 725–728 (2012)
9. M. Bansal, K. Gimpel, K. Livescu, Tailoring continuous word representations for dependency parsing. Proc. Annu. Meet. Assoc. Comput. Linguist. (2014)
10. Q. Liu, H. Jiang, S. Wei, Z.-H. Ling, Y. Hu, Learning semantic word embeddings based on ordinal knowledge constraints. Proc. of ACL (2015)
11. K. Kawakami, C. Dyer, Learning to represent words in context with multilingual supervision. Workshop in ICLR (2016)
12. R. Socher, J. Pennington, E.H. Huang, A.Y. Ng, C.D. Manning, Semi-supervised recursive autoencoders for predicting sentiment distributions. Proc. of EMNLP (2011)
13. A. Mundra, A. Soni, S.K. Sharma, P. Kumar, D.S. Chauhan, Decision support system for determining: right education career choice. ICC 2014-Comput. Netw. Secur., 8–17 (2014)
14. T. Mikolov, S. Kombrink, L. Burget, J. Černocký, S. Khudanpur, Extensions of recurrent neural network language model. IEEE Trans. Acoust. Speech Signal Process. (ICASSP), 5528–5531 (2011)
15. H. Arif, K. Munir, A.S. Danyal, A. Salman, M.M. Fraz, Sentiment analysis of roman Urdu/Hindi using supervised methods. Proc. of ICICC**8**, 48–53 (2016)
16. Z. Huang, W. Xu, K. Yu, Bidirectional LSTM-CRF models for sequence tagging, arXiv preprint arXiv:1508.01991 (2015)
17. C. McCormick, Word2vec tutorial-the skip-gram model, (2016)
18. A. Nenkova, A survey of text summarization techniques, (2012)
19. H. Somers, Example-based machine learning, (1999)
20. K. Cho, Y. Bengio, Neural machine translation by jointly learning to align and translate (2014)
21. X. Rong, Word2vec parameter learning explained,arXiv preprint arXiv:1411.2738 (2014)
22. R. Sharma, S. Morwal, B. Agarwal, R. Chandra, M.S. Khan, A deep neural network-based model for named entity recognition for Hindi language. Neural Comp. Appl., 1–13 (2020)
23. A. Mridula, C.R. Kavitha, Opinion mining and sentiment study of tweets polarity using machine learning. IEEE Second Int. Conf. Inventive Commun. Comput. Technol. (ICICCT), 621–626 (2018)
24. S. Mundra, A. Dhingra, A. Kapur, D. Joshi, Prediction of a movie's success using data mining techniques. Inf. Commun. Technol. Intell. Syst., 219–227 (Springer, Singapore, 2019)
25. S. Kabra, R. Agarwal, Auto spell suggestion for high quality speech synthesis in Hindi, arXiv preprint arXiv:1402.3648 (2014)

The manufacturer's authorised representative in the EU is Springer Nature Customer Service Centre GmbH, Europaplatz 3, 69115 Heidelberg, Germany. If you have any concerns regarding our products, please contact ProductSafety@springernature.com

Printed and bound by CPI Group (UK) Ltd, Croydon, CR0 4YY
30/09/2026
02225291-0011